NASA SP—4019

# ASTRONAUTICS AND AERONAUTICS, 1974

A Chronology

by
Nancy L. Brun

The NASA History Series

*Scientific and Technical Information Office*     1977
NATIONAL AERONAUTICS AND SPACE ADMINISTRATION
*Washington, D.C.*

# Foreword

New technology in a rapidly advancing space age remains a challenge, but this chronological collection of aeronautical and space events of 1974 shows that the emphasis in technology has shifted from the problems of how to operate in air and space to the practical use of those environments to meet human needs.

Of NASA's 16 payloads launched during the year, 7 were operational satellites, expected to produce profits for their owners. Three of the nine experimental spacecraft were launched to develop technology for applications. Although the immediate use of technology was emphasized in 1974, NASA also looked to the future. A little more than a third—six spacecraft—were launched for scientific investigations, an investment in tomorrow.

Other nations and commercial interests increasingly joined us in efforts to explore and use space. In NASA's 1974 total, 11 spacecraft were paid for by non-NASA users; 10 were international. In addition, two Italian launches used NASA launch vehicles.

Operational satellites—all of whose costs, including launch costs, were paid for by others—included six for communications, two of them the first U.S. domestic communications satellites and four for other nations or international groups. And a new operational weather satellite joined the network that reports data from pole to pole. Meanwhile, new technology for improved services was pursued: NASA launched the first Synchronous Meteorological Satellite, for continuous day-and-night weather monitoring, and orbited an experimental communications satellite for France and West Germany.

NASA's *Ats 6* Applications Technology Satellite demonstrated a new use for powerful communications satellites, transmitting educational courses and health services to small low-cost receivers in remote areas. And remote sensing by satellite and aircraft, though still experimental, found increasing use around the globe for monitoring the earth's geology, ecology, resources, and pollution.

NASA continued the systematic exploration of the solar system and the observation of the universe in 1974. Studies of the sun, the planets, and the stars added to knowledge of atmospheric processes, geological formations, energy sources, and physical laws that affect the earth. The German-American *Helios 1* probed interplanetary space to within 45 million kilometers of the sun, closer than any previous spacecraft had flown. Three planetary probes launched in previous years gave us the first close look at Mercury, new clues to the origin and evolution of Venus, and new information on the weather, atmosphere, and radiation belts of Jupiter. Jupiter was found to be a ball of liquid hydrogen, its great red spot a gigantic cluster of storms at least 400 years old. These Mariner and Pioneer probes were sweeping on toward further planetary investigations even as preparations continued for future probes to softland on Mars and to fly past Jupiter, Saturn, and Venus. Scientific satellites of the earth, sounding rockets, balloons, and aircraft were used to study spectra of the stars, celestial x-ray and gamma ray sources, and the earth's own atmosphere and magnetic field and interaction with the solar wind.

Manned space flight continued to demonstrate man's capability to live and work in space. *Skylab 4*, man's longest mission to date, extended into 1974, gathering data for 84 days on his ability to work in space, his physiology, the sun, and the earth. The Apollo-Soyuz Test Project neared readiness for its mid-1975 U.S.–U.S.S.R. flight to test compatible docking systems and conduct joint experiments in space. Development of the first true space transportation system moved nearer its goal of a reusable space shuttle and reusable space laboratory, as the shuttle reached test and fabrication stages and the European Space Research Organization awarded the prime contract for its contribution, Spacelab, to be carried into orbit and back in the shuttle.

Aeronautical research was reoriented, with more emphasis placed on reducing both the amount of energy required for transportation and the pollution produced by transportation. NASA sought new solutions for the problems of noise, pollution, and safety, while experimenting with alternate fuels and composite materials. Flight tests began on a new general-aviation wing, the GAW-1, and the X-24B lifting body tested maneuverability and landing abilities of a vehicle designed for reentry from space.

NASA made advances toward low-cost production of solar cells to convert sunlight into electricity and demonstrated the use of solar energy to cool and heat houses. Wind-driven electric generators and ways to reduce fuel consumption and pollution by cars were other targets of research during the year.

Thus this nation, in cooperation with others, turned capabilities that man has never before possessed toward solution of today's problems, while investing significant resources to increase our knowledge and technology so that in future decades we will be in a better position to help meet humanity's cultural and economic needs.

August 1976

John E. Naugle
*Associate Administrator*

# Contents

| | PAGE |
|---|---|
| Foreword.................................................... | iii |
| John E. Naugle, Associate Administrator | |
| Preface..................................................... | vii |
| January.................................................... | 1 |
| February................................................... | 19 |
| March...................................................... | 51 |
| April....................................................... | 75 |
| May........................................................ | 97 |
| June........................................................ | 111 |
| July........................................................ | 129 |
| August..................................................... | 143 |
| September.................................................. | 163 |
| October.................................................... | 181 |
| November.................................................. | 197 |
| December.................................................. | 207 |
| 1974 in Summary.......................................... | 221 |
| Appendix A: Satellites, Space Probes, and Manned Space Flights, 1974.................................................. | 227 |
| Appendix B: Major NASA Launches, 1974................... | 257 |
| Appendix C: Manned Space Flights, 1974................... | 261 |
| Appendix D: NASA Sounding Rocket Launches, 1974......... | 263 |
| Appendix E: X–24B Lifting-Body Flights, 1974............... | 273 |
| Appendix F: Abbreviations of References..................... | 275 |
| Index and List of Abbreviations and Acronyms................ | 279 |
| Errata in Earlier Volumes................................... | 317 |

# Preface

The 14th volume in the NASA series of day-by-day records of aeronautical and space events has somewhat narrowed its scope and selectivity in its brief accounts from immediately available, open sources. This year the emphasis is even more directly focused on concrete air and space activities. More coverage has, perforce, always been given the national space agency's own activities, plans, and achievements, and that is true of this volume, but the text continues to reflect some events in other agencies and countries.

The 1974 Chronology includes fewer examples of public comment and reaction and less technical detail in more condensed summaries. New technological advances in other fields—such as medicine and ground and sea transportation, except when related to NASA's programs—have been dropped, as have most death notices. Air and space highlights of the President's budget are given in tables in the text. The more routine notices of satellite launches, when little information is available beyond the fact of launch, are now given in Appendix A only, not in the text. Most sounding rocket launches and lifting-body test flights also have been dropped from the text; two new appendixes, Appendixes D and E, present the data in table format. Routine missile and nuclear tests, defense contracts, and disarmament items are no longer covered.

Corrections for errors found in previous volumes of this series are carried on Errata pages following the index. This innovation will be continued in later volumes.

The sources, identified by abbreviations that are explained in Appendix F, were those immediately available in NASA and other Government agencies, the Congress, and the professional societies, as well as the press. Contradictory accounts have been resolved and doubtful ones verified whenever possible by querying participants. Cross-references are given in the text, and the detailed index will aid in tracing related events through the year. The index also serves as a glossary of acronyms and abbreviations.

*Astronautics and Aeronautics, 1974* was written by Nancy L. Brun of the NASA History Office, with contributions from other staff members and the History Office Summer Seminar. General editor of the volume was Frank W. Anderson, Jr., Publications Manager of the Office, and technical editor was Carrie E. Karegeannes. Archivist Lee D. Saegesser collected and verified documentation and provided research assistance.

The assistance of many persons throughout NASA and other Federal agencies has contributed to reliability and comprehensiveness of the volume. Comments, additions, and corrections are always welcomed by the NASA History Office.

Monte D. Wright
*Director, NASA History Office*

## January 1974

*1 January:* Appointment of Gerald J. Mossinghoff as Assistant General Counsel of NASA became effective. Mossinghoff had been Deputy Assistant Administrator for Legislative Affairs since 1971, after earlier serving as Director of Legislative Liaison. He would continue to serve as Acting Deputy Assistant until a successor was appointed. (NASA Ann, 3 Jan 74; NASA Release 74–1)

- Josef Boehm, rocket pioneer from Peenemuende, Germany, and Chief of Marshall Space Flight Center's Electromechanical Engineering Div., died at the age of 65. An associate professor of kinematics before joining Dr. Wernher von Braun in 1939 at the Rocket Research Center in Peenemuende, Boehm came to the United States in November 1945 with Dr. von Braun under U.S. contract. He was instrumental in designing and engineering the first U.S. spacecraft for the Army Ballistic Missile Agency, beginning with *Explorer 1* (launched 31 Jan. 1958), and continued with MSFC after it was formed in 1960. He had a major role in developing the Skylab Workshop's Apollo Telescope Mount. (*Marshall Star*, 9 Jan 74, 4; *Huntsville Times*, 1 Jan 74)

*2 January:* Skylab 4 Astronauts Gerald P. Carr, Dr. Edward G. Gibson, and William R. Pogue, launched 16 Nov. 1973, discussed their changed attitudes during an inflight press conference held on their 48th day aboard the Orbital Workshop. Science Pilot Gibson said that from space "you see the Earth as one unit, you see the Sun as a star, and you can see all the other stars out there . . . and just the number of possible combinations . . . which could create life, . . . makes [life elsewhere in the universe] seem very much more likely." Pogue said that the mission had had a great spiritual impact on him. "At the beginning of the mission, I tried to operate like a machine. . . . Now I'm trying to operate as a human being within the limitations I possess." Now he would "feel much more inclined toward humanistic feeling toward other people." (Transcript)

- The Air Force announced award of a $75 318 000 fixed-price-incentive, firm contract to Martin Marietta Corp. for Titan III launch services at Waterton, Colo.; Vandenberg, Calif.; and Cape Canaveral, Fla. (DOD Release 2–74)

*3 January:* Johnson Space Center announced extension of two contracts for program support to the U.S.–U.S.S.R. Apollo-Soyuz Test Project.

A $695 077 cost-plus-fixed-fee contract extension covering 1 Jan. 1974 through 30 Sept. 1975 had been given General Electric Co. to continue support of all Apollo hardware-related programs including automatic checkout equipment.

A $2 424 250 cost-plus-fixed-fee extension for 1 Jan. 1974 through September 1975 had been made in a Boeing Co. contract to provide systems and project engineering support to the ASTP and Skylab program offices. (JSC Release 74–001)

*4 January:* Twenty-seven spacecraft on twenty-six vehicles—eleven for NASA and fifteen cost-reimbursable launches for other organizations—were scheduled for launch by NASA during 1974. For the first time, NASA would launch more spacecraft for other organizations than for itself.

Included in the NASA launches were two Synchronous Meteorological Satellites (SMS) to be launched in cooperation with the National Oceanic and Atmospheric Administration; a proof flight of the Titan-Centaur booster carrying a mass model of Viking spacecraft and a Space Plasma High Voltage Interaction Experiment (SPHINX); Helios-A, a NASA and West German cooperative satellite to study the solar environment; GEOS-C Geodetic Explorer; Nimbus-F experimental weather satellite; San Marco C2, a NASA–Italian cooperative satellite; UK–5, a NASA and United Kingdom cooperative satellite; ATS–F Applications Technology Satellite; a Hawkeye Explorer scientific satellite; and ANS–A, a NASA–Netherlands astronomical satellite.

The 15 NASA launches that would be paid for by domestic and foreign corporations and governments included 2 Skynet communications satellites, the UK–X4 scientific satellite for the United Kingdom, 3 Intelsat and 2 Marisat comsats for Communications Satellite Corp., the first 3 Westar domestic comsats for Western Union Telegraph Co., Aeros-B scientific satellite for West Germany, ITOS–G operational weather satellite and GOES–A (SMS–C) meteorological satellite for NOAA; and Symphonie A experimental communications satellite for a West German and French consortium. (NASA Release 74–2; NASA Gen Mgt Rev Rpts, 14 Jan 74, A66; 11 Feb 74, A64)

- Goddard Space Flight Center launched an Aerobee 200 sounding rocket from White Sands Missile Range to study Comet Kohoutek's ultraviolet rays. The rocket carried cameras, spectrometers, and other instruments to 232.6-km altitude to determine amounts of oxygen, hydrogen, and carbon dioxide in the comet. The flight was "highly successful," Dr. Stephen P. Maran, GSFC Manager of Operation Kohoutek, said.

  The comet, discovered 5 May 1973 by Czechoslovak astronomer Dr. Lubos Kohoutek, was once predicted to be the "comet of the century" as it passed near the earth on its 75 000-yr journey around the sun. It was only twice as bright as nearby stars but reflected enough light to permit instruments to be aimed at it precisely. (Maran, interview, July 74; Watson, *W Post*, 5 Jan 74; Dilts, B *Sun*, 5 Jan 74, A1)

- *Pioneer 10*, launched 2 March 1972, was 26 070 000 km beyond Jupiter after passing within 130 000 km of the giant planet 3 Dec. 1973 on a course that would carry it past Pluto and out of the solar system in 1987. *Pioneer 11*, launched 5 April 1973, had been in flight across the Asteroid Belt since August 1973, on its way toward a Jupiter encounter in December 1974. (KSC Release 2–74)

- The Shuttle Avionic Integration Laboratory (SAIL) was being established at Johnson Space Center to provide a central facility for testing and evaluating avionics, flight software, flight procedures, and ground equipment for the space shuttle. Testing in SAIL would provide information on system operation before each major step in the shuttle flight tests. (JSC *Roundup*, 4 Jan 74)

- The North American Air Defense Command's count of objects in orbit around the earth totaled 624 payloads and 2349 pieces of debris, left from 7038 objects placed in space by man since the beginning of the space age in 1957. Space debris was falling back into the atmosphere at a rate of one piece a day, but NORAD's tracking and impact prediction program analyzed the decay trajectory of each satellite that might survive reentry and predicted the time and area of impact, issuing warnings if any were deemed hazardous. Most returning objects were burned up by friction heat on reentry and most of the rest fell into oceans and seas. To date, the only fatality caused by space debris was a cow killed nearly 13 yrs ago when an 88-kg chunk of metal from a U.S. space probe struck the earth near a Cuban village.

Eventually, however, increasing space debris and reentries might require traffic control assistance for supersonic aircraft and spacecraft. "Short of sending up some sort of a celestial vacuum cleaner . . . the best approach . . . appears to be tracking and impact prediction," NORAD spokesmen said. (Miles, *LA Times*, 7 Jan 74)

- Marquis Childs commented on *Pioneer 10*—which had passed within 130 000 km of the planet Jupiter on 3 Dec. 1973—in the Baltimore *Sun:* "The scientific brains that have gone into the Pioneer project are one of America's greatest resources. They underwrite the technological-scientific lead that is perhaps this country's greatest asset. The momentum must be maintained if our rivals—the Russians, the Germans, the Japanese—are not to pass us." (B *Sun*, 4 Jan 74)

*6 January:* The Aerospace Industries Assn. of America, Inc., predicted "a slight decline" in sales in 1974 after a jump of more than 10% in 1973, the *New York Times* reported. Industry sales were expected to total $24.7 billion in 1974, 1% lower than the 1973 boom that had pulled the industry out of a five-year slump. AIA President Karl G. Harr, Jr., had said, however, that the energy crisis could bring a greater decline. Fuel shortages already had softened markets for commercial aircraft, and jet transport orders could taper off, although the Department of Defense was beginning a new generation of tactical and strategic weapons.

Aerospace industry business with the space program was expected to continue its five-year decline, dropping by about 2.7% to sales of $2.3 billion in 1974.

Profits of the industry had risen in 1973, to an average of 2.7% of sales after taxes, but employment dropped 1% to 935 000 persons in 1973, and was expected to fall to 915 000 in 1974. (Finney, *NYT*, 6 Jan 74)

*7 January:* A 5-million-km hydrogen cloud surrounding Comet Kohoutek was photographed by a Naval Research Laboratory camera aboard a NASA Aerobee 200 sounding rocket. The rocket, launched from White Sands Missile Range, reached an altitude of 193.1 km. The image of the cloud, invisible from the earth because it glowed at an ultraviolet wavelength that did not penetrate the atmosphere, was obtained by photographing the comet in the light of the Lyman-alpha line of atomic hydrogen using a special electronic UV camera. Scientists believed the comet consisted largely of water ice which evaporated as it neared the sun, breaking up into component hydrogen and oxygen atoms. The lighter hydrogen atoms escaped rapidly, producing the cloud. (*Naval Research Reviews*, Jan 74, 29)

- Communications Satellite Corp., on behalf of itself and other joint owners—including American Telephone & Telegraph Co., ITT World Communications, Inc., RCA Global Communications, Inc., and Western Union International Corp.—announced intention to award a $3.16-million contract to E-Systems Inc. for construction of two 32-m-dia antennas for international satellite communications on the existing earth station sites of Andover, Me., and Etam, W. Va. (ComSatCorp Release 74-2)

*8 January:* The first domestic communications satellite service, connecting East and West Coasts, was officially inaugurated by RCA Corp. from New York City at 2:40 pm EDT. Within minutes, conversations were relayed by Canada's *Anik 2* satellite (launched by NASA 20 April 1973) on a circuit leased from Telesat Canada. Sen. Jacob K. Javits (R-N.Y.) in New York, Federal Communications Chairman Dean Burch in Washington, D.C., Alaska Gov. William A. Egan in Juneau, Sen. Mike Gravel (D-Alaska) in Anchorage, and Sen. Ted Stevens (R-Alaska) in Nome exchanged congratulations.

Operated by RCA Global Communications, Inc., and RCA Alaska Communications, Inc., for the parent RCA Corp., the system used earth relay stations near New York, San Francisco, Juneau, and Anchorage. These were to be joined by others in 1976, when RCA hoped to have three satellites of its own. Transmission charges for coast-to-coast messages were less via satellite than by earthbound systems; a telephone circuit rented for $1700 a month rather than the $2298 charged by American Telephone & Telegraph Co. for a similar land-based private circuit. Quality also was better because a connection could be completed with fewer relay points, one relay to the satellite and another from the satellite to the ground. (RCA Release, 8 Jan 74; *W Star-News,* 9 Jan 74; *WSJ,* 9 Jan 74, 34)

• The Library of Congress Congressional Research Service published *United States and Soviet Progress in Space: Summary Data through 1972 and a Forward Look* (74-35SP), prepared by Dr. Charles S. Sheldon II, Chief of the Science Policy Research Div. In its manned program the U.S.S.R. had launched a *Salyut 2* space station 3 April 1973 but a failure—possibly a wildly firing thruster—had damaged the station, which reentered the earth's atmosphere May 28. Other later attempts to launch and man a space station also appeared to have failed. Two manned launches, *Soyuz 12* and *Soyuz 13,* had been successful in September and December, the second spacecraft remaining in space eight days to conduct astrophysical, biological, and earth resources experiments.

By 30 June 1974 the United States expected to have spent $75 billion on combined military and civilian space programs since the beginning of the space program, spending about $4.7 billion in FY 1974. The Soviet Union did not publish space budget data, but its program was thought to be "at least of the same magnitude as that of the United States at its former peak" of $7.7 billion a year.

The U.S. had several major space research and development centers operated by the military and private industry as well as NASA. There was no reason to believe the Soviet aerospace industry was as fully equipped but missiles of similar capabilities and different design philosophies which had appeared in Moscow parades "suggest the existence of more than one design and development team for space work."

The U.S. had launch vehicles scaled to lift payloads weighing from 9 kg to 136 000 kg. The Soviet Union had started to use in 1957 its original intercontinental ballistic missile and still used the same basic vehicle with stages added to improve performance. The very large Soviet vehicle predicted by NASA officials had not been seen.

A Soviet lunar landing did not seem imminent but was still expected, as part of Soviet long-range plans, as soon as the problems of unreliable hardware were solved. U.S.–U.S.S.R. cooperation in space efforts was "the hope of well-intentioned people everywhere." Two motives for cooperation were to lessen tension and to save money. Although there was already exchange of information between scientists and engineers, "no assessment of the overall prospects can be made successfully without forecasting the future political climate." (Text)

*8–31 January:* NASA's *Mariner 10* probe, launched 3 Nov. 1973, met a number of problems but continued its flight millions of kilometers from the earth toward a 5 Feb. flyby of Venus and 29 March flyby of Mercury.

Despite the loss of the prime power chain, causing automatic transfer to the backup system, *Mariner 10*'s instruments continued scientific studies. Comet Kohoutek's tail was observed in the extreme ultraviolet range. On 13 Jan. hydrogen Lyman-alpha intensity, beginning 20° from the comet's nucleus, climbed rapidly as the nucleus region drifted toward the UV airglow spectrometer's field of view. The nucleus was scanned 16 Jan.

On 17 Jan., while the spacecraft was powered down for its second trajectory correction maneuver, shutting off several instrument heaters allowed the TV optics heaters, which had failed to function earlier, to turn on. Scientists had feared that below-freezing temperature might prevent camera operation during the Venus encounter. On 21 Jan., instruments for Comet Kohoutek and solar wind experiments were turned on.

On 22 Jan. the UV spectrometer measured hydrogen Lyman-alpha intensities out to 17° to map the comet's hydrogen cloud. Comet Kohoutek was scanned seven hours 24 Jan. by the UV spectrometer—the only instrument of its kind to observe the comet from outside the earth's own hydrogen cloud.

During the seventh and final roll calibration maneuver 28 Jan., telemetry indicated a rapid increase in the use of attitude control gas, the nitrogen gas supply dropping from 2.7 to 2.1 kg. The gyros were turned off and the spacecraft put into celestial control, with the sun and the star Canopus as references. Nitrogen-gas-use rate returned to normal and estimates of the supply indicated a comfortable margin for reaching Mercury and executing experiments. The backup electric power system was still in use, as studies of that problem continued.

On 31 Jan. *Mariner 10* was 3 200 000 km from Venus and 38 800 000 km from the earth, traveling at 30 238 km per hr. (JPL Mariner Venus/Mercury 1973 Bull Nos 14, 15; NASA PAO press briefing transcript, 31 Jan 74; *W Post*, 10 Jan 74, A5)

*9 January:* Lockheed Propulsion Co. asked the General Accounting Office to set aside NASA's 20 Nov. 1973 selection of Thiokol Chemical Corp. for negotiation of a $106-million contract to design and develop the space shuttle's solid-fueled rocket motor. Lockheed charged that it had an edge in technical evaluation of the engine and that it had been the low bidder before NASA had made "improper adjustments" in its bid. Lockheed also maintained Thiokol's transportation costs from Utah to Florida would be higher and that Thiokol's proposed design would not meet program objectives without major revision. (LATNS, *Today*, 10 Jan 74; NASA Release 74–130)

- A photograph of the U.S.S.R.'s space launching complex at Baykonur Cosmodrome taken by NASA's *Erts 1* Earth Resources Technology Satellite (launched 23 July 1972) from 900 km altitude was published in the Washington newsletter *Defense/Space Business Daily*. The photo, taken 16 March 1973, showed that the complex stretched 60 km east and west, 24 km north of Tyuratam. The newsletter published a closer, 5 Oct. 1972 photo on 17 Jan. (*SBD*, 9, 10, 17 Jan 74, 41–42, 49–50, 89; NASA photos 74-H-6, 74-H-11, 74-H-12, 74-H-13)

*10 January:* All spacecraft and experiment checks on *Explorer 51* (Atmosphere Explorer-C), launched 15 Dec. 1973, were completed. Spacecraft systems and all 14 experiments were operational. Spacecraft performance was nominal except for Programmer 2, which had experienced some degraded load capability. The orbital-adjust propulsion system had lowered the perigee to 152 km in two stages. The first "excursion" to a lower altitude—125 km—was planned for February. (NASA MOR, 10 Jan 74)

- Ames Research Center's C-141 AIRO (Airborne Infrared Observatory) aircraft was officially transferred to operational status for research development. A total of 26 research programs had been approved for C-141 participation, using its 91-cm IR telescope—18 from universities, 5 from NASA or other Government agencies, and 3 from foreign countries. AIRO also was to participate in studies of Comet Kohoutek. (ARC *Astrogram*, 18 Jan 74, 1,4)

*11 January:* Solar energy would be used by Marshall Space Flight Center engineers to heat and cool three house trailers parked together to simulate a residence. The system would take from the sun three fourths of the thermal energy needed for the simulated living quarters and furnish 2.7 metric tons of air conditioning in the summer and 20.5 kw of heat during winter, with surplus energy in storage for operation at night and through three cloudy days. A conventional system would still be necessary to make up for undercapacity and provide energy during extended cloudy periods.

The system, a project sponsored by the NASA Office of Applications to develop solar energy technology, would use a 120-sq-m solar-collector roof, a storage tank, pumps, coils, a modified air conditioner, plumbing, an auxiliary heater, and automatic controls. (MSFC Release 74-6)

- The National Science Foundation was spending more than $1 million in FY 1974 on wind research programs and to build a large windmill, the *Wall Street Journal* reported. Windmills, for thousands of years one of man's basic machines, were again being taken very seriously in the worldwide energy crisis. Researchers believed windmills could be grouped to form power stations and provide a significant portion of U.S. energy needs, becoming competitive with other energy sources. The $200 000 NSF windmill would be built by NASA at Lewis Research Center's Plum Brook Station to capture the maximum possible power from the wind with the minimum construction costs. In addition, the NASA team would look into methods of storing electricity for periods when the wind was not blowing and of generating current at a constant 60 cps, the rate all large appliances were designed to use. Other researchers were working on windmill complexes based at sea, where powerful ocean winds could provide a regular flow of power. (Brand, *WSJ*, 11 Jan 74, 1; NASA Release 74-33)

- NASA's shrinking work force and the relatively few available women and minority engineers and scientists were responsible for the agency's admittedly poor record on equal employment, Dr. Dudley G. McConnell,

NASA Assistant Administrator for Equal Opportunity Programs, told the Senate Committee on Appropriations' Subcommittee on Housing and Urban Development, Space, Science, and Veterans. The hearing was investigating the October 1973 dismissal of a NASA official, Mrs. Ruth Bates Harris, after she and two coauthors asserted in a report that the agency was not moving fast enough to hire women and minority workers and that NASA's proportion of minority employees was the lowest in the Government. Despite the difficulties, NASA in the past had hired well above the agency's current percentage of minorities: 10.3% in 1971; 13.2% in 1972; and 16.3% in 1973.

Associate Deputy Administrator Willis H. Shapley denied the report was the reason for the Harris termination but conceded NASA's poor record on minority hiring. Subcommittee Chairman William Proxmire (D–Wis.) ordered the agency to report to him quarterly on progress toward its own equal opportunity goals. (Transcript)

- The Air Force's first flight-test B–1 strategic bomber had been successfully mated with its wings in Palmdale, Calif., the Air Force Systems Command announced. The B–1 was being developed by prime contractor Rockwell International Corp. to provide a modern supersonic strategic manned bomber to replace the aging B–52. The B–1, designed to carry twice the payload of the B–52 at three times the speed, was equipped with movable fore and aft wings to carry out a variety of missions from both normal and short runways, in addition to low-level operation at high subsonic speeds and high-altitude supersonic flight. (AFSC Release OIP 001.74)

- The U.S. would sell fighter aircraft to Iran and Saudi Arabia, the *Washington Post* reported. Iran would buy 30 Grumman Aerospace Corp. F–14 Tomcat swing-wing fighters equipped with the advanced, long-range Phoenix missile system, with deliveries to begin January 1976. Saudi Arabia would buy 30 new Northrop Corp. F–5E Tiger II lightweight fighters and 20 of the earlier two-place F–5B trainers. (Miles, *W Post*, 11 Jan 74, A12)

*14 January:* *Skylab 4* Astronauts Gerald P. Carr, Dr. Edward G. Gibson, and William R. Pogue (launched 16 Nov. 1973 in the final mission to crew the Orbital Workshop) set a new record for the longest manned flight time in space, completing 59 days 11 hrs 9 min at 10:10 pm EDT and surpassing the time set by *Skylab 3* on its 28 July–25 Sept. 1973 mission. *Skylab 4* had been given a go-ahead 10 Jan. for a seven-day extension, with weekly reviews through the end of a planned 84-day mission. (*A&A 73*)

- Three U.S.–U.S.S.R. Apollo-Soyuz Test Project working groups began technical meetings at Johnson Space Center on mission plans and experiments, communications and tracking, and life support and crew transfer. Thirty-five Soviet scientists and engineers would work at JSC for periods from 16 to 90 days in preparation for the joint, manned, earth-orbital mission in July 1975 to test compatible rendezvous and docking systems and techniques. (NASA Release 74–9; NASA OMSF, interview, 9 July 74)

*14–18 January:* Final reports on six space tug system studies were presented to NASA at Marshall Space Flight Center by contractors Grumman Aircraft Corp., Martin Marietta Co., McDonnell Douglas Astronautics Co., General Dynamics Corp., and Lockheed Corp. The reports—covering a storable propellant version, a cryogenics propellant version, and a growth stage version—would become the basis for the completion of Phase A studies of the space tug program and recommendations for Phase B definition studies. (*Marshall Star*, 16 Jan 74)

*15 January:* A new, portable, battery-powered spotlight, with a peak capability of 1-million candlepower and 50 times brighter than high-beam automobile headlights, had been developed from NASA arc light research, NASA announced. The 3-kg spotlight, called the Stream Lite-1 Million and marketed by Streamlight, Inc., used a unique xenon lamp with an operating lifetime of 200 hrs at maximum intensity. (NASA Release 74-6)

*16 January:* Plans for an Arab communications satellite network were presented by Arab League engineer Salah 'Amir during a meeting with League Secretary General Mahmud Riyad in Cairo. The plans—prepared with the assistance of the United Nations Educational, Scientific, and Cultural Organization (UNESCO) and the International Communications Union—had been completed in 1972. The network would establish radio, TV, and news agency links among the Arab states to transmit standard educational, cultural, and development programs. Fifty ground stations would be established with the costs of each—between $500 000 and $1 000 000—defrayed by the country of its location. The costs of the satellites and launch facilities would be paid by member states. (Cairo MENA, FBIS–Inter-Arab, 17 Jan 74, A6)

- The Air Force announced the award of a $1 553 500 cost-plus-incentive-fee contract to Avco Corp. for design, development, and testing of a space reentry vehicle for the reentry vehicle nosetip test program. (DOD Release 18–74)

- A 12-story addition to the RCA Corp. building in Rockefeller Center in New York City would be constructed to use solar energy heat, RCA announced. The $6-million addition, a management conference center, would use panels on its solid exterior surfaces to capture solar energy in the form of radiant heat, which would be stored and substituted for other forms of energy. (RCA Release, 16 Jan 74)

- A *Huntsville Times* editorial commented on the "lengthening space-flight endurance record" that *Skylab 4* (launched 16 Nov. 1973 to man the Orbital Workshop) "sets with each successive turn it now makes around the old planet," with a February target splashdown date. "America's heart doubtlessly goes out to them, for that's an awfully long time for anyone to have to be away from earth's fuel shortages, inflation, rainy weather, dishwater-dull Super Bowls, Washington politics, air pollution, and traffic bottlenecks." (*Huntsville Times*, 16 Jan 74)

*17 January:* A NASA plan for energy conservation and management, effective immediately, was outlined in a memo from Dr. George M. Low, NASA Deputy Administrator, to heads of all program and staff offices and Center Directors. NASA expected to exceed by a wide margin the nationwide goal of a 5% reduction in energy consumption for the following 12-mo period as established in a 29 June 1973 Presidential Memorandum. Management actions included scheduling all energy resources, using fewer cars and aircraft, encouraging employee carpools, and examining alternative power sources in large energy consumers such as wind tunnels. NASA would also play a key role in the research, development, and technological application for the longer term solution of energy requirements and efficiencies. A Special Task Force on Energy Conservation, activated 10 Dec. 1973, would review and recommend policies and guidelines on energy conservation, consider NASA's response to external energy directives, develop suggestions for conserving energy, and review NASA's progress in conservation. (Text)

- Johnson Space Center had issued requests to 11 interested aerospace companies to submit contract proposals for computing and data processing for space program software systems, JSC announced. Primary objective of the contract would be to define, design, develop, and maintain software systems (for space shuttle, earth resources, and other science programs) that had maximum flexibility for supporting a wide variety of missions and future requirements. The cost-plus-award-fee contract would run from 1 July 1974 through 31 Dec. 1979. (JSC Release 74–30)
- A $70-million to $100-million cutback in development of the complex $4.4-billion surface-to-air (SAM–D) air defense missile system had been ordered by Dr. James R. Schlesinger, Secretary of Defense, the *Washington Post* reported. The cutback had been ordered to speed efforts to develop a lower cost, highly mobile SAM missile in addition to the production of an already developed foreign-designed missile effective against low-flying aircraft over a battlefield. (Getler, *W Post*, 17 Jan 74, A3)
- President Nixon issued Executive Order 11760 designating the European Space Research Organization (ESRO) a public international organization entitled to privileges, exemptions, and immunities provided for by the International Organizations Immunities Act. (*PD*, 21 Jan 74, 40)

*18–27 January:* NASA launched *Skynet IIA*, first of two second-generation United Kingdom communications satellites on the 100th Thor-Delta—a three-stage, thrust-augmented booster—from Kennedy Space Center Launch Complex 17 at 9:38 pm EDT. The satellite was placed in an extremely eccentric orbit when the attitude control system of the rocket's 2nd stage failed. The apogee was 3406 km, perigee 95 km, period 121.6 min., and inclination 37.6°. Tracking stations were unable to pick up signals from the spacecraft.

On 23 Jan. the satellite was found by the Air Force Satellite Control Facility and attempts were made 24 Jan. to raise the orbit by firing the spacecraft's onboard thrusters. Although the firing was succcessful, the orbital attitude was not raised enough to extend the life of the satellite. *Skynet IIA* reentered the earth's atmosphere and burned up 25 Jan. over the southwestern Pacific Ocean.

The Skynet II communications satellite program was being carried out under a U.K.–U.S. agreement signed 1 April 1970 and a USAF–NASA agreement 1 May 1972 to provide the U.K. in-orbit, X-band military communications between designated earth stations and replace *Skynet A* and *B*, launched by NASA 21 Nov. 1969 and 18 Aug. 1970. NASA responsibilities included integration of the U.K.-built spacecraft with the launch vehicle and provision of the vehicle and related services including network tracking communications and telemetry during ascent and transfer orbit phase. Another Skynet II satellite was to be launched in the fall. (NASA MORs, 8 Jan 74, 4 Sept 74; AP, *NYT*, 30 Jan 74; AP, *W Star-News*, 19 Jan 74, A4)

*18 January:* First evidence of water molecules in a comet had been identified in the tail of Comet Kohoutek by Canadian scientists Dr. Gerhard Herzberg and Dr. Hin Lew, both of Canada's National Research Council, NASA announced. Dr. Herzberg and Dr. Lew identified the molecules by analyzing emissions of light at five wavelengths in the red region of the comet's spectrum, where the positively charged molecules, ionized by solar radiation, were pushed by the solar wind. Data had been collected by telescopes at Asiago Astrophysical Observatory in Italy and Univ. of California's Lick Observatory in October and November 1973.

The discovery of hydrogen in the comet supported the "dirty snowball" theory that comets might be composed of an icy nucleus, various other frozen gases, and tiny dust particles as proposed by Dr. Fred L. Whipple, retired Director of the Smithsonian Astrophysical Observatory and principal consultant to NASA on comets. (NASA Release 73 [74]–13)

- Hazard analyses and quality control techniques developed by NASA in the Apollo program would improve the reliability of safety and antipollution equipment used in the offshore drilling of oil wells, NASA announced. At the request of the U.S. Geological Survey, NASA teams at Marshall Space Flight Center, the Mississippi Test Facility, and the Michoud Assembly Facility had recommended development of subsurface safety valves, activated either automatically after sensing increased velocity in the oil line or by remote control from a surface location, to protect the marine and coastal environment. (NASA Release 74–10)
- Development of a two-man version of the F–5E International Fighter had been initiated by the Air Force, the Air Force Systems Command announced. Designated the F–5F and to be built by Northrop Corp., the two-seater would have the same performance and weapon capability as the single-seat F–5E but was to provide the capability of training fighter pilots in all aspects of fighter aircraft operations. The first test aircraft was expected to fly in September 1974. (AFSC Release OIP 002.74)

*21 January:* An all-time record number of helicopters—413—had been exported by U.S. manufacturers in 1973, Aerospace Industries Association of America, Inc., announced. The exports totaled $84 775 500, up from $73 797 500 in 1972, when 260 helicopters had been exported. (AIA Release 74–1)

- Lockheed Aircraft Corp. would need a maximum of $75 million to overcome the threat that it would run out of cash, the New York Times News Service quoted Lockheed's Chairman of the Board Daniel J. Haughton as saying in a London news conference. Haughton had expressed confidence that Lockheed's bankers would provide the necessary funds. In addition he had expected that a long-term solution—possibly a merger—would be found by June. Lockheed's financial problems had begun with severe technical setbacks during development of the L–1011 TriStar airliner in 1971. Disaster had been averted by congressional approval of a Government guarantee of up to $250 million in loans. (*W Star-News*, 21 Jan 74, A13)

*22 January:* The U.S.S.R. proposed that the United Nations establish a center for collection and distribution of data from earth-orbiting satellites. The proposal was made in response to a U.S. offer to supply the U.N. with a master copy of all information obtained from *Erts 1* Earth Resources Technology Satellite, launched 23 July 1972, if a facility would be available to process the data. The Soviet proposal suggested a single international center, under the U.N. or another international organization and financed by voluntary contributions. The U.S.S.R. previously had been unwilling to bring in any international authority. U.S. space experts later suggested the new proposal was a significant development in closer Soviet-U.S. cooperation, reached by the scientific exchanges between the two countries. Another view was that the U.S.S.R. was trying to head off a possibility that the U.S. might turn over its expertise to private enterprise.

Brazil had earlier proposed a treaty that would obligate space powers to obtain consent of governments for remote-sensing activities over their countries and for release of the data obtained. (UN Doc A/AC.105/C.1/WG.4/L.6/Add 3; *NYT*, 10 Feb 74, 20)

*23 January:* President Nixon transmitted a message to Congress outlining legislative proposals and executive actions to deal with the energy crisis. Proposals to meet the short-term emergency included a special act to permit restrictions on public consumption of energy and temporarily relax pollution control requirements for power plants and automotive emissions, a windfall profits tax to prevent private profiteering at public expense, unemployment insurance to help those who lost their jobs because of the energy crisis, and establishment of a Federal Energy Administration. To meet the long-range goal of energy self-sufficiency, proposals would allow market pricing of new natural gas, permit surface mining of coal in an environmentally safe manner, permit development of deepwater port facilities off shore, eliminate depletion allowances for foreign oil and gas production, and accelerate licensing of construction of nuclear facilities.

The President concluded that, although shortages had been long in appearing, the oil embargo—imposed on the U.S. by the Arab oil producing countries because of American support of Israel during the October 1973 war—"opened our eyes to the short-sighted policy we had been pursuing." The U.S. must never again allow itself to be overly dependent on foreign supplies of a vital good. "By 1980 . . . I believe we can . . . be essentially independent of foreign energy producers." (*PD*, 28 Jan 74, 72–87)

- The Air Force announced the award of a $1 348 414 contract to General Electric Co. for the design, development, and testing of a space reentry vehicle for the reentry vehicle nosetip test program. (DOD Release 31–74)

*24 January:* Ames Research Center scientists had found more evidence that life on primitive earth might have been triggered by chemical evolution of nonliving matter, NASA announced. Seventeen varieties of fatty acids, similar to those used by plants and animals to produce more complex biological molecules, had been discovered in two meteorites. The family of simple carbon-hydrogen acid molecules found in the meteorites had counterparts in biologically formed earth materials, but had not been found in nonorganic sources before. (NASA Release 74–16)

- NASA's equal employment opportunity program was reviewed during a hearing before the Senate Committee on Aeronautical and Space Sciences. Dr. George M. Low, NASA Deputy Administrator, said that during the 1970s NASA faced the challenge of catching up in a most important area of human need. The challenge would be difficult because the NASA work force had decreased by nearly one third. In addition, 47% of the NASA work force were scientists and engineers. Of these, 3.4% were minorities, while the nationwide percentage of minorities in this category was 3.5%, and these persons were in high demand in private industry, where starting salaries were $1200 to $1700 per year higher.

Dr. Dudley G. McConnell, NASA Assistant Administrator for Equal Opportunity Programs, said NASA would during 1974 achieve a minority complement of 6.1% of the total permanent work force, hiring 80 minority members and 80 women into professional positions. In addition NASA would help increase the availability of technically trained women and minority members by NASA's cooperative education program and by initiating an aerospace fellowship program.

Mrs. Ruth Bates Harris, former NASA Deputy Assistant Administrator for Equal Opportunity Programs, testified that NASA had turned away qualified women and minority applicants. She said NASA ranked 22 out of 26 agencies in percentage of female attorneys. Also, although NASA had a 47% scientific and engineering work force, in the 53% nonscientific and technical category NASA had a 3.7% minority participation, not significantly better than the 3.6% in the scientific and engineering category. (Transcript)

- A space shuttle launch processing system (LPS) was being developed at Kennedy Space Center to provide a flexible, reliable, and cost-effective method of testing systems, controlling launch operations, and monitoring status of vehicle, ground support equipment and facilities during ground operations. A single automatic system, replacing the multiplicity of systems used in previous programs, would support every technical and management activity required to process the launch vehicles, their payloads, and all ground support equipment. (KSC *Spaceport News*, 24 Jan 74, 4–5)

- Jet aircraft pollution would be measured under a $1.1-million NASA contract awarded to United Air Lines, Inc., NASA announced. Under the cost-plus-fixed-fee contract, United would design, develop, test, and fly a Boeing 747—equipped to measure dust particles and gases—to determine the effects of jet exhaust on the natural environment, weather, and human health and to measure changes in the level of the ozone shielding people on the earth from solar ultraviolet radiation. Flights would be made over the continental U.S. and Hawaii in addition to international flights by another 747 under a subcontract with Pan American World Airways, Inc. The instruments were expected to be installed by the end of 1975. NASA would work closely with the National Oceanic and Atmospheric Administration to correlate study results. (NASA Release 74-20)

- The Anglo-French Concorde supersonic transport would not go into transatlantic service until 1975 and would not fly commercially until 1976, French Minister of the Armed Forces Robert Galley said in Paris. The press reported sources in the French aviation industry as saying that increased fuel prices and possible shortages were responsible for the delay. (UPI, *NYT*, 25 Jan 74, 53)

- A new Air Force air-traffic-control radar system that could track aircraft approaching a runway in heavy rain would be put into full production, the Air Force Systems Command announced. The AN/TPN-19, to be produced by Raytheon Co., consisted of an operational shelter, an airport surveillance radar to detect aircraft within 97 km and direct them to final approach entry positions, and a precision-approach radar to guide aircraft along the 32-km final approach. The all-weather, $2 617 000 system could be transported anywhere in the world and be operational within two hours. (AFSC Release OIP 181.73)

*25 January:* Results from NASA's *Pioneer 10* mission (launched 2 March 1972 toward a 3 Dec. 1973 encounter with Jupiter) were published in a series of articles in *Science*. Plasma analyzer data had indicated that Jupiter had a detached bow shock and magnetopause similar to the earth's but with a much larger spatial extent and with the size highly responsive to changes in solar wind dynamic pressure. The thermal structure of Jupiter was found to be closely related to the visual appearance of the planet, with illuminated and nonilluminated sides having the same brightness temperatures. The net thermal energy of the planet was twice the solar energy input.

Magnetic field measurements showed a very strong magnetic field and implied the dipole was inclined at about 15° to Jupiter's axis of rotation. The satellite Io was found to have an ionosphere extending from near the surface to about 1000 km. A new density of 3.5 g per cc—close to the density of the moon and Mars—was established for Io. Densities of the other satellites of Jupiter were progressively lower in proportion to their distance from the planet.

More comprehensive reports of *Pioneer 10* experiments were published in the *Journal of Geophysical Research* of 1 Sept. (Opp, Wolfe, et al., Smith et al., Kliore et al., *Science*, 25 Jan 74, 302–324; *Journ Geophysical Research*, *Pioneer 10* issue, 1 Sept 74)

- A *Science* editorial commented on *Pioneer 10*'s 3 Dec. 1973 flyby of Jupiter: "One of the impressive features of the *Pioneer 10* mission was the performance of the spacecraft and its scientific equipment. . . . The craft, with its transmitter of only eight watts, has been storing and sending tremendous quantities of data to Earth. Even after their radiation exposure during flyby, the electronics components continue to function well as *Pioneer 10* proceeds on its way out of the solar system." (Abelson, *Science*, 25 Jan 74, 261)

- A new medical device, the mobile automatic metabolic analyzer (MAMA) developed from Apollo and Skylab technology to measure energy expended by ambulatory patients, was presented to the Univ. of Alabama Medical School by Dr. Rocco A. Petrone, Director of Marshall Space Flight Center. The instrument—primarily a mass spectrometer—would accurately measure metabolic activity, recording oxygen consumed, carbon dioxide produced, and nitrogen and moisture exchanged to gauge the progress of severely disabled persons through phases of rehabilitation. MAMA was developed and manufactured at MSFC under NASA's technology utilization program. (NASA Release 74–18)

- Fraudulent transmissions to Johnson Space Center purporting to be from *Skylab 4* astronauts, working in the Orbital Workshop since their 16 Nov. 1973 launch, were heard during a telephone call to New York City from Unifi, Inc., in North Carolina. The supposed astronauts described "an approximately 10-megaton explosion" on the spacecraft as they took military photos of silos over Moscow. After the "astronauts" said they were sending scrambled messages "on channel 5 and channel B," they were heard supposedly speaking to President Nixon at the White House. They acknowledged that their wives had been notified and were being flown to Houston. The voices then reported that their "secret documents and equipment" had been jettisoned. The transmissions then abruptly ended. The press later reported that 12 other persons had heard similar "space messages." A Houston area telephone company employee was later taken into custody. (Anderson et al., *W Post*, 19 Feb 74, B17; JSC Hist Off, interview, 9 Dec 75)

- Julian Nott and Felix Pole broke the world altitude record for a hot-air balloon larger than 4000 cu m, ascending to 13 961 m in the 10 620-cu-m *Daffodil II*. The crew—which lifted off from Bhopal, India—was protected by a pressurized cabin and special suits from the United Kingdom Royal Air Force. (NAA *News*, Aug 74; NAA Record Book)

- A. Ernest Fitzgerald, a financial analyst who had been fired from the Air Force after he exposed a $2-billion cost overrun in the development of the C–5A jet transport, filed a $3.5-million damage suit against Dept. of Defense and Air Force officials. The Civil Service Commission had ordered the Air Force to rehire Fitzgerald 19 Sept. 1973. (*W Star-News*, 26 Jan 74, A3)
- A Soviet civil aviation magazine, apparently alarmed at an estimated 10 crashes of Soviet-built aircraft in 16 mos, with a death toll of about 600 persons, had implored aviation workers to be more attentive to their work, United Press International reported. (*NYT*, 25 Jan 74, 47)

*25–26 January:* The U.S.S.R. conducted the first long-range tests of its new SS–19 intercontinental missile with multiple independently targetable reentry vehicle (MIRV) warheads. Two missiles were test-fired 7200 km from the Soviet missile research center at Tyuratam to a target area in the Pacific Ocean 1370 km northwest of Midway Island. Chief Pentagon spokesman Jerry W. Friedheim later described the tests to the press as a "significant milestone" in the Soviet program to develop MIRVs but said that the U.S.S.R. would not have a missile force with MIRV warheads in operation before 1976. (Finney, *NYT*, 29 Jan 74, 1)

*27 January:* Hills 1000 m high on the planet Mercury, as well as valleys 700 m deep and craterlike surface features 50 km across, had been found by Jet Propulsion Laboratory radar astronomers during high-resolution scans, NASA announced. The features were charted by directing high-powered beams at Mercury in 14 probes between 15 July and 8 Dec. 1972 from Goldstone Deep Space Station's 64-m dish antenna. Months of computer work had analyzed the data and formulated a map of radar findings, which might be verified 29 March when *Mariner 10* (launched 3 Nov. 1973) was scheduled to fly past and take the first closeup photos of Mercury. (NASA Release 74–12)

*28 January:* Three Intelsat IV communications satellites transmitted the Muhammad Ali–Joe Frazier boxing match in New York City to 17 countries. The three-hour match, won by Ali by a decision, was transmitted to Argentina, Australia, Barbados, Brazil, Chile, Colombia, Jamaica, Japan, Korea, Mexico, Nicaragua, Philippines, Puerto Rico, Thailand, United Kingdom, Venezuela, and Zaire. (ComSatCorp Release 74–6; *W Star-News* Library)

*28 January–1 February:* The American Institute of Aeronautics and Astronautics held its 10th Annual Meeting and Technical Display 28–30 Jan. and 12th Aerospace Sciences Meeting 30 Jan.–1 Feb. in Washington, D.C. "Spaceship Earth—A New Perspective" was the theme of the meeting, chaired by Dr. Wernher von Braun, Vice President of Fairchild Industries, Inc., and former Marshall Space Flight Center Director and NASA Deputy Associate Administrator for Planning.

At the honors banquet 30 Jan. Dr. Kurt H. Debus, Director of Kennedy Space Center, received the Louis W. Hill Space Transportation Award for 1973 for scientific, engineering, and organizational contributions to manned and unmanned space missions, beginning with the first U.S. satellite, *Explorer 1*, in 1958. Dr. John H. Wolfe, Chief of Ames Research Center's Space Physics Branch, was given the AIAA Space Science Award for 1973 for his leading role in the Pioneer missions and his contribution to man's understanding of interplanetary space.

The Goddard Award was shared by Paul D. Castenholtz, Rockwell International Corp.; Richard C. Mulready, United Aircraft Corp.; and John L. Sloop, former NASA Assistant Associate Administrator for Advanced Research and Technology—for significant contributions to the development of the practical liquid oxygen and hydrogen rocket engine.

Winner of the AIAA History Manuscript Award for 1973 was Dr. William M. Leary, Jr., Univ. of Georgia historian, for "The Dragon's Wings: The China National Aviation Corporation and the Development of Commercial Aviation in China." Carroll H. Woodling, Chief of Johnson Space Center Crew Training and Simulation Div., received the de Florez Training Award for advancing development of space flight simulators and for flight training programs.

The G. Edward Pendray Award was awarded to Frederick I. Ordway III, Univ. of Alabama, for his contributions to the literature recording the history and benefits of the space program. Willis H. Hawkins, Lockheed Aircraft Corp., was awarded the Sylvanus Albert Reed Award for contributions to the design and development of aerospace vehicles—including vertical take-off, transport, interceptor, and supersonic fighter aircraft—and to missile and space programs including ramjet and reentry test vehicles. The Lawrence Sperry Award went to Dino A. Lorenzini, Air Force Academy, for developing a modern inertial guidance test facility incorporating computer-control, data-acquisition, sensor-excitation, and seismic-isolation technology.

Dr. Harold A. Rosen, Hughes Aircraft Co., received the Spacecraft Design Award for contributions to the development of satellite communication systems, including the spin-stabilized synchronous-orbit spacecraft concept and commercial systems.

Dr. Alan M. Lovelace of the Air Force Systems Command delivered the von Kármán Lecture, "Advanced Composites," pointing to a need for competitive cost and innovative design to realize the advantages of composites in material, fabrication, and configuration. The Dryden Research Lecture was delivered by Herbert F. Hardrath, Langley Research Center, who reviewed the use of fracture mechanics to choose materials, configure safe and efficient structures, specify inspection procedures, predict lives of flawed structures, and develop the reliability of current and future airframes.

Daniel J. Fink, Vice President of General Electric Co. and General Manager of the GE Space Div., was installed as 12th president of AIAA. The 19 new Fellows elected included Dr. Rocco A. Petrone, MSFC Director; John P. Campbell and John P. Reeder, LaRC; and Astronaut Charles Conrad, Jr. Inducted as Honorary Members were Sen. Frank E. Moss (D–Utah) and Rep. Olin E. Teague (D–Tex.). (A&A, Mar 74; AIAA Bull, Jan 74; AIAA Releases, 21 Dec 73; AIAA History Newsletter, 24 May 74; KSC Release 14–74; ARC Astrogram, 18 Jan 74, 1; Marshall Star, 30 Jan 74, 1)

*29 January:* Dr. Harrison H. Schmitt, Chief of the Astronaut Office at Johnson Space Center and *Apollo 17* astronaut, began a 90-day assignment as special assistant to the NASA Administrator to coordinate NASA's effort in energy research and development and facilitate NASA aid to other agencies working on energy research. (NASA Release 74–37; Off Energy Prog, interview, Aug 74)

*30 January:* Lockheed Aircraft Corp. had signed a protocol agreement in Moscow with the U.S.S.R. on possible future cooperation, including

**30 January**

development of civilian transports and helicopters, air traffic control, and navigation and communications systems, Lockheed announced. The protocol was one of 20 agreements the U.S.S.R. signed with U.S. companies. (Lockheed Release, 30 Jan 74)

**30 January–4 February:** Two representatives of the 10-nation European Space Research Organization that would develop and build Spacelab for use with NASA's space shuttle accompanied U.S. scientists on the Ames Research Center's *Galileo II* Convair 990 aircraft on flights evaluating electronic instruments designed to measure sea conditions. Dr. D. J. Shapland of the United Kingdom and J. De Waard of The Netherlands observed the flying laboratory, which would serve as a pattern for shuttle sortie missions carrying Spacelab. Spacelab was to have two elements: a manned laboratory module permitting scientists and engineers to work in a shirt-sleeve environment and an instrument platform, or pallet, for telescopes, antenna, and other equipment requiring space exposure. (NASA Release 74–31; ARC Aerospace Sci Div, interview, 17 July 74)

**31 January:** A background science press briefing on *Mariner 10*'s approaching 5 Feb. encounter with the planet Venus was held at NASA Hq. Launched by NASA 3 Nov. 1973 toward Venus and Mercury, *Mariner 10* would be the first two-planet mission, the first spacecraft to use gravity-assist from one planet to reach another, and the first to take closeup photos of Venus.

Dr. S. Ichtiaque Rasool, NASA Deputy Director for Planetary Programs, said Venus was the planet most similar to the earth in size, density, and distance from the sun. If its atmosphere were like the earth's, Venus should also have similar temperature, but the temperature had been found to be very high, 755 K (900°F). And the atmosphere was 100 times denser than the earth's, in one theory acting as a blanket, absorbing the radiation that tried to leave the surface and giving a "greenhouse effect." Further, the Venus atmosphere had been determined by measurements from the earth and by U.S.S.R. spacecraft to be mostly carbon dioxide.

Chief questions were: Why such a high temperature? Why such a heavy atmosphere? Why did the earth get all the liquid water and on Venus everything stayed in the atmosphere? The earth had life because it had oceans. Why did it happen on earth and why did it not happen on Venus? And what could happen to the earth if pollution or other causes should increase its temperature? "How much increase in temperature do you need to have the whole thing go this way, and you become like Venus, not next week, but [in] millions of years. Is a process like that feasible at all on the earth?" (Transcript)

• Chairman Don Fuqua (D–Fla.) and members of the Subcommittee on Manned Space Flight of the House Committee on Science and Astronautics received an annual program review at Kennedy Space Center, preliminary to FY 1975 congressional space budget hearings. The Subcommittee was briefed on preparations for the space shuttle, manpower, and plans for the Apollo-Soyuz Test Project. Center personnel would be further reduced in the next five months, with the closeout of the Skylab program. Runway construction was to begin in the spring for the shuttle. (KSC Release 18–74)

• Kennedy Space Center had extended its contract with Chrysler Corp. Space Division for the period 1 Jan. 1974 through 31 July 1975, KSC announced. The $7 092 547 extension, which brought total value of the contract to $36 467 389, provided for prelaunch, launch, and postlaunch support of the Saturn IB 1st stage for a Skylab rescue vehicle, if required, and for the launch of an Apollo spacecraft in the 1975 U.S.-U.S.S.R. Apollo-Soyuz Test Project. (KSC Release 17-74)

*During January:* A 65% noise reduction during landing had been demonstrated by 38 United Air Lines, Inc., flight crews in an in-service evaluation of a NASA-developed landing technique under visual and instrument flight conditions, NASA announced. Using the NASA technique, a Boeing 727 aircraft approaching on a two-segment path exposed only a 50-sq-km area to greater than 90 effective perceived decibels of noise. The program was closely coordinated with the Department of Transportation, the Federal Aviation Administration, and the Environmental Protection Agency. Program guidance was provided by a Panel on Noise Abatement Flight Procedures with members from FAA, DOT, aircraft and avionics manufacturers, airlines, and pilots' associations. (NASA Aeronautics Update, Jan 74)

- Dr. James C. Fletcher, NASA Administrator, said in his article "Are Skylab and the Space Shuttle Worth the Investment?" in *Government Executive* that everything done in the Skylab program—with total runout costs of $2.6 billion through FY 1974—had been necessary for future progress in space. He believed the Skylab team had "taken another giant leap for mankind comparable to the first step upon the Moon or the first satellite in Earth orbit."

  And even if the shuttle were not expected to save $1 billion a year in launch and payload costs, as it was—after an $8-billion 1972-1991 investment—the U.S. "should still build it. We cannot run Spaceship Earth without it." We could not "begin to think of obtaining solar power from collectors in space without the Shuttle; we cannot hope to unlock the still hidden energy secrets of the Sun and stars without improved space observatories launched by the Shuttle; we cannot hope to develop the tools for management of natural resources and protection of our environment on a global scale without manned and unmanned Earth observatories launched and serviced by the advantages of a 'hydrogen economy' in the future unless we are willing to support such important steps toward that future as a hydrogen-fueled Shuttle. In short, there is no new frontier in space for America and mankind without the Shuttle." (*Govt Exec*, Jan 74, 38-42)

## February 1974

*1 February:* Selection of Hughes Aircraft Co. for negotiation of a cost-plus-award-fee contract for continuing design of the Pioneer Venus spacecraft system was announced by NASA. Proposed cost of design work was $3 million, with an option for final design, development, fabrication, and testing of two flight spacecraft and launch support at $55 million. Two Pioneer Venus missions, planned for 1978, would launch one spacecraft to orbit Venus, transmitting data for nearly eight months, and another to explore the planet's atmosphere down to the surface by ejecting four probes transmitting data during their hour-long descent. (NASA Release 74–34)

- A $24 475 000 cost-plus-award-fee contract extension awarded to Boeing Co. for support services at Kennedy Space Center became effective. Services included test support management, plant engineering and maintenance, supply and transportation operations, security services, and fire protection and rescue through 31 Jan. 1975, bringing the total contract amount to an estimated $98 952 758. (KSC Release 21–74)

*1–2 February:* Congressmen and staff of the House Committee on Science and Astronautics were briefed at Marshall Space Flight Center and Michoud Assembly Facility in preparation for hearings on NASA's FY 1975 budget. Committee members were briefed on Skylab, space shuttle, tug, Spacelab, concept verification test, and use of space transportation systems at MSFC. On 2 Feb. they were briefed on the space shuttle external tank at Michoud. (*Marshall Star*, 6 Feb 74, 2)

*2 February:* The U.S. exported a record $5.125 billion worth of aerospace equipment during 1973 according to Dept. of Commerce figures, the Aerospace Industries Association of America announced. Aerospace exports were 6.4% of total U.S. exports and had increased 34.1% over 1972. Much of the increase was from exports of complete aircraft, which totaled $3.096 billion in 1973, up from $2.028 billion in 1972. (AIA Release 74–3)

- Four U.S. schools would be heated by solar power in a National Science Foundation project, the *Washington Post* reported. The project would evaluate solar energy systems in practical situations. All four systems—in Osseo, Minn.; Boston, Mass.; Baltimore, Md.; and Warrenton, Va.—would augment rather than replace the existing heating systems. The system in the Warrenton school was expected to cut heating costs in the five affected classrooms from $1600 to $10 a year. Built by InterTechnology Corp. under a $173 000 NSF contract, the Warrenton system consisted of a 230-sq-m glass, plastic, and coated metal solar collector through which ran 1-cm pipes. Water heated in the pipes was pumped from the collector into underground storage tanks and from there into conventional hot water radiators. The tanks would provide heat through five straight sunless days. The only outside energy required would be electricity to run the pumps, less than 1% of that required to power a baseboard system. (Ringle, *W Post*, 2 Feb 74, D5)

*4 February:* President Nixon, in his message transmitting the U.S. budget for FY 1975 to Congress, said the budget emphasized fiscal balance and continued "moderate restraint" on the economy to curb inflation, a strong defense force, a comprehensive energy program to "reestablish our ability to be self-sufficient in energy," the "New Federalism philosophy of strengthening the role of State and local governments," and a "more intensive focus on the tangible results that programs achieve."

The $304.4-billion budget proposed an increase for defense from $79.5 billion in FY 1974 to $85.8 billion, "so that we can increase our defense preparedness and preserve present force levels in the face of rising costs." Despite dollar increases caused by rising costs and by pay raises accompanying transition to an all-volunteer armed force, the President said, the proportion of defense costs to the total budget had fallen from 44% in 1969 to an estimated 29% in 1975, after the end of U.S. combat in Vietnam.

Project Independence, to reestablish U.S. self-sufficiency in energy, would require about $10 billion for accelerated energy research and development over the next five years and would encourage an even larger private R&D investment. Higher prices would be necessary to stimulate development of fuel supplies, but the emergency windfall profits tax had been proposed to curb excessive profits. The national energy policy included reorganization of Federal administrative machinery, stringent energy conservation measures, mandatory allocation of petroleum products, mandatory reporting on oil production and inventories, modernization of railroad regulations, accelerated development of domestic oil and gas reserves, increased use of coal reserves, development of a fast-breeder nuclear reactor, faster approval of energy facility sites, faster construction of nuclear power plants, and increased research into advanced energy sources, including fusion power and geothermal and solar energy.

Shortages of clean fuels would mean "some temporary variances from air quality plans" would be necessary to meet high-priority energy needs. "The progress we have made in pollution control in recent years, however, along with reductions in energy consumption, should insure that overall air quality will continue to improve." (*CR,* 4 Feb 74, H407-14; *PD,* 11 Feb 74, 161-70)

- President Nixon sent a $304.4-billion FY 1975 U.S. budget request to Congress, an increase of $29.8 billion over FY 1974. With 90% of the increase mandatory under existing law, the budget included few new programs and few cuts. Federal funds for research and development would increase by 9% from FY 1974, to $19.8 billion total obligations, "reflecting recognition of the larger potential contribution of science and technology to the solution of critical national problems." A substantial increase was requested for energy R&D, $2.05 billion in FY 1975 from $942 million in FY 1974.

The FY 1975 R&D request proposed investment in priority national needs, but included basic research as well as efforts to speed the use of federally sponsored R&D. The non-Federal investment in R&D was expected to be $16 billion, 45% of a national total of $36 billion.

NASA's total budget request of $3.247 billion was $245 million above the FY 1974 appropriation and $96.2 million above the 1974 budget plan. If approved, the increase would be NASA's first since 1965, except for a $100-million FY 1973 increase that the agency had not been permitted to spend. Significant budget items are shown in the table.

The Dept. of Defense FY 1975 budget request of $85.8 billion included $9.3 billion for research, development, test, and evaluation, an increase

## NASA FY 1975 Budget Highlights

(in thousands; items do not add to program totals because only selected items are listed)

| Budget Item | FY 1974 Authorization | FY 1975 Request | FY 1975 Difference |
|---|---|---|---|
| Total budget authority | $3 054 500 (3 002 100 appropriation) | $3 247 129 | +$192 629 (+245 029) |
| *Research and development* | 2 245 500 (2 194 000 appropriation) | 2 346 015 | +100 515 (+152 015) |
| Manned space flight | 1 032 000 | 1 124 800 | +92 800 |
| Space shuttle | 475 000 | 800 000 | +325 000 |
| Space flight operations | 555 500 | 323 300 | −232 200 |
| Skylab | 233 800 | — | −233 800 |
| ASTP | 90 000 | 114 600 | +24 600 |
| Development, test & operations | 220 200 | 175 200 | −45 800 |
| Space life sciences | 21 000 | 18 000 | −3 000 |
| Space sciences | 552 000 | 547 015 | −4 985 |
| Physics & astronomy | 63 600 | 140 515 | +76 915 |
| OSO | (un- | 7 630 | |
| OAO | dis- | 2 380 | |
| HEAO (suspended 1974) | trib- | 40 400 | |
| Orbiting Explorers | uted) | 33 000 | |
| Large Space Telescope | — | 6 200 | +6 200 |
| Lunar & planetary | 311 000 | 266 000 | −45 000 |
| Viking | (un- | 89 016 | |
| Mariner Jupiter-Saturn 1977 | dis- | 69 761 | |
| Pioneer Venus (new start 1975) | trib- | 27 100 | |
| Helios | uted) | 1 400 | |
| Lunar sample analysis | | 5 798 | |
| Lunar science operations | | 4 333 | |
| Launch vehicle procurement | 177 400 | 140 500 | −36 900 |
| Space applications | $161 000 | $177 500 | +$16 500 |
| Weather and climate | 51 100 | 35 000 | −16 100 |
| Nimbus 5 & F | (un- | 7 000 | |
| Tiros-N | dis- | 9 000 | |
| SMS-A & -B | trib- | 1 400 | |
| GARP | uted) | 7 400 | |
| Pollution monitoring | 13 900 | 29 800 | +15 900 |
| Earth resources survey | 49 600 | 58 600 | +9 000 |
| ERTS-B | (undis- | 11 100 | |
| Aircraft survey | tributed) | 17 300 | |
| Earth and ocean physics | 10 700 | 18 500 | +7 800 |
| SEASAT-A | — | 8 000 | +8 000 |
| Space processing | 3 100 | 3 500 | +400 |
| Energy applications | 2 000 | 2 000 | — |
| Communications satellites | 22 100 | 8 300 | −13 800 |
| Heat Capacity Mapping Mission | — | 2 600 | +2 600 |
| Aeronautics & space technology | 252 000 | 241 200 | −10 800 |
| Aeronautical | 180 000 | 166 400 | −13 600 |
| Space & nuclear | 72 000 | 74 800 | +2 800 |
| Tracking & data acquisition | 244 000 | 250 000 | +6 000 |
| Technology utilization | 4 500 | 5 500 | +1 000 |
| *Construction of facilities* | 112 000 (101 100 appropriation) | 151 490 | +39 490 (+50 390) |
| Infrared telescope facility, Hawaii | — | 6 040 | +6 040 |
| X-ray telescope facility, MSFC | — | 4 060 | +4 060 |
| Space shuttle facilities | 67 200 | 86 020 | +18 820 |
| *Research & program management* | 707 000 (707 000 appropriation) | 749 624 | +42 624 (+42 624) |

## DOD FY 1975 RDT&E Budget Highlights

(in thousands; items do not add to program totals because only selected items are listed)

| Budget Item | FY 1974 Budget Plan | FY 1975 Budget Plan | FY 1975 Difference |
|---|---|---|---|
| Research, development, test, & evaluation | $8 333 009 | $9 322 469 | +$989 460 |
| *Military astronautics* | 593 926 | 527 248 | −66 678 |
| *Aircraft* | 1 682 381 | 1 829 318 | +146 937 |
| B-1 advanced strategic bomber | 449 000 | 499 000 | +50 000 |
| A-10 close-air-support aircraft | 107 000 | 268 000 | +161 000 |
| AF air combat fighter | — | 36 000 | +36 000 |
| USN VFX fighter prototype | — | 34 000 | +34 000 |
| Tanker/missile-carrier aircraft | — | 20 000 | +20 000 |
| Advanced medium STOL transport (AMST) | 25 000 | 56 000 | +31 000 |
| F-14 interceptor aircraft | 737 000 | 756 000 | +19 000 |
| F-15 air superiority fighter | 1 129 000 | 1 076 000 | −53 000 |
| *Missiles* | 2 124 762 | 2,352 993 | +228 231 |
| Trident missile | 528 000 | 649 000 | +121 000 |
| Air-launched cruise missile | 11 000 | 80 000 | +69 000 |
| Submarine-launched cruise missile | 3 000 | 45 000 | +42 000 |
| Advanced ICBM technology | 4 000 | 37 000 | +33 000 |
| Minuteman-site-defense prototype development | 110 000 | 160 000 | +50 000 |
| Terminally guided maneuvering reentry vehicle (MARV) | — | 20 000 | +20 000 |

of $989.5 million from the FY 1974 budget plan. Increases in military astronautics would improve the satellite warning system and the Air Force Satellite Communications System (AFSATCOM). NAVSTAR Global Positioning System development would continue. The planning effort toward use of the NASA shuttle for military payloads would increase.

The National Oceanic and Atmospheric Administration's FY 1975 budget request of $467.1 million, a $77.8-million increase, included space and environmental programs.

The Federal Aviation Administration requested $845.4 million within a Department of Transportation total of $2.2 billion. Although the DOT total was a decrease from FY 1974, it included an increase for R&D.

The Atomic Energy Commission's $3.06-billion request included a 30% increase in energy activities and more funds for civilian efforts than for defense for the first time.

## NOAA FY 1975 Budget Highlights

(in thousands)

| Budget Item | FY 1974 Estimate | FY 1975 Request | FY 1975 Difference |
|---|---|---|---|
| NOAA total budget authority | $389 308 | $467 096 | +$77 788 |
| Environmental satellite services (including polar orbiting & geosynchronous systems; next generation polar-orbiting model) | 62 356 | 61 475 | −881 |
| Public forecast & warning services | 45 620 | 47 875 | +2 255 |
| Global monitoring of climate | 594 | 1 112 | +518 |
| International projects (including Global Atmospheric Research Program's GATE) | 9 723 | 9 148 | −575 |

### DOT FY 1975 R&D Budget Highlights
(in thousands)

| Budget Item | FY 1974 Estimate | FY 1975 Estimate | FY 1975 Difference |
|---|---|---|---|
| DOT total budget authority | $5 520 477 | $2 195 656 | −$3 324 821 |
| Research & development | 390 000 | 417 000 | +27 000 |
| FAA budget authority | 1 311 820 | 845 383 | −466 437 |
| Engineering & development (aircraft safety, environment, aviation medicine, noise reduction, STOL, VTOL) | 11 500 | 11 241 | −259 |

### AEC FY 1975 Energy and Space Budget Highlights
(in thousands)

| Budget Item | FY 1974 Estimate | FY 1975 Estimate | FY 1975 Difference |
|---|---|---|---|
| AEC total budget authority | $2 388 914 | $3 057 648 | +$668 734 |
| Space nuclear systems (radioisotope thermoelectric generators for NASA & DOD missions, especially Viking, Mariner, Jupiter-Saturn, LES; also technology improvements & extension to energy applications) | 26 100 | 27 000 | +900 |
| Enriched uranium production | 331 615 | 426 970 | +95 355 |
| Civilian reactor R&D (including fast-breeder reactor) | 290 390 | 349 750 | +59 360 |
| Fusion research | 53 000 | 82 000 | +29 000 |

The National Science Foundation's $672.1-million request included increases in every energy-related category; national and special research program funding decreased slightly. And the Environmental Protection Agency's $695.2 million would increase funding for R&D and pollution control.

### NSF FY 1975 R&D Budget Highlights
(in thousands)

| Budget Item | 1974 Estimate | 1975 Estimate | 1975 Difference |
|---|---|---|---|
| NSF total budget authority | $579 260 | $672 100 | +$92 840 |
| National & special research programs (including troposphere & atmosphere research in GARP; also science & energy R&D policy support to Presidential Adviser) | 91 600 | 82 300 | −9 300 |
| Research Applied to National Needs (RANN) | 75 100 | 84 000 | +8 900 |
| National research centers (including very-large-array radio astronomy facility) | 42 500 | 52 500 | +10 000 |

### EPA FY 1975 Budget Highlights
(in thousands)

| Budget Item | 1974 Estimate | 1975 Estimate | 1975 Difference |
|---|---|---|---|
| EPA total budget authority | $4 628 864 | $695 200 | −$3 933 664 |
| Research & development | 168 916 | 171 068 | +2 152 |
| Pollution abatement & control | 356 094 | 407 976 | +51 882 |

**4 February**

(NASA Release, FY 75 Budget Briefing, Background Material; NASA, FY 75 Budget Chronological History; OMB, *Budget of US Govt, FY 75;* OMB *Special Analyses,* FY 75; NASA, Extracts Budget US Govt, FY 75; DOD Release 43–74; *Annual DOD Rpt, FY 75; W Post,* 5 Feb 74, A1, A10–12)

- Dr. James C. Fletcher, NASA Administrator, released the budget statement he had given at a 2 Feb. press briefing. The FY 1975 NASA budget plan for new authorization of $3.247 billion and spending of $3.273 billion, although a "constrained budget," would carry forward the programs planned in FY 1974. Goals and major programs were the same, with three new projects in space science and applications, but the space shuttle had been slipped four to six months.

  NASA's FY 1974 budget had "reflected a sharp temporary reduction, because of Government-wide fiscal problems . . . , below the $3.4 billion level previously planned as the long-term level." Increases in FY 1975 and later years were therefore "in order to approach again the level required to maintain . . . advances in space and aeronautics for the rest of the decade." The FY 1975 increase of "about $100 million," however, was less than required to return to NASA's complete program as planned, and adjustments had been necessary. Principally, the first manned orbital flight of the shuttle was rescheduled to the second quarter of 1979, instead of the end of 1978, but NASA had a "firm commitment now" that there would be no more slips for budgetary reasons, and none was expected for technical reasons. NASA Comptroller William E. Lilly said at the briefing the stretchout was expected to increase shuttle costs about $50 million "in 1971 dollars." The $800 million requested for the shuttle was $89 million less than NASA had submitted to the Office of Management and Budget.

  In the three new space flight projects, Pioneer Venus had top scientific priority as the next step in planetary exploration. Two missions launched in 1978 would study the Venus atmosphere for a better understanding of that planet and of the earth's atmosphere, meteorology, and climatology.

  In the second new start, SEASAT, an experimental applications satellite, would measure ocean characteristics to aid scientific understanding of ocean dynamics and determine if a forecasting system could be developed. SEASAT–A was to be launched in 1978. The third start, the Heat Capacity Mapping Mission (HCMM), would launch an Explorer satellite in 1977 to make thermal measurements of the earth's surface to help find mineral resources and potential geothermal energy sources and to discriminate rock structures for planning civil works such as highways.

  NASA was asking $6 million to build the world's first large infrared telescope, on Mauna Kea, Hawaii, for planetary research to complement planetary space flights, especially to provide data for the 1977 launch of the Mariner Jupiter-Saturn probe.

  Another new thrust, still in the planning stage, was the Tracking and Data Relay Satellite System (TDRSS), to return the large amounts of data expected from the shuttle in the 1980s and permit phase-out of some tracking stations. NASA would not develop the spacecraft for the TDRSS but would lease satellite services from a private organization, beginning in CY 1979.

Analysis of data from the Skylab missions that were ending with the 8 Feb. splashdown of *Skylab 4* would continue for many years. Apollo-Soyuz Test Project joint engineering and training activities with the U.S.S.R. were on schedule. All shuttle hardware prime contractors and most subcontractors had been selected, and the Dept. of Defense tentatively planned to modify an existing upper stage for use as an interim space tug with the shuttle. *Erts 1*, launched 23 July 1972, was still working well, exceeding its design life by more than six months. ERTS-B, previously planned for 1976 launch, had been moved to 1975. A number of NASA and 11 commercial applications launches were scheduled during the year. A nominal communications satellite research program would be continued. In space astronomy, OSO-I was scheduled for 1975 and the HEAO series would be resumed, with launches planned for 1977, 1978, and 1979. Astronautics programs would continue to develop technology to improve fuel economy, reduce noise, and improve performance of U.S.-built civil transport aircraft.

NASA's Civil Service employment was to be stabilized at the end of FY 1974 levels except for a further reduction of 354 persons at Marshall Space Flight Center following completion of Skylab. NASA's FY 1975 total would be 24 616, about 2200 less than in FY 1973 and 354 less than in FY 1974. Support service contractors at the Centers had reductions planned at about the same level, the 25 000 FY 1973 total dropping to 24 000 in FY 1974 and to 22 000 in FY 1975. (Text; NASA, FY 75 Budget Briefing transcript & background material)

*4–6 February:* The U.S.–U.S.S.R. Joint Committee on Cooperation in Atomic Energy reached accord in Washington, D.C., on implementation of the Agreement on Scientific and Technical Cooperation for the Peaceful Uses of Atomic Energy, signed 21 June 1973 by President Nixon and Soviet Communist Party General Secretary Leonid I. Brezhnev. A Protocol on Cooperation in Controlled Thermonuclear Fusion and Plasma Physics Research was signed by Atomic Energy Commission Commissioner Clarence E. Larson and U.S.S.R. State Committee Chairman Andronik M. Petrosyants. Joint coordinating committees were established for fusion and fast-breeder reactor cooperation. A program of cooperation in research on fundamental properties of matter also was agreed on. (AEC Release, 15 Feb 74)

- Research into automotive propulsion systems was reviewed by NASA, Dept. of Transportation, and Environmental Protection Agency officials in hearings before the House Committee on Science and Astronautics' Subcommittee on Space Science and Applications. H.R. 10392 had been introduced 19 Sept. 1973 to authorize NASA research toward development of economical, energy-conserving systems with clean emissions and improved performance. Dr. George M. Low, NASA Deputy Administrator, testified 4 Feb. that Lewis Research Center was testing EPA's experimental engine, under a June 1973 agreement with EPA on work toward a low-pollution gas-turbine-powered vehicle. Upgraded engine work would be completed by mid-1975 and vehicle performance evaluation before the end of CY 1975. First results of Jet Propulsion Laboratory investigations of hydrogen injection into fuel to improve efficiency and reduce pollution in conventional automobile engines promised achievement of 1977 Federal emission standards with 25% improvement in fuel economy.

LeRC Director Bruce T. Lundin testified that, in addition to the gas turbine engine, a broader, longer range, advanced technology program was under way, with 13 industry contracts for such improvements as a low-emission combustor, higher-temperature heat exchangers, lower-cost turbine discs, turbine wheel cooling, and low-cost fuel controls. Dr. William H. Pickering, JPL Director, reported a test automobile using bottled gas to inject hydrogen into the fuel had already met the carbon monoxide emission standard and had almost met the nitrogen oxide standard, although the hydrocarbon emissions remained a problem. With engine modifications to produce hydrogen from gasoline, instead of using bottled gas, performance per gallon of fuel was expected to improve 25% over the conventional engine. (Transcript)

*4–8 February:* The first meeting of the International Telecommunications Satellite Organization's Assembly of Parties was held in Washington, D.C. Seventy-two of the eighty-four member governments were represented. Raymond J. Waldmann (U.S.) was elected Chairman and Dr. Osama Anani (Jordan) was elected Deputy Chairman. Elected as Vice-chairmen were Ambassador Alejandro Orfila (Argentina), Ambassador Adolfo Alessandrini (Italy), Ambassador Leonard O. Kibinge (Kenya), and Ambassador Motoo Ogiso (Japan).

The Assembly of Parties found that the proposed U.S. Geostationary Operational Environmental Satellite (GOES) and maritime satellite system would not interfere with the INTELSAT system. GOES was technically compatible with the use of the radio-frequency spectrum and orbital space by the existing or planned satellites. The maritime satellite system would not prejudice direct satellite telecommunication links. The Assembly refused a United Nations request for free use of INTELSAT satellites, although a priority access arrangement, for periods not exceeding 90 days during an emergency, would be worked out. (INTELSAT Release 74–12).

*5–22 February: Mariner 10,* launched 3 Nov. 1973, passed within 5800 km of Venus—within 15 km of the planned aim point—at 1:01 pm EDT 5 Feb. The encounter sequence was accomplished according to plan and data were transmitted over the 45 million km between the earth and Venus at a rate of 117.6 kilobits per sec. Excellent ultraviolet cloud-structure pictures and new data on particle environment and mass, density, and shape of the planet were returned.

Because *Mariner 10* approached Venus on the dark side, the TV sequence was not begun until shortly before the encounter, at 12:21 pm EDT. The spacecraft flew behind Venus at 1:07 pm EDT for 20 min; TV signals were stored on tape and recovered without difficulty after *Mariner 10* emerged from the occultation zone. During occultation, the spacecraft's radio antenna was aimed toward the edge of the planet so that the radio signals were bent around Venus by the extremely dense atmosphere and sent on to the earth.

All science instruments functioned normally except the scanning electrostatic analyzer. The TV sequence was completed 13 Feb. after some 4165 pictures were returned to the earth. During the flyby, the 503-kg spacecraft used the gravitational field of Venus to slow its speed and direct the flight path toward a 29 March encounter with Mercury. By 22 Feb. *Mariner 10*—its speed increasing as it came nearer the sun—was traveling 3 million km per day toward the planet.

Preliminary science results were presented at a press briefing 7 Feb. at Jet Propulsion Laboratory. Dr. Herbert S. Bridge, Massachusetts Institute of Technology scientist, said the solar wind was interacting with the ionosphere of Venus, producing a magnetic tail that could be detected as a modified plasma flow as far out from the planet as 500 Venus radii. Dr. James A. Dunne, Mariner project scientist, said the magnetic field doubled just before radio occultation but that no evidence of an intrinsic planetary magnetic field was found. Preliminary radio science results indicated that Venus was very round, 100 times less oblate than the earth, and confirmed that Venus had a dayside ionosphere. The mass of the planet was slightly less than that derived from previous data. Hydrogen was a major element of the Venus cloud deck, possibly present in droplets of sulfuric acid as well as water vapor. *Mariner 10* found no evidence of deuterium on the planet, indicating that the hydrogen had originated from the solar wind. Helium, carbon, and atomic oxygen were found in the atmosphere, with possible traces of argon and neon. Outer atmosphere temperature of 590 K (600°F) was suggested.

TV data indicated a uniform main cloud deck with three or four layers and a stratified haze layer. TV pictures in UV showed a classical "Y" feature that originated in the equatorial region and spiraled toward the pole. Clear circulation patterns could be seen at the poles, caused by rising gas transporting excess solar heat from the equatorial zone by convection. The atmosphere was smooth, with no swirling storms. (NASA MOR, 11 March 74; Briefing transcript; Sullivan, *NYT*, 5 Feb. 74, 18; 6 Feb. 74, 1; Miles, *LA Times*, 6 Feb 74)

*5 February:* Dr. James C. Fletcher, NASA Administrator, presented NASA's $3.247-billion FY 1975 budget request' and program [see 4 Feb.] to the Senate Committee on Aeronautical and Space Sciences to open hearings on the NASA authorization bill. In a review of FY 1974 accomplishments, Dr. Fletcher pointed out that as the Skylab program came to an end the three crews had traveled some 113 million km and orbited the earth 2475 times in a working scientific laboratory—chalking up "about 100 miles per gallon of fuel." Skylab had focused on the sun, the earth, and man. "Apollo extended man's reach to the moon; Skylab added near-Earth to man's domain."

In answer to questions Dr. Fletcher said the principal difference between NASA's minimum request to the Office of Management and Budget and the final figures sent to Congress by the President was the $89 million subtracted from the shuttle program, causing a four- to six-month delay. Other differences brought a total decrease of $124 million below "what we could have gotten by with without any major changes" in NASA's program. The proposed minimum budget would not have restored any of the programs NASA had had to drop because of FY 1974 budget cuts and therefore had been "well below the constant level" envisaged for NASA in FY 1973. Of $7.9 million cut from aeronautics programs, the chief reduction was in the quiet propulsion lift technology program, deleting the system integration model effort. Deletion of $1 million from the requested technology utilization increase canceled proposed increases in numbers of dissemination centers and reduced buildup of support to state and local governments. A $5.7-million cut in the amount requested for earth observations instruments had deferred their development.

Dr. George M. Low, Deputy Administrator, said there had been "some discussions during the past several months" of an OMB proposal to consolidate weather satellite programs under the Department of Defense. NASA, DOD, OMB, and the National Oceanic and Atmospheric Administration had decided "to keep separate systems for the DOD and NASA [with NOAA], but we will make maximum use" of hardware commonality in building satellites for them. (Transcript)

- Dr. James R. Schlesinger, Secretary of Defense, said in FY 1975 budget hearings before the Senate Armed Services Committee that, although it was the first time in more than a decade the budget did not reflect support of U.S. forces in combat, it had to support a minimum peacetime defense and deterrence posture. The $85.5-billion budget appeared large because of inflation, increased military and civilian pay, and increased retirement benefits.

An FY 1974 supplemental request of $6.2 billion included $480 million in increased fuel prices and $231 million for extra costs of arms for Israel during the October 1973 Mid-East war.

The FY 1975 budget included programs to maintain and improve ballistic missile systems; continue strategic systems research; increase the active air defense of the U.S.; and increase the total airlift and sealift capability. (Testimony)

*6 February:* *Skylab 3* Astronauts Alan L. Bean and Jack R. Lousma, launched 28 July 1973 to spend 59 days aboard the Orbital Workshop, were honored by Secretary of the Navy John W. Warner in a Pentagon ceremony. Bean received his second Navy Distinguished Service Medal. Lousma received the Navy Distinguished Service Medal and Navy astronaut wings. (DOD Release 54–74)

- The *Apollo 16* command module (launched 16 April 1972) went on public display at the Alabama Space and Rocket Center in Huntsville. The CM, called *Casper*, had carried Astronauts Thomas K. Mattingly II, Charles M. Duke, Jr., and John W. Young to the moon and back during the fifth successful lunar landing mission. (*Marshall Star*, 6 Feb 74, 1)
- Dr. Karl G. Henize, scientist-astronaut, suffered a "landing mishap" while on instrument flight plan after flying his T–38 trainer aircraft from Ellington Air Force Base to Bergstrom AFB, Tex. Dr. Henize was not injured. A wheel on the T–38 collapsed and the aircraft was damaged. (JSC Release 74–50; JSC Hist Off, interview)

*7 February:* Rep. Olin E. Teague (D–Tex.) praised the *Skylab 4* astronauts (launched 16 Nov. 1973) on the eve of their splashdown after a record 84-day mission, concluding the Sylab program. On the floor of the House of Representatives Rep. Teague called the Skylab record "unprecedented." After damage to the Orbital Worshop during its 14 May 1973 launch, the three crews had repaired and altered the systems, exceeded their expected workloads, and returned vast amounts of useful information on earth rseources, solar phenomena, physical science, life sciences, and technology. The crews had spent 41 hrs in extravehicular activity and 171½ days in earth orbit, demonstrating that man could live indefinitely and work productively in space.

To date the U.S. manned space program had amassed 30 flights with some 21 850 hrs of active operation in space. "These programs— Mercury, Gemini, Apollo, and Skylab—have made direct contributions to our daily lives as well as new scientific knowledge. . . . Skylab has amply demonstrated that space can be used for practical purposes." (CR, 7 Feb 74, E533)

- Prof. Nikolay A. Kozyrev, Soviet astronomer, had obtained convincing proof that Mercury had an atmosphere, *Tass* reported. A glowing halo around the planet, indicating the existence of an atmosphere, was seen in a spectrogram made while Prof. Kozyrev tracked the planet's path as it crossed the sun's disc. He calculated the refraction of light in the atmosphere to be 0.002 of the earth's refraction and the pressure 0.02 of the earth's atmospheric pressure. The thickness of the atmosphere, established from the brightness of the halo, was 600 km. Kozyrev had said the thickness indicated the presence of a light gas, probably hydrogen. He believed that Mercury's atmosphere was continuously replenished by the flow of protons from the sun. (Tass, FBIS–Sov, 11 Feb 74, U1)

- Francis C. Schwenk, Director of the Research Div. in NASA's Office of Aeronautics and Space Technology, testified on four NASA projects in hearings on space nuclear propulsion and power before the House Committee on Science and Astronautics' Subcommittee on Aeronautics and Space Technology. Early research had shown that high-temperature thermionic converters combined with a compact nuclear reactor could provide an efficient power plant for advanced planetary missions. NASA technology would now focus on thermionic energy conversion for achieving highly efficient operation at lower temperatures. During the Apollo and Pioneer programs, NASA had proved the advantages of a small (40- to 70-w) radioisotope-thermoelectric-generator power system, which had operated at 6% efficiency at a cost of $20 million per kilowatt. NASA was developing technology for a 0.5- to 2.0-kw gas-turbine conversion system (Brayton energy conversion system) to achieve efficiencies of 25% and reduce costs to one fourth.

  Gaseous-core nuclear rocket research had led to plasma-core reactors as advanced energy sources for terrestrial as well as space propulsion uses; use of uranium hexafluoride as plasma fuel was expected to achieve efficiencies of more than 50%. Lewis Research Center research into superconducting magnets and plasma confinement and heating in advanced steady-state schemes was expected to show within a year whether a superconductive magnetic apparatus (SUMMA) could produce fusion-like plasmas as predicted. (Text)

- NASA's report on emergency preparedness and defense mobilization activities for FY 1973 was issued by the Senate in the 23rd annual report of activities of the Joint Committee on Defense Production. NASA had revised emergency plans at field installations for increased readiness for a broader range of emergencies. The agency's contract policies had contributed to maintenance of a broad industrial base of small businesses to meet conditions of a national emergency. Research and development programs contributing to energy solutions included studies of new aeronautical technology, solar and wind energy, clean fuel production, improved use of coal, and more efficient fuels. Technology utilization teams were applying aerospace technology to public problems in medicine, urban construction, law enforcement, mine safety, air and water pollution, and transportation. (Sen Rpt 93–683)

- The New Mexico House of Representatives approved a bill to appropriate $1.8 million for a proposed International Space Hall of Fame in Alamogordo. (UPI, *NYT*, 8 Feb 74)

*8 February:* President Nixon wired salutations to *Skylab* 4 Astronauts Gerald P. Carr, Dr. Edward G. Gibson, and William R. Pogue aboard the recovery ship U.S.S. *New Orleans* after their successful splashdown at 11:17 am EDT. The astronauts, launched 16 Nov. 1973, had set an 84-day space flight record in the Orbital Workshop.

The President said: "On behalf of the American people, I salute Skylab's third crew of astronauts on their safe return to earth. They have successfully completed man's longest space journey and brought to an end one of the most scientifically productive endeavors in the history of human exploration. From the Skylab program, we have learned that we can live and work in space for long periods of time. And we have found that the results of these efforts can be of enormous practical value to life on earth.

"Skylab now joins the ranks of the *Santa Maria*, the H.M.S. *Beagle*, the *Spirit of St. Louis*, and the *Eagle*. Each of these great vehicles has carried us beyond the contemporary limits of human knowledge into a new comprehension of our own possibilities and a new definition of our own destiny. We welcome the men of Skylab home, and we salute them and all their predecessors who have launched us on this great adventure." (*PD*, 11 Feb 74, 193; *A&A 1973*)

- "Skylab . . . has demonstrated that this nation is capable of conducting broader and more useful beneficial activities in space that directly relate to our own planet Earth," Dr. James C. Fletcher, NASA Administrator, said at a postrecovery press briefing at Johnson Space Center. It had "moved the space program from the realm of the spectacular into a new phase that can be characterized . . . as almost businesslike if not yet quite routine." The Skylab experience was necessary for the future in space and had confirmed that NASA was proceeding in the right direction with development of the space shuttle.

  Skylab Program Director William C. Schneider said 30 earth resources passes had been planned and 39 accomplished during the *Skylab 4* mission. There had been 350 hrs of Apollo Telescope Mount observations planned and 338 hrs completed. Observations of Comet Kohoutek included 13 by the Apollo Telescope Mount and 111 by other instruments. All 28 planned corollary experiments and all major medical and student experiments had been completed. Original planning had called for 140 days in orbit for the three manned missions; 172 were accomplished. Skylab's greatest achievement "was to prove . . . there is no limit in our space research"; the limit "is only our resolve, not our ability to do work, not the ability of men to work and not our technical knowledge." (Transcript)

- *Skylab 4* Astronauts Gerald P. Carr, Dr. Edward G. Gibson, and William R. Pogue had experienced a harrowing last hour in space when they could not get commands to the secondary control system, which pointed the spacecraft heat shield at the correct angle during reentry. After discovering a leak in the prime reaction control system, the astronauts tried and failed to activate the secondary system. "Our hearts fell and our eyeballs popped," Carr said at a 22 Feb. Johnson Space Center press conference. Carr, commander of the mission, instead had used a manual procedure to bypass the computer controlling the engine's firing and guided the spacecraft to a safe on-target splashdown. The press later quoted a NASA spokesman as saying that the failure to activate the backup control system was "just a matter of a switch not being thrown." (O'Toole, *W Post*, 15 Feb

74, A1; UPI, *Richmond Times-Dispatch,* 16 Feb 74; Maloney, *H Post,* 23 Feb 74)

- The Army was issuing a $1 988 000 firm-fixed-price contract to Algernon Blair Industrial Contractors, Inc., for modifications to orbiter propulsion system test facilities for space shuttle engine testing at NASA's Mississippi Test Facility, the Dept. of Defense announced. (DOD Release 59-74)
- The Dept. of Defense announced the Air Force was issuing a $1 400 000 fixed-price-incentive contract to General Electric Co. for engine refurbishment and increase in test hours for eight YJ-101 test engines planned for use in the YF-17 aircraft. (DOD Release 59-74)
- Dr. Fritz Zwicky, California Institute of Technology astronomer and a leading expert on jet propulsion, died of a heart attack in Pasadena, Calif., at the age of 76. Dr. Zwicky's inventions included the aeropulse, hydropulse, hydroturbojet, monopropellants, and coruscatives. An early member of Aerojet Engineering Corp. (forerunner of Aerojet-General Corp.), Dr. Zwicky was Director of Research at Aerojet from 1943 to 1949. Under his leadership the corporation developed JATO (jet-assisted takeoff) for getting heavy-laden bombers into the air and in 1949 he was awarded the Medal of Freedom for his work in jet propulsion. In 1934, in a report coauthored with Dr. Walter Baade, Dr. Zwicky had drawn the first clear-cut professional distinction between common novas and supernovas. The report said a supernova blew off considerable mass and could leave a pure neutron core, a theory which had gained increasing support. He discovered his 100th supernova in 1973. Only 378 supernovas had been discovered throughout history. (*NYT,* 11 Feb 74, 36; *Who's Who*)

*9 February:* Medical tests of *Skylab 4* Astronauts Gerald P. Carr, Dr. Edward G. Gibson, and William R. Pogue—begun after their 8 Feb. splashdown—continued on board the U.S.S. *New Orleans.* All three astronauts reached their maximum preflight work rates on the bicycle ergometer. Each astronaut completed the full run on a whirling chair, reaching a rate of 30 rpm while making 150 head movements with no problems. A test of muscle function showed no significant decrease. As soon as the men had landed, the pull of gravity had compressed the vertebrae of their spinal columns, removing the inch or more of height gained in weightlessness. Carr had gained slightly in weight; Pogue had lost one kilogram and Dr. Gibson a little more than two. (McElheny, *NYT,* 11 Feb 74, 7)

- A Baltimore *Sun* editorial commented on NASA's space program and the 8 Feb. splashdown of *Skylab 4:* "The Apollo moon missions and Skylab have moved NASA to a point where few if any scientists any longer condemn the space program as 'more circus than science,' and Skylab, especially, has given manned spaceflight an unassailable new reputation as a superb scientific instrument. . . . It is unfortunate that the space program entered its most useful and significant period in conjunction with the loss of interest in space on the part of a public obsessed with Watergate and the energy crisis" but "there are no indications the American people wish to junk the space program." NASA's FY 1975 budget included promising projects. "The evolution of man into spaceman has just begun and there are wondrous things ahead." (B *Sun,* 9 Feb 74)

*10 February: Skylab 4* Astronauts Gerald P. Carr, Dr. Edward G. Gibson, and William R. Pogue, who splashed down 8 Feb. after 84 days aboard the Orbital Workshop, completed a third day of post-splashdown tests before a brief welcoming ceremony at dockside in San Diego, Calif. The astronauts then flew to Ellington Air Force Base, Tex., for a reunion with their families. Medical tests would continue throughout the week and the three men would continue to eat the precisely controlled diet begun three weeks before their 16 Nov. 1973 launch. (McElheny, *NYT*, 11 Feb 74, 7)

- A *New York Times* editorial noted that the 8 Feb. safe return of the *Skylab 4* astronauts "was considered so certain that the television networks had decided well in advance not to broadcast the splashdown." It "really was a public tribute to the high technical achievements of the Skylab program and of previous manned American space projects." With Skylab's record and dividends, "the case for a permanent manned station in space can now be regarded as proved." (*NYT*, 10 Feb 74)

- The success of the Skylab program had made the idea of continuous occupation of an orbital base more concrete to the directors of the American space efforts, a *New York Times* article reported. Such a base could be assembled by flights of the space shuttle being developed by NASA and could be used to observe the sun and the earth and for manufacturing special materials under weightless conditions. Skylab Program Director William C. Schneider had said in an interview, "We have shown that there is no man or machine limitation on whatever we want to do in space." The confidence of space scientists had been strengthened by the apparent adaptation of the astronauts to space, although the *Times* reported *Skylab 2* Astronaut Joseph P. Kerwin and Johnson Space Center Director Christopher C. Kraft as cautioning that nine men "is a very small statistical sample." More long test flights would have to be made. However, if the medical data held true, the design of future space stations could be made much simpler and the number of space shuttle flights to establish it would be fewer. Schneider had noted that the conflict between the Workshop's need for power and its program for observation was still to be solved. The current design called for Workshop solar panels to point toward the sun rather than the earth. In a future space station, observatories might be placed on long booms which could turn on their own, minimizing the station's need to maneuver. (McElheny, *NYT*, 10 Feb 74, 40)

- Computer enhancement of multispectral photographs from *Erts 1* satellite (circling the earth since 23 July 1972) had been proved useful for finding mineral and water deposits on the earth. Computer image enhancement, first used at Jet Propulsion Laboratory to improve early spacecraft pictures of the moon and Mars, had identified surface signs of known mineral deposits for JPL and U.S. Geological Survey investigators. Twenty ERTS-identified points had been verified in a test area, exploratory drilling had uncovered water-bearing rock 12 m below an ERTS-indicated water hole, and fractures indicating other drilling sites had been revealed. (NASA 74-27)

- Noting the space program's emphasis had shifted earthward now that space flight was an accepted and almost routine part of life, John N. Wilford commented in the *New York Times* that "it is only natural that after surveying the canyons of Mars, the clouds and haze of Venus and the radiation belts of Jupiter, men turn back and take a new look at the earth and, in the words of T. S. Eliot, 'know the place for the first time.'" (*NYT*, 10 Feb 74, 7)

*10–12 February:* Soviet space probes *Mars 4* and *5*, launched 22 and 25 July 1973, neared Mars after a journey of 460 million km. *Mars 4* approached the planet 10 Feb. Because of faulty functioning of one onboard system, the braking engine was not fired and the spacecraft flew by Mars at a distance of 2200 km. TV pictures were taken during the flyby and transmittal of information on outer space characteristics was to continue.

*Mars 5* was put into orbit around Mars at 6:45 pm Moscow time (11:45 am EDT) 12 Feb., with a 32 500-km apogee, 1760-km perigee, 25-hr period, and 35° inclination.

*Mars 6* and *7*, launched 5 and 9 Aug. 1973, would approach Mars in March. During a 22 Sept. 1973 interview with Tass, Director Roald Sagdeyev of the Space Research Institute of the Soviet Academy of Sciences had said one of the four Mars probes would softland near the Martian south polar cap to test the physical properties of the soil and surface rocks and check the possibilities of transmitting TV pictures of the surrounding terrain to earth. (Tass, FBIS-Sov, 14 Feb. 74, Ul; *A&A 1973*)

*11 February:* The Titan IIIE-Centaur proof-flight vehicle (TC–1), launched by NASA from Eastern Test Range Complex 41 at 9:48 am EDT carrying a Viking spacecraft model and the SPHINX satellite, was destroyed after the Centaur engine failed to start. At 8.1 min from launch, after normal separation from the Titan booster, the Centaur main-engine start failed and the vehicle automatically went into the restart sequence. A second attempt to start the engine failed at 9.2 min after liftoff. Since the vehicle had not achieved orbital velocity the Range Safety Officer at the Antigua, West Indies, tracking station transmitted destruction commands and the Centaur was destroyed at 12.5 min after the liftoff. The vehicle fell into the Atlantic Ocean 3530 km downrange. Inflight data indicated that the liquid-oxygen boost pump had failed to operate during both attempted engine starts. A committee was formed to investigate the failure and recommend corrective action.

Although the primary objective of demonstrating the capability of the Titan-Centaur launch vehicle was not accomplished, some objectives were fulfilled: demonstration of the capability of the launch facility to support a Titan-Centaur launch, structural integrity of the vehicle, Centaur guidance and control of the Titan, Titan-Centaur separation, and Centaur standard shroud capability.

The Viking Dynamic Simulator had been instrumented to verify flight loads for the two planned 1975 missions to Mars. Remaining attached to the Centaur stage, the VDS was to have flown a typical Viking launch trajectory with a shortened burn to limit the spacecraft to a low earth orbit, to test the guidance system, and estimate injection accuracy.

The SPHINX Space Plasma High Voltage Interaction Experiment spacecraft had been intended for launch into an elliptical orbit to investigate the effect of charged particles in space on high-voltage solar cells, insulators, and conductors.

The launch vehicle combined NASA's versatile high-energy liquid-hydrogen and liquid-oxygen Centaur upper stage with the Air Force-developed Titan III booster, a two-stage liquid-propellant core rocket augmented by two strap-on solid-fueled rocket motors. It was designed to fill the performance and cost gap between the Atlas-Centaur and the Saturn vehicles and boost NASA's heaviest unmanned payloads into orbit or interplanetary trajectories. The first of six scheduled operational missions—the U.S. and West German Helios solar probe scheduled for the fall—would be configured to a two-burn mission to obtain data lost on the proof flight.

The Titan-Centaur program was managed by NASA's Office of Space Science. Lewis Research Center was responsible for the Titan-Centaur system. The Air Force Space and Missile Systems Organization was responsible for the Titan IIIE booster, including aerospace ground equipment and Launch Complex 41 site activation. Kennedy Space Center directed launch operations at ETR. (NASA MORs, 25 Jan, 24 June 74; NASA Release 74-25; *UPI, W Post,* 12 Feb 74, A2; Robinson, *Today,* 13 Feb 74)

- A standardized solar-electric propulsion stage (SEPS) that could become part of a space transportation system for payloads in planetary and earth-orbital missions was under study at Marshall Space Flight Center. SEPS, one of several concepts under study, would be able to accept modules such as a docking subsystem for earth-orbital missions or a science package for planetary missions. Another concept, for an attachable SEP module, would provide propulsion only and depend on spacecraft subsystems for all other functions. SEPS would carry a cluster of six to nine thrusters, fueled by ionized mercury, and two lightweight 27-m solar-array wings to supply 25 kw of power to the thrusters and payloads. It was being designed for use with the space shuttle and space tug, and application of an attachable SEP module to a Mariner spacecraft was being studied. The tug, supplemented by SEPS, would be able to deploy and retrieve a 2500-kg geosynchronous equatorial payload in 100 days, tripling the amount the unaided tug could deliver. After one delivery, the SEPS could continue to operate between the shuttle orbit and intermediate orbits, delivering and retrieving 20 more payloads before the thruster life was used up. (NASA Release 74-38; NASA OMSF, interview)

- A vertical-axis windmill was under study at Langley Research Center as a potential source of nonpolluting energy. The windmill, based on a 1927 principle, had two four-meter curved blades attached at the top and bottom to a five-meter vertical shaft, set on a simple gear system, and a generator that converted wind power to electricity. The airfoil-shaped blades rotated in almost any wind and were expected to achieve enough revolutions per minute to provide the energy requirements of a single-family house at an estimated construction and installation cost of $500 to $1000. A much larger windmill project at Lewis Research Center, for the National Science Foundation, was studying large concentrations of windmills for industrial use [see 11 Jan.]. (NASA Release 74-33)

- Dr. Harriet G. Jenkins assumed her duties as Deputy Assistant Administrator for Equal Opportunity Programs at NASA Hq. Dr. Jenkins would assist Dr. Dudley G. McConnell, Assistant Administrator for Equal Opportunity Programs, in agency-wide direction of both employee and contractor aspects of NASA equal opportunity programs. She had been an educational consultant for the Response to Educational Needs project of the Anacostia District of Washington, D.C., public schools and had served in the Berkeley, Calif., school system 20 yrs. (NASA Release 74-40)

- Dr. Charles A. Berry, NASA Director of Life Sciences, was quoted as saying in an interview: "From what we know today, there is no medical reason to bar a two-year mission to Mars.... We still need more data, but I don't think the medical findings from a six-month or a year-long mission would differ appreciably from the experience with our three Skylab missions." On a two-year mission to Mars (not likely to be considered before the early 1990s), "we're going to have a hard time keeping people completely busy," leaving astronauts more time to think of their isolation. "That's another problem we'll have to deal with." Meanwhile, all the data indicate "that man is really adapting to the zero-gravity environment of space. He does that in a way that causes some very large fluctuations initially in various body systems" but, "with time, it appears that these fluctuations settle down." (*US News*, 11 Feb 74, 62–64)

- An *Aviation Week & Space Technology* editorial said the FY 1975 aerospace budget request offered the aerospace industry "its best prospect in a decade." The dollar increases were "significant because they were aimed at alleviating problems caused by a decade of neglect in basic weapon system development and diplomatic euphoria" and recognized that aerospace was an "important sinew of national power and economic strength." It was the first budget in many years that had no major program cancellations or cuts. The worst cut, $89 million, meant a six-month delay in the first shuttle flight but was accompanied by a written guarantee of future support for full funding of the program. The military budget was well balanced between research, development, and procurement. While the NASA program would not satisfy the space buffs, it was a strong continuing effort supplemented by $500 million in military space activities and the growing communications and applications space activity in other agencies. The aerospace industry must now spend extra effort in delivering the hardware and performance to meet the challenge required by this budget. (Holz, *Av Wk*, 11 Feb 74, 7)

- The TV film "Houston, We Have a Problem" was criticized by former astronaut James A. Lovell, commander of the aborted *Apollo 13* moon-landing mission (launched 11 April 1970), in a letter to the NASA Administrator, Dr. James C. Fletcher. The fictitious film focused on the personal lives of the Mission Control team that brought the crippled *Apollo 13* spacecraft back to earth. Lovell said the film was "in poor taste"; the safe return of *Apollo 13* was one of NASA's finest hours and "it is not necessary to resort to soap opera plots to enliven the . . . story." (Carmody, *W Post*, 1 March 74, C7)

- Economy moves left Kennedy Space Center dangerously short on fire protection, according to Jack N. Anderson and Les Whitten in a *Washington Post* article. The Center was being protected between 3:00 pm and 7:00 am by eight on-duty fire fighters, not enough to man the $750 000 worth of fire equipment or fight a conflagration, they said. The 3 237 500-sq-km facility contained 3747 cu m of liquid oxygen, 3028 cu m of liquid hydrogen, and rocket fuel "to stoke a flash fire." Boeing Co., which held the KSC fire protection contract, planned to lay off 21 firemen and shift others to lower paying jobs. A Boeing spokesman had said no manned missions were planned until 1975 and the base was adequately protected. The *Post* quoted a NASA spokesman as saying that the cuts were necessary and that additional fire fighting help was only five minutes away. (*W Post*, 11 Feb 74, B13)

*12 February:* With the 8 Feb. splashdown of the *Skylab 4* astronauts, 1816 employees at Kennedy Space Center saw their jobs come to an end, the Baltimore *Sun* reported. The *Sun* quoted a NASA spokesman as saying that by 30 June the contractor and Civil Service work force at

KSC would be 9450. The latest group to be laid off included mostly contract employees for Boeing Co., General Electric Co., and International Business Machines, Inc. (AP, B *Sun*, 12 Feb 74, A9)
- Tass commentator Georgy Sergeyev commented on the U.S.–U.S.S.R. Apollo-Soyuz Test Project: The success of Skylab and the first flight of Yuri A. Gagarin, the space walk of Aleksey A. Leonov, and the exploration of the moon by American astronauts, were only "milestones along the road of exploration of outer space." ASTP would be another step. The joint Soviet-American mission had become possible, not only because of the efforts of scientists, but also because of the "conditions of a favorable political climate established in Soviet-American relations as a result of the fruitful . . . summit meetings." The main task was to see that "these changes become irreversible." ASTP would go a long way toward "making space exploration really internationalist." (Tass, FBIS–Sov, 14 Feb 74, B6)

*13 February:* The Air Force launched an unidentified satellite from Vandenberg Air Force Base on a Titan IIIB-Agena vehicle. The satellite entered orbit with a 404-km apogee, 128-km perigee, 89.8-min period, and 110.4° inclination. The press later reported the satellite would support the Navy's surface surveillance program. It reentered 17 March. (Pres Rpt 74; *Av Wk*, 4 March 74, 24)
- The House of Representatives by a 253-to-2 vote passed H.R. 11864, the Solar Heating and Cooling Demonstration Act of 1974. The bill would authorize $50 million over five years for the early commercial demonstration of solar heating technology by NASA and the Dept. of Housing and Urban Development, in cooperation with the National Bureau of Standards, the National Science Foundation, the General Services Administration, and other Federal agencies, and early development and commercial demonstration of technology for combined solar heating and cooling. H.R. 11864 was the clean bill that resulted from the 19 Dec. 1973 recommendation of the House Committee on Science and Astronautics' Subcommittee on Energy to revise, amend, and combine H.R. 10952, introduced 16 Oct. 1973, and 18 similar bills. The bill was sent to the Senate. (*CR*, 13 Feb 74, H774; *Background & Legis Hist*, Com Print, Feb 74)
- Marshall Space Flight Center awarded a 90-day $950 000 cost-plus-fixed-fee contract to Thiokol Chemical Corp. for studies, analysis, planning, and design required to define the interface and performance relationships of the solid-fueled rocket motor to the solid-fueled rocket boosters, external tank, and orbiter of the space shuttle. Award of the 90-day contract was necessary to continue the overall development schedule of the total space shuttle system. A contract award to Thiokol in November 1973 to develop the motors for the shuttle had been delayed pending the resolution of a protest filed by Lockheed Propulsion Co. (NASA Release 74–130; *Huntsville Times*, 17 Feb 74)
- NASA and European Space Research Organization officials met in Paris to discuss future cooperative missions to the planets. Joint mission possibilities included an out-of-the-ecliptic flight to study the sun, missions to Jupiter and Mercury, and experiments on missions to Venus and Mars. NASA would receive suggestions for the 1980s from European nations during the next few months. NASA Director of Planetary Programs Robert S. Kraemer had said that international cooperation would

be necessary for one third to one fourth of the payloads the agency envisioned for the next 15 yrs. (NASA, Spec Asst to Deputy Assoc Admin, interviews, 20–21 Jan 75; Couvault, *Av Wk*, 1 April 74, 38–9)

- Concorde 202, the second production Anglo-French supersonic transport, made its maiden flight from British Aircraft Corp.'s airfield at Filton, England. The aircraft, developed jointly by BAC and Aérospatiale France, was airborne 1 hr 16 min, including 30 min of supersonic flight, reaching a speed of mach 1.4 and an altitude of 12 800 m. On arrival at Fairford, the Concorde made two low-level passes over the runway for instrument calibration tests before making a perfect touchdown, piloted by BAC's Director of Flight Testing M. Brian Trubshaw. (BAC-Aérospatiale Release 2c/74)

- A *New York Times* editorial commented on *Mariner 10*'s 5 Feb. encounter with Venus: In less than 15 yrs since a Soviet spacecraft had radioed back to earth the first pictures of the far side of the moon, American and Soviet spacecraft had provided picture coverage of the moon, Mars, and Jupiter. "Now Venus . . . and, if all goes well, Mercury," too, would be photographed. "Man's understanding of the solar system, its nature, origin, and history has been profoundly transformed by these remarkable feats, all carried out in so brief a period . . . at a rather small cost. . . . What are in effect telescopes in space rockets have now realized achievements that can surpass the possibilities of even the largest earth-space telescopes in giving us information about our neighbors in the solar system." (*NYT*, 13 Feb 74, 36)

*14 February:* Tentative NASA plans for launching 810 payloads into space in the next two decades were outlined by Dr. James C. Fletcher, NASA Administrator, in an address before the National Space Club in Washington, D.C. The total, for late 1973 to 1991, included non-NASA payloads for Government agencies (other than the Dept. of Defense), foreign agencies, and private industry. The tentative payload model—presented to Congress in October 1973 to show "what *could* be done in the 1980s" when the space shuttle was in use—included as many as 10 missions to the outer planets Uranus, Neptune, Jupiter, and Saturn. In 1990 and 1991 two very heavy payloads might be sent to orbit one of Jupiter's moons and land an instrument package. Tentative missions to the inner planets included five missions to Venus, one of them with a lander, and two spacecraft to orbit Mercury. Tentative Mars missions included one spacecraft to orbit the planet, two landers, and two to return samples from the two moons of Mars. Lunar missions under consideration included a lunar polar orbiter in 1979, two other orbiters in the 1980s, two lunar rovers in the 1980s, a lunar halo satellite to ensure communications with the hidden side of the moon, and two lunar rovers in 1990 and 1991 that could return samples to the earth from any point on the moon. Other plans under consideration were a flyby and rendezvous with Comet Encke and a landing on the nucleus, a flyby of Haley's Comet in 1985, and two missions to asteroids in 1986.

Of the 810 total payloads, 57 were designated for planetary exploration and 753 for earth orbit. Scientific satellites for earth orbit were Large Space Telescopes, High Energy Astronomical Observatories, Large Solar Observatories, Large Radio Astronomy Observatories, and x-ray telescopes. The large spacecraft would be unmanned but visited regularly by space shuttle crews and brought back to earth for refurbishing. In the applications field were large Earth Observatory Satellites. And NASA expected to orbit 120 communications and navigation satellites for other agencies or private industry through 1991, not including those for DOD.

Of some 300 payloads to be flown in the Spacelab, most would be NASA payloads in the fields of astronomy, physics, earth observations, earth and ocean physics, communication and navigation, life science, and space technology. Of non-NASA payloads, 10 would be for private industry space processing, beginning in 1985. Some 10% were expected to be flown for foreign users.

The large number of contemplated missions, even with the increasing number of U.S.S.R. launches, was expected to have very little impact on the environment. Effects of the major constituents of the solid-fueled rocket booster exhaust—water vapor, carbon dioxide, hydrogen chloride, and aluminum oxide—would be negligible. (Text)

- France reluctantly decided to cut back construction of the Anglo-French Concorde supersonic airliner and to scrap production of the Mercure short-range twin-jet airliner, at a special meeting of Premier Pierre Messmer and five ministers. France had been demanding production of eight Concorde aircraft but would now accept as few as six in the knowledge that, with a Labor Party victory in the 28 Feb. British elections, the U.K. might be willing to scrap the entire project. The press said the French decision came after a report from France's rocket testing center suggesting a $90-million fuel-tank modification to boost range at the expense of reduced transatlantic loads. A suggested 10% increase in engine thrust, requiring 8% more wing area at a cost of as much as $400 million, was believed rejected at the meeting.

  The decision to scrap the 120-seat, 1200-km range Mercure was taken after an initial production run of 10 aircraft and writeoff costs reported by the press at $120 million. Competition from the already established DC-9s and Boeing 727s and 737s was too great for the Mercure, whose one buyer was Air Inter, the internal French airline. (Randal, *W Post*, 17 Feb 74, A1)

- NASA Deputy Associate Administrator for Applications Leonard Jaffe—U.S. Representative to the United Nations Committee on the Peaceful Uses of Outer Space Working Group and Task Force on Remote Sensing—reviewed U.S. environmental and earth resources programs before the Working Group and Task Force. The U.S. had designed tape recorders and studies on *Erts 1* Earth Resources Technology Satellite (launched 23 July 1972) for foreign use of the data, but the "extent and vigor of foreign interest was perhaps greater than we expected." Brazil and Canada had built ground stations to receive data directly and, by agreement with the U.S., were disseminating data. Italy also would establish a ground station, and nine other countries had expressed interest.

  *Erts 1* performance had shown that data collection with a spatial resolution of 100 m was economical and adequate for a wide range of significant applications. Although total discontinuance of photographic imagery was not planned, the speed, reliability, and flexibility found in computer analysis indicated that future systems would increasingly use automated methods.

  The mapping accuracy of the satellite's instruments was adequate for 1:1 000 000 and even 1:250 000 scale when system errors were removed by computer processing. Processing and disseminating data was a key factor in the cost of an earth observation program. No one data-processing and dissemination facility would ever be able to meet the needs of the world efficiently. Estimates put the cost of an additional system similar to the *Erts 1* data-collection system—aside from spacecraft costs—at $11 million for the facility and $9.9 million for five years of data processing. A minimal system with only multispectral scanner and real-time observations would cost $2.5 million for the facility and $3.6 million for five years of data processing. (Text; NASA OA, interview)

- French, Canadian, and U.S. scientists were participating in projects to expand Univ. of Hawaii facilities on top of 5000-m Mauna Kea into one of the world's greatest centers for celestial observations. In addition to NASA's planned 3-m infrared telescope announced 4 Feb., Hawaii had recently concluded an agreement with the National Research Council of France and the National Research Council of Canada for a $20-million enterprise to operate a new 3.8-m telescope for general astronomical observations. The university already had three telescopes on Mauna Kea. (Durdin, *NYT*, 14 Feb 74, 33; NASA Release 74–23)

- A *Washington Star-News* editorial commented on the triumph of the Skylab program: Although a "fellow wondering where his next tank of gas is coming from has difficulty getting excited about the cosmos," the country "could take much encouragement from the successful completion of the $2.6-billion Skylab program." We could expect that this had been well worth the effort and that a similar crash program on energy could meet with the same success. Americans would not venture into space again until the space shuttle project at the end of the decade, except for the 1975 linkup of U.S. and Soviet spacecraft. If all went well, that "could be the most fascinating and heartening adventure since that first landing on the moon." (*W Star-News*, 14 Feb 74, A18)

*15 February:* NASA Associate Administrator for Manned Space Flight Dale D. Myers had resigned effective 15 March to return to private industry, NASA announced. Myers would rejoin Rockwell International Corp. as president of a new aircraft group with headquarters in Los Angeles. Myers, who had headed all NASA manned space flight programs since January 1970, had received the NASA Certificate of Appreciation in 1969 for his contributions to the *Apollo 8* moon-orbiting mission, the NASA Public Service Award in 1969 for his contributions to the *Apollo 11* landing mission, and the NASA Distinguished Service Medal in 1971 for his contributions to the continued success of the Apollo program.

Skylab Program Director William C. Schneider had been named Acting Associate Administrator for Manned Space Flight. (NASA Release 74–42)

- A one-fifth scale model of a large portion of the Air Force B–1 swing-wing strategic bomber had been tested in transonic and supersonic wind tunnels at the Air Force Systems Command's Arnold Engineering Development Center, AFSC announced. The tests verified inputs to the computer control system that adjusted the B–1's double inlets for an adequate air supply to the engines during flight maneuvers. Test data also would be used to optimize the inlets' boundary-layer control. (AFSC Release OIP 196.73)

*16 February:* Japan successfully launched *Tansei 2* experimental MS–T2 satellite from Kagoshima Space Center, Uchinoura, on a three-stage MU–3C booster at 2:00 pm local time (1:00 am EDT) after a two-day postponement because of bad weather. A thrust-vector-control system had been installed on the 2nd stage of the booster to stabilize its flight.

**16 February**

The improved version of the four-stage MU–4S was the first guided booster Japan had used. The satellite entered orbit with a 3229-km apogee, 283-km perigee, 121.7-min period, and 31.2° inclination. KYODO news service reported the objective was "to test the control devices and not for scientific observation."

*Tansei 2*, named "Light Blue" for the Univ. of Tokyo colors, was Japan's fifth satellite and third test satellite. *Ohsumi* (launched 11 Feb. 1970) and *Tansei 1* (launched 16 Feb. 1971) were test satellites returning performance data. *Shinsei* (launched 28 Sept. 1971) had carried instruments to study cosmic rays and electric waves, and *Denpa* (Radio Explorer Satellite launched 19 Aug. 1972) had carried instruments to probe the ionosphere. (GSFC *SSR*, 28 Feb 74; FBIS–Japan, 22 Feb 74, 13–14; *A&A 1971*)

*18–26 February:* The cooperative Italian-NASA *San Marco 4* (San Marco C2) Explorer satellite was successfully launched by an Italian crew from the San Marco launch platform off the coast of Kenya on a NASA Scout booster. Liftoff was at 1:05 pm local time (6:05 am EDT), 2 hrs 50 min late because of low cloud cover. The satellite entered orbit with a 930.5-km apogee, 234.5-km perigee, 96-min period, and 2.9° inclination to measure diurnal variations of the equatorial neutral atmosphere's density, composition, and temperature. The data would be correlated with data from *Explorer 51* (launched 15 Dec. 1973) for studies of the physics and dynamics of the thermosphere. The launch had been delayed from 6 Feb. to allow NASA's Spaceflight Tracking and Data Network (STDN) to give better support for the 8 Feb. *Skylab 4* splashdown.

The Italian-built 170-kg *San Marco 4* carried three experiments: an Italian density-drag accelerometer to measure atmospheric drag, a U.S. omegatron to measure the temperature and density of thermosphere molecular nitrogen, and a U.S. neutral mass spectrometer to measure primary constituents of the neutral atmosphere. Data gathered by *San Marco 4* in the equatorial zone and *Explorer 51* in the auroral zone would provide a good picture of the effects of magnetic storms on the thermosphere and differences in the thermosphere's response to energy coming from the sun to the poles and the equator.

The omegatron experiment was activated 19 Feb. and the neutral mass spectrometer 22 Feb. By 26 Feb. both instruments were acquiring worthwhile scientific data. The Italian density-drag accelerometer, activated shortly after launch but not performing properly, was being investigated. All spacecraft systems were functioning as planned.

*San Marco 4* was the fourth satellite launched under cooperative agreements between NASA and the Italian Space Commission. *San Marco 1* was launched 15 Dec. 1964; *San Marco 2*, 26 April 1967; and *San Marco 3*, 24 April 1971. Under the latest agreement, signed 6 Aug. 1973, Italy had designed and built the spacecraft, provided one experiment, and conducted launch operations. The U.S. provided the Scout booster, two experiments, technical consultation, launch crew training, and spacecraft tracking and data acquisition. (NASA MORs, 15, 26 Feb 74; NASA Releases 74–30, 74–39)

*18 February–6 March:* Thomas L. Gatch was lost at sea while attempting the first transatlantic balloon flight. Gatch lifted off from Harrisburg, Pa., 18 Feb. in a two-meter sealed gondola, the *Light Heart*, suspended from 10 helium-filled balloons. He hoped to reach 12 000-m altitude and ride the jet-stream winds across the Atlantic to Europe, but changes in air currents pushed him south of his planned course. Radio contact was made with Gatch 19 Feb. by an airliner 1600 km east northeast of Puerto Rico. He was spotted at 305-m altitude 21 Feb. by a Liberian ship 1670 km off the coast of Africa. Further inconsistent and unconfirmed radio contacts and sightings made tracking Gatch's balloon difficult.

After reports of a sighting on the Spanish Sahara, the Spanish Foreign Legion searched the 480-km Moroccan-Spanish Sahara border for two days, while U.S. military aircraft and ships were put on lookout. The Dept. of Defense ordered an intensive visual radar and radio search 1 March of the Atlantic Ocean 1530 km west of Africa, but the search was given up 6 March after exploration of 578 000 sq km of ocean. (*W Star-News*, 18, 22, 26 Feb. 74; *W Post*, 25 Feb, 2, 7 March 74; *B Sun*, 21, 25 Feb 74)

*19 February:* NASA Associate Administrator for Manned Space Flight Dale D. Myers testified on the manned space program before the House Committee on Science and Astronautics' Subcommittee on Manned Space Flight during FY 1975 authorization hearings. Myers said NASA's ability to hold close to the original cost target for the space shuttle in spite of schedule adjustments in the last two years was "due to bringing on our contractors below our cost estimates" and the ability of the management structure to remain dynamic and react to changes and "surprise problems." He was confident that the average-cost-per-flight commitment of $10.5 million in 1971 dollars would be maintained. However, resolution of the protested selection of Thiokol Corp. for the solid-fueled rocket motor contract was necessary before work on the motor could begin; "a timely start of this . . . activity is very important to orderly progress in the shuttle development schedule."

In the Apollo-Soyuz Test Project, the ASTP spacecraft was to be delivered to Kennedy Space Center in the fall, joint docking system qualification testing and preflight docking system compatibility tests were to be completed during the winter, experiment hardware was to be qualified by early spring, and flight readiness activities were to be completed by spring, leading to the July launch. NASA had allocated $13 million for experiments for ASTP. More than 140 proposals from the scientific community had been evaluated by eminent scientists and medical doctors. The final, "first-rate" experiment package had been given final approval by the Administrator.

Following ASTP, two Saturn Vs, two Saturn IBs, one complete and one partially complete command and service module, the unused Skylab backup Workshop cluster, and one ASTP docking module and system would remain unused. Acquisition cost of the hardware was $870 million. With the conclusion of the programs, the hardware would be stored. Total storage costs per year would be $100 000. After FY 1975, disposition of the hardware would have to be considered. (Transcript)

- A nine-month, $321 394 study contract to provide NASA with engineering concepts, requirements, and design trade-offs for a zero-gravity atmospheric-cloud-physics-experiment laboratory to fly on the space shuttle had been awarded to McDonnell Douglas Astronautics Co., Marshall Space Flight Center announced. The proposed laboratory would permit study of "trigger" actions in clouds, such as the formation of ice crystals or water droplets, electrical charges on droplets, and water-droplet–ice-crystal interactions to aid in the ultimate control and modification of hazardous weather conditions. The laboratory could be reused on at least 20 Spacelab missions. (MSFC Release 74-27; MSFC PAO)

- Jean-Pierre Causse, Head of the European Space Research Organization's Spacelab Program, announced his resignation effective in early April. He would join the French company Saint-Gobain-Pont-à-Mousson as Director of Research. (*Spacelab Newsletter*, 74-2; NASA prog off, interview)
- Communications Satellite Corp. reported 1973 earnings of $36 299 000, equal to $3.63 per share, an increase from $24 967 000, or $2.50 per share, for 1972. Net operating income for 1973 totaled $29 424 000 and revenues $119 291 000, up from income of $21 428 000 and revenues of $105 965 000 for 1972. Leased full-time half circuits had totaled 3583 on 31 Dec. 1973, 21% more than the 2971 at the end of 1972. (ComSatCorp Release 74-11)

*19 February–1 March:* The U.S.S.R. launched a series of test missiles into the north Pacific Ocean, the tests coinciding with resumption of the Strategic Arms Limitation Talks in Geneva. Pentagon spokesman Jerry W. Friedheim said an SSX-18 carrying several multiple independently targetable reentry vehicles (MIRVs) was launched from Tyuratam 19 Feb., and the press reported 5550-km flight tests of the SSX-19 and SSX-16 missiles were made before the conclusion of the tests. (Tass, FBIS-Sov, 19 Feb 74, V1; 1 March 74, V7; Hoffman, *W Post*, 21 Feb 74 A3)

*20 February:* Pioneer 10, launched by NASA 2 March 1972 toward a 3 Dec. 1973 encounter with Jupiter, was 60 million km beyond Jupiter and 913 million km from the earth, traveling at 80 000 km per hr on its flight path out of the solar system. Communications with the spacecraft, which took 1 hr 42 min, indicated *Pioneer 10*'s operational systems were functioning well. Minor changes in operational systems during passage through Jupiter's intense radiation belts had disappeared or had no effect on spacecraft performance, but the cosmic ray experiment had suffered some loss of function. Experimenters were assessing the problem.

*Pioneer 11*, launched 5 April 1973, was 660 million km from the earth and traveling through the Asteroid Belt at 58 000 km per hr toward a December 1974 encounter with Jupiter. Its meteoroid detector had recorded eight hits in the belt, about the same rate of penetrations as for *Pioneer 10* when taking into account design differences in the meteoroid detectors. (NASA Release 74-43)

- Skylab Program Director William C. Schneider testified before the House Committee on Science and Astronautics' Subcommittee on Manned Space Flight in NASA's FY 1975 authorization hearings on early results from Skylab's earth resources data. Perhaps the most interesting so far was the unconfirmed indication of mineral deposits near Ely, Nev. The Skylab earth resources experiment package had also identified several areas of citrus fruit fly infestation on the Mexican side of the Rio Grande, demonstrated the use of remote sensing for snow mapping, identified new urbanization patterns in Phoenix, Ariz., and inventoried vegetation patterns in California test sites. A Baltimore-Washington area photo and similar imagery for 12 other cities were being used by the Dept. of Interior to test their use for updating the 1970 census.

Dr. Joseph P. Kerwin, astronaut on the 25 May–22 June 1973 mission to man the Orbital Workshop, testified that Skylab astronauts were healthy, happy, and not under significant stress during their long flights. After initial decreased appetite, an astronaut had a feeling of well-being with normal thirst and appetite, although he had a continued feeling of fullness in the head and tended to assume a peculiar posture with neck extended and moving backwards and elbows, hips, and knees flexed at 20° to 30°. He tended to become sleepy and required frequent exercise to restore vigor and alertness. His voice had a peculiar nasal quality. After the third day he was virtually immune to motion sickness. Without gravity, there was no automatic sensation of up and down and without vision he could quickly get lost.

Dr. Harry C. Gatos, Massachusetts Institute of Technology scientist and Skylab materials processing investigator, testified Skylab experiments had shown that segments of crystals partially regrown in space exhibited none of the inhomogeneities from dopants (electronically active chemical elements) that characterized earth-grown crystals and inhibited their efficiency. In addition to preparation of special materials for electronic and medical uses, materials processing studies in space would provide an understanding of the structures of materials and permit preparation of materials to perform closer to their theoretical limits. Crystal-growing technology was developing new means to create and store power and could, within five years, provide a practical economic basis for using solar power. (Transcript)

- The first two Lockheed S-3A Viking carrier-based jet aircraft were delivered to the Pacific fleet in San Diego. The $10-million swept-wing submarine-hunters carried 60 acoustic listening devices that could be dropped to the sea within 10 sec to transmit underwater sounds to the Viking. The Viking, which could fly 800 km per hr and drop from a cruising altitude of 12 000 m to sea level in two minutes, would replace the S-2, the mainstay of the Navy's antisubmarine defenses for 20 yrs. The Navy wanted delivery of 187 Vikings through 1975. (Holles, *NYT*, 21 Feb 74, 13)

- Dr. James C. Fletcher, NASA Administrator, said in a speech at Utah State Univ. that he was "certain there is intelligent life somewhere in this universe, but probably not in this solar system." Primary life would probably be found in our solar system, with Mars as the most likely candidate. The U.S.S.R. was expected to land a spacecraft on Mars within the next few weeks, but in an area not likely to have any life form. A NASA spacecraft would land on Mars in 1976 in the area most likely to have water. NASA intended to continue systematic exploration of the solar system and to search for extraterrestrial life. (Bean, *Logan* [Utah] *Herald Journal*, 20 Feb 74)

*21 February:* Dr. Goetz K. Oertel, NASA Chief of Solar Physics, testified during NASA FY 1975 authorization hearings before the House Committee on Science and Astronautics' Subcommittee on Manned Space Flight that Skylab data had changed the understanding of the solar wind. Scientists had thought the solar wind reached the earth because of an expansion of the sun's corona, but Skylab data indicated the sun's magnetic fields confined the corona material and pulled it inward. The corona expanded and escaped in a significant amount only through coronal holes in the magnetic field.

Dr. Charles A. Berry, Director of Life Sciences, testified that the *Skylab 2* and *Skylab 3* crews (launched 25 May and 28 July 1973) had experienced similar reductions in their blood cell mass, with a red blood cell reduction. However, after the 59-day *Skylab 3* mission the compensating response of the blood-forming elements was faster, returning the mass almost to the preflight level within two weeks after splashdown—rather than the six weeks required after the 28-day *Skylab 2* mission. The *Skylab 4* crew (which splashed down Feb. 8 after 16 Nov. 1973 launch) showed much less red blood cell loss and a much faster return to preflight levels after the 84-day mission, indicating adaptive changes to the space environment. The space life sciences program sought to understand the mechanisms of adaptive changes and would "undoubtedly assist our understanding of basic body physiology of man here on earth."

Capt. Chester M. Lee, Apollo-Soyuz Test Project Program Director, said that during Skylab missions the orbital altitude and the limited number of ground stations had permitted communications coverage only 38% of the time. The lower orbital altitude for ASTP would reduce communications coverage to 17%. To alleviate the problem, the Apollo command module was being modified for communications via the ATS–F Applications Technology Satellite, scheduled for 1974 launch. Communications would move from a manned spacecraft through an unmanned communications satellite to the ground for the first time. Coverage would be increased to 50%. The present configuration of the communications network would furnish minimum real-time data to meet experimental requirements. (Transcript)

- Preliminary results of the 18–26 Dec. 1973 *Soyuz 13* mission were published in *Izvestiya*. The two crew members of the "orbiting astrophysical observatory" had made 10 000 spectrograms with the Orion 2 telescope, recording emissions of stars in ultraviolet. Remote luminaries up to 12th stellar magnitude had been recorded. The biomass of the microbe culture in the Oasis 2 experimental closed-cycle biological system had increased more than 35 times during the mission. (*Izvestiya*, FBIS–Sov, 5 March 74, U1)
- Sen. Sam J. Ervin, Jr. (D–N.C.), introduced S. 3034, a bill to prohibit the reservation of appropriated funds except to provide for contingencies or to effect savings. (*CR*, 21 Feb 74, S1955)

*22 February:* NASA Deputy Associate Administrator for Programs George W. Cherry testified in hearings on technology for subsonic aircraft before the House Committee on Science and Astronautics' Subcommittee on Aeronautics and Space Technology. NASA aeronautical programs were structured to provide the airframe industry and its airline customers an efficient and environmentally benign technology base for a new-generation subsonic aircraft. The supercritical aerodynamics program included studies to improve structural efficiency without increasing drag, thrust, or fuel consumption. F–8 and F–111 aircraft flight tests with the supercritical wing had shown that using supercritical aerodynamics on passenger transports would increase profits 2.5% over those of conventional aircraft, or $78 million a year, on a fleet of 280 two-hundred-passenger aircraft.

Advanced transport technology studies showed that aircraft extensively employing composite materials would show a yearly profit of 2.7%, or $100 million, on a fleet of 280 two-hundred-passenger aircraft.

NASA propulsion technology had initiated an advanced multistage axial-flow experimental compressor weighing 60% less than conventional compressors. This, with technology advances in other components, could reduce gross aircraft weights and fuel consumption 10%. (Transcript)

- Large underwater oscillations, called "internal waves" and usually invisible at the surface, had been detected by National Oceanic and Atmospheric Administration scientists Dr. John R. Apel and Robert Charnell, using images from NASA's *Erts 1* Earth Resources Technology Satellite (launched 23 July 1972), NOAA announced. The scientists speculated that internal-wave formation by lunar and solar tides and their subsequent breaking and dissipation of energy were important processes in the gradual lengthening of the day over millions of years. The waves also affected the propagation of sound in the ocean and contributed to the mixing processes in its upper layers. Internal waves changed the way sunlight was reflected from the ocean surface overlying them, and the change was detected in the satellite images. (NOAA Release 74-32)

*23 February:* The U.S. Postal Service announced the design of a 10-cent stamp to commemorate the Skylab program. To be issued 14 May on the first anniversary of the launch of the *Skylab 1* Orbital Workshop, the stamp would depict the 7 June 1973 two-man spacewalk that freed the jammed solar panel during the *Skylab 2* mission (25 May to 22 June 1973). The horizontal stamp was designed by Robert T. McCall of Arizona. (USPS Release 11)

- A *Washington Post* editorial commented on solar heating legislation: There was "enough sunshine in this country . . . to supply twice the energy needed to heat and cool all our buildings." Technically, solar heating was simple. Solar cooling was more complicated, but within the grasp of a determined engineering effort. The probem was psycho-economic rather than technical. Industry did not want to produce hardware because they were not sure people would buy it. A bill [H.R. 11864] passed in the House by a 253-to-2 vote [see 13 Feb.] would make $50 million available for a national demonstration of solar heating in 4000 homes in various parts of the country. The Senate would begin hearings on a similar measure [S. 2658], and companion measures would encourage solar heating and cooling with tax incentives and mortgage rule changes. We need "a wide-open, long-range determined public energy policy . . . and we need it in all areas that promise to help free us of our dependence on shrinking oil resources. The new solar-heating legislation is a welcome step in this direction." (*W Post*, 23 Feb 74, A16)

- Dr. George A. Van Biesbroeck, Univ. of Arizona astronomer noted for his work with twin stars, asteroids, and comets, died in Arizona at the age of 94. Dr. Van Biesbroeck had charted the orbit of Neptune's second satellite, published catalogs of double stars, and discovered 11 asteroids and 2 comets. In the 1930s he had been instrumental in establishing the McDonald Observatory in Fort Davis, Tex. Dr. Van Biesbroeck had won the Franklin L. Burr Award from the National Geographic Society in 1952 for finding new proof of the bending of light from stars in support of the Einstein theory of relativity. (AP, *NYT*, 26 Feb 74, 36)

*24 February–1 March:* This would be the age we could contact other civilizations, Dr. Frank D. Drake, Cornell Univ. astronomer, told the annual meeting of the American Assn. for the Advancement of Science in San Francisco. The problem was finding ways to detect them. Analysis of data from a 1973 search of 500 stars by Univ. of Maryland and Univ. of Chicago scientists had failed to reveal radio emissions suggestive of other technological civilizations. Possibly "no one is seeking to make radio transmissions" and eavesdropping would be the only method of detection. Project Cyclops, a NASA-supported study proposing construction of a 16-km-wide field of 1400, 91-m antennas, would have a high probability of success if civilization existed within 200 light years. The ultimate $5-billion cost would be less than the Apollo moon landings and would give man a chance to contact the "galactic community." (*NYT*, 14 March 74, 15; Salisbury, *CSM*, 7 March 74)

*25 February:* NASA Deputy Associate Administrator for Applications Leonard Jaffe—U.S. Representative to the United States Committee on the Peaceful Uses of Outer Space Working Group and Task Force on Remote Sensing—told a session of the Working Group and Task Force the U.S. believed its policy of openness and benefit-sharing in earth resources and environmental remote-sensing programs was consistent with the Outer Space Act of 1967, which called for use of outer space for the benefit of all countries. The U.S. could see no justification in the suggestion that sovereignty over natural resources included control over all information about these resources. The U.S. would regret any setback to the principle of open and unimpeded international exchange of information and would continue its open distribution policy. A restrictive policy would slow the development of remote-sensing systems and cut off developing countries from beneficial data. Limiting data to conform to national boundaries would require costly techniques and would destroy the most useful functions of remote sensing; most ecological systems had to be studied globally. (Text)

- Hearings on S. 2658 and H.R. 11864, to enact the Solar Heating and Cooling Demonstration Act of 1974, were held by the Senate Committee on Aeronautical and Space Sciences. Rep. Mike McCormack (D–Wash.) said the legislation—which called for NASA, with other agencies, to develop and demonstrate solar technology—would set up a program that the National Science Foundation was not authorized to do and therefore would create no conflict with NSF's ongoing research and development program in solar energy.

Dr. H. Guyford Stever, NSF Director, said that the bill should be considered in the context of the "intensive and coordinated R&D effort" already under way in the Executive Branch. The NSF Act had authorized NSF as lead agency in basic and applied solar research and, with interagency cooperation, NSF could bring "practical solar energy systems into widespread use." Dr. Stever expressed the Nixon Administration view that the Energy Research and Development Administration (ERDA), proposed in other legislation, would be the preferred agency to accelerate applied solar technology when programs reached the stage of large demonstration and development. However, if that legislation were not approved, the Dept. of Housing and Urban Development would be preferable over NASA as lead agency, because of HUD's experience with the local groups that would carry out the projects.

Dr. James C. Fletcher, NASA Administrator, also expressed the Administration's support of ERDA and of continuing NSF as the lead agency until ERDA was established. Dr. Fletcher suggested as an alternative plan, if ERDA were not established, that NASA serve as lead agency in phase one, the development of the technology and construction of solar units. Phase two, building the houses and installing the units, could be best handled by HUD. There would be "no big problem" in transferring the responsibility to ERDA should it be established. (Transcript)

- John H. Disher, Skylab Program Deputy Director since August 1965, had been named Director of Advanced Programs in NASA's Office of Manned Space Flight, NASA announced. Skylab had concluded operations with the 8 Feb. splashdown of its third manned mission to visit the Nation's first experimental space station. In the Advanced Program Office, Disher would head evaluations of concepts for manned space activities within changing national needs and priorities, to guide transition from present to future programs. From 1959 to 1961 Disher had headed advanced manned missions. He had also served as Apollo Test Director and Assistant Director for Apollo Spacecraft Development. (NASA Release 74-48)
- Marshall Space Flight Center had requested quotations from industry on a contract to study a two-way data link using laser beams and operating between ground and payloads in low earth orbit, MSFC announced. The study, which could lead to a flight system on the space shuttle, would indicate the most effective lasers and frequencies and analyze effects of orbital dynamics on acquisition, tracking, and pointing systems. (MSFC Release 74-28)

*26 February:* NASA Associate Administrator for Manned Space Flight Dale D. Myers testified during the Senate Committee on Aeronautical and Space Sciences hearings on the FY 1975 NASA authorization that space shuttle subcontractors, vendors, and suppliers had been selected in 32 states. Of the major subcontractors, 30 had already begun work. By June 1974 the total major subcontractor team would be at almost 100% of the required level. Shuttle employment was 16 000 and would reach 27 000 by June 1974 and 37 000 by June 1975. The first major subsystem test of the shuttle main engine would be made before the end of June 1974. Fabrication of Orbiter 1 would begin by the end of 1974, with the first horizontal flight test scheduled for the second quarter of 1977, the first manned orbital flight for the second quarter of 1979, and initial operational shuttle capability for 1980. (Transcript)

- Dr. Philip E. Culbertson, Director of Mission and Payload Integration in NASA's Office of Manned Space Flight, testified on manned space systems in development, before the House Committee on Science and Astronautics' Subcommittee on Manned Space Flight during FY 1975 authorization hearings. Dr. Culbertson said 986 payloads were projected for the 12-yr period beginning in 1980, with one third each for applications, science, and the Dept. of Defense. Duplication of the accomplishments expected from the space shuttle sortie (Spacelab) program would require automated spacecraft, sounding rockets, a manned space station, and a system for manned transport. The shuttle, with effective integration of payloads and the ability to service in-orbit spacecraft, offered a potential savings of $14 billion over expendable systems from 1980 to 1991. Studies had shown that 725 flights were required to fly the projected 986 payloads, an average of 60 flights a year. At a 10% rate of discount and a 10% rate of return, the space transportation system would break even if the shuttle were flown only 25 times a year. (Transcript)

*26-28 February:* NASA Associate Administrator for Applications Charles W. Mathews—testifying during the House Committee on Science and Astronautics' Subcommittee on Space Science and Applications hearings on NASA's FY 1975 authorization—said the 1974 decision to phase out NASA communications activity had been reconsidered because of industry unwillingness to risk capital on advanced communications research and development. Private industry could meet short-term commercial needs but "continued NASA activity is essential to identify and meet long term national needs." The investment level in advanced communications R&D would increase from $1.0 million in FY 1974 to $1.4 million in FY 1975. R&D included investigations into advanced antenna techniques to define radiation pattern contours to fit geographical boundaries and eliminate unwanted spillover; technology to open near- and far-infrared wavelengths for communications; new techniques in solid-state, high-power devices to replace traveling wave tubes; and use of millimeter-wave regions to expand the spectrum for satellite communications.

The ATS–F Applications Technology Satellite scheduled for 1974 launch would conduct health and education telecommunications and radio beacon experiments and measure charged particles and magnetic fields over the U.S. for one year. The satellite would then be moved over India to beam educational TV to 4000 villages. ATS–F (to be Ats 6 after launch) also would be used to extend coverage during the 1975 Apollo-Soyuz Test Project docking mission. Following completion of the Indian experiments, ATS–F would be moved back over to the U.S. to continue experiments there. (Transcript)

27 *February:* The House of Representatives approved the conference report on the Energy Emergency Act, S. 2589, clearing it for the White House. The act would declare a nationwide emergency and authorize special Presidential actions to conserve fuels and increase supply. (NASA *LAR*, XIII/20; *CR*, 22 Jan 74, H63)

28 *February:* NASA's Acting Associate Administrator for Aeronautics and Space Technology Edwin C. Kilgore—testifying during the Senate Committee on Aeronautics and Space Sciences hearings on NASA's FY 1975 authorization—said that NASA aeronautics and space technology activities, budgeted at $241.2 million, were designed to be highly responsive to national needs. To provide organizational focus, a Research Division had been reestablished, to which the OAST Research Council would report its evaluation of NASA's basic research. New research included an acoustic composite nacelle program to reduce noise and fuel consumption and a program to develop new higher performance materials for advanced turbine engines. New starts in long-haul technology, including the active control aircraft program and a remotely piloted research vehicle, would help increase productivity and efficiency, environmental acceptability, and reduced fuel consumption.

A new Office of General Aviation Technology had been organized in FY 1974 to improve aircraft and increase safety. Research using a new low-speed airfoil, a derivative of NASA's supercritical wing research, showed a 30% increase in maximum lift and a 50% increase in the lift-to-drag ratio over conventional airfoils. New programs in advanced avionics technology, including microelectronics and digital circuitry, would provide practical low-cost systems for general aviation. (Transcript)

• The number of Federal scientists and engineers had increased to 166 700 by October 1972, an increase of 6% from 1967 and 1% from 1971, the National Science Foundation reported. The Depts. of Defense and Agriculture had continued to employ the largest number, although both agencies had decreased since October 1971, by less than 0.05% and by 2%.

NASA's scientists and engineers decreased 3% from 1971. Nonprofessional scientific and technical personnel grew 1% Government-wide from 1967 to 1972. The yearly rate of salary increase for Federal scientists and engineers decreased from 9% for 1967–1970 to 5% for 1970–1972. (*NSF Highlights*, 28 Feb 74)

*During February:* The European Space Research Organization awarded study and advance procurement contracts totaling $2.8 million for its maritime satellite MAROTS. GEC/Marconi of the United Kingdom received a six-month $1.16-million contract for Phase B detailed design and definition work on the communications payload. Thomson-CSF of France was awarded $100 000 to continue critical development of an L-band transistor amplifier. Hawker Siddeley Dynamics Ltd., also of U.K., was awarded a six-month $1.13-million contract for a Phase B detailed design study to adapt the Orbital Test Satellite platform to a maritime mission and for associated advance procurement. MAROTS would be launched into geostationary orbit in August 1977 to acquire experimental data and preoperational experience. (ESRO *Newsletter*, June 74, 4)

## March 1974

*1 March:* The refanned JT8D jet engine successfully completed its first test at full power, at Lewis Research Center. The engine, part of NASA's program to reduce jet engine noise, was run through its complete operating speed range and achieved a thrust value above takeoff thrust. Built by United Aircraft Corp. Pratt & Whitney Div., the JT8D was used in a major portion of the U.S. narrow-body commercial air fleet. Modified by replacing the two-stage fan with a larger, single-stage fan, the engine could reduce the noise footprint areas of these aircraft by 75%. (NASA Release 74–57; LeRC PIO, interview, Oct 75)

- The YCH–53E helicopter, largest and most powerful helicopter in the Western world, made its first flight, lifting off from the Stratford, Conn., plant of United Aircraft Corp. Sikorsky Aircraft Div. The three-engine helicopter had a rotor blade assembly 24 m in dia and fuselage 23 m long and could carry a 14 500-kg external payload. (*W Post*, 12 March 74, D7)

*2 March:* American Broadcasting Co. Science Editor Jules Bergman recalled tales of astronauts in a *TV Guide* article and noted that, with the splashdown of *Skylab 4*, an incredible era of space flight had come to an end. In 30 manned flights the U.S. had moved from the first 15-min suborbital flight to 12 Americans walking on the moon and then finally to the near-routine Skylab endurance missions. It would be a long while before someone set out for the planets, a trip which would require far more advanced propulsion systems than now available. But "we *will* go to the planets. . . . We will go beyond Jupiter. We will go beyond this solar system and eventually build starcraft to go so far that man's only limit will be his imagination and ingenuity. But the mission clearly for the next decade is this Earth: it is the only home man has at the moment." (*TV Guide*, 2 March 74)

*3 March:* A Turkish Airlines, Inc., DC–10 jumbo jet airliner carrying 335 passengers and 11 crew members crashed in the Forest of Ermenonville 42 km northeast of Paris, France, killing all aboard in the worst air disaster in history. The crash—the second by a jumbo jet—occurred shortly after takeoff for London from the Paris Orly Airport. Aviation experts from the Federal Aviation Administration, National Transportation Safety Board, manufacturer McDonnell Douglas Corp., and Turkish government were sent to the scene to help French officials determine the cause. Eyewitness reports were conflicting but initial evidence pointed to a mid-air explosion; six bodies and one section of seats were found 10 km from the main point of impact. Later findings indicated that the rear cargo door had flown off. The sudden loss of pressure in the cargo hold had collapsed the cabin floor, severing the aircraft control cables, and sucked out the six passengers before the plane plunged steeply earthward. (Robertson, *NYT*, 4 March 74, 1; 6 March 74, 5; AP, *B Sun*, 7 March 74, A2; Randal, *W Post*, 4 March 74, A1)

*4 March:* Dr. James B. Schlesinger, Secretary of Defense, said in his annual Defense Posture Statement submitted to Congress that the U.S. would not insist on strict numerical missile equality with the U.S.S.R. in a future arms agreement. Rather, he wanted an "essential equivalence" in a diversified strategic force "to make available to the President a reasonable range of strategic options as USSR and PRC capabilities evolve." The U.S. might be willing to agree to a reduction of its land-based missile force within an overall balance.

Dr. Schlesinger also said he had directed the Navy to concentrate on smaller aircraft carriers rather than the large, $1-billion, nuclear-powered ships. He argued that four nuclear-powered carriers—one already in operation and three under construction—would meet "high-threat" needs but were not needed for the usual "low-threat situations." The carrier fleet was to remain at 15 ships through FY 1975 and be reduced by 1980 to 12 large carriers including the 4 nuclear vessels. (Text; Finney, *NYT*, 4 March 74, 14; 5 March 74, 7)

- Selection of the official emblem of the joint U.S.–U.S.S.R. Apollo-Soyuz Test Project mission, scheduled for July 1975, was announced by NASA and the Soviet Academy of Sciences. The circular emblem displayed the English word "Apollo" and the Russian "Soyuz" around a center disc depicting the two spacecraft docked in earth orbit. (NASA Release 74-49)
- Widespread material and component shortages—increased by the oil embargo imposed by oil producing countries during the October 1972 Middle East war—were causing severe problems for the aerospace industry, *Aviation Week & Space Technology* reported. Suppliers found it easier and more profitable to supply material such as aluminum to non-aerospace industry, which placed quantity orders without specific deadlines and rigid specifications. Government price controls had encouraged some suppliers to halt production or divert supplies to export markets, where profits could be two to three times above the domestic.

  A Hughes Aircraft Co. survey showed that reasons for shortages included a diminished supply of petrochemicals, which were the basis of most plastics and synthetic rubbers, and the curtailed production of materials requiring large amounts of energy for fabrication. (*Av Wk*, 4 March 74, 12-13)
- Prof. Mikhail Tikhonravov, Soviet rocket pioneer, died in Moscow at the age of 73. Prof. Tikhonravov had worked on the development of the first man-made satellites, pilot-operated spacecraft, and automatic space stations. With Sergey P. Korolev, he had worked on development of early liquid-fuel rockets in the 1930s. He had been an early expert in glider and aircraft engineering. Tikhonravov had been awarded the Orders of Lenin and the Red Banner of Labor, the title of Hero of Socialist Labor, and the Lenin Prize. (*NYT*, 7 March 74, 40; *Av Wk*, 11 March 74)

*5 March:* The U.S.S.R. launched *Meteor 16* meteorological satellite from Plesetsk to collect information for weather forecasts. Orbital parameters were 892-km apogee, 830-km perigee, 102.2-min period, and 81.2° inclination. The satellite carried instruments to photograph the clouds and snow cover on day and night sides of the globe and to collect data on the heat reflected by the earth and its atmosphere. (GSFC *Wkly SSR*, 28 Feb–6 March 74; Tass. FBIS–Sov, 6 March 74, Ul; *SF*, Sept 74, 355)

- The X-24B lifting body successfully completed its first supersonic flight near Flight Research Center, in the joint NASA–Air Force program to develop a hypersonic vehicle for reentry from space and technology for future aircraft capable of sustained cruise at hypersonic speeds. Piloted by John A. Manke and launched from a B-52 aircraft at 13 700 m, the research vehicle reached an altitude of 18 400 m and speed of mach 1.09. The powered flight satisfied the primary objectives of obtaining stability and control data at mach 1.1 for power-off and power-on conditions, performance and longitudinal trim data at mach 0.8 and 0.9, and pressure data on both fins for identification of asymmetric flow. (NASA, X-24B Flash Rpt, 5 March 74; NASA Release 73-130; Dryden FRC Release, "NASA Facts," March 1976)

- Major reorganizations were announced at NASA Hq. and Marshall Space Flight Center by Dr. James C. Fletcher, NASA Administrator, and Dr. Rocco A. Petrone, MSFC Director.

Dr. Fletcher said the Headquarters reorganization and key personnel appointments would consolidate planning and direction of research and development programs under the Associate Administrator and overall planning and direction of the field center operations under the newly created position of Associate Administrator for Center Operations. Dr. Petrone was named Associate Administrator. He would remain as MSFC Director until early summer when he would be replaced by Dr. William R. Lucas, MSFC Deputy Director. Dr. John E. Naugle, Associate Administrator for Space Science, was appointed Deputy Associate Administrator. He would also continue as Acting Associate Administrator for Space Science until a successor was named. Dr. George M. Low, in addition to his duties as Deputy Administrator, would serve as Acting Associate Administrator for Center Operations until the new position was filled. He would be assisted by Edwin C. Kilgore, Deputy Associate Administrator for Management in the Office of Aeronautics and Space Technology. Gen. Bruce K. Holloway (USAF, Ret.), who would also continue as the Assistant Administrator for Dept. of Defense and Interagency Affairs, was named Acting Associate Administrator for Aeronautics and Space Technology. Bernard Moritz was appointed Associate Administrator for Organization and Management.

The reorganization, effective 15 March, followed completion of the Apollo and Skylab programs and would provide mechanisms for the phaseover from conventional launch vehicles to the space shuttle and new programs, Dr. Fletcher said.

At MSFC, Dr. Petrone said that the reorganization, effective 30 May, would enable the Center to fulfill requirements of its varied assigned programs and improve its competitive position to obtain new assignments. The major changes were in the Science and Engineering Directorate. A Deputy for Operations and a Deputy for Systems had been established. Business operations and control of all resources had been consolidated into a centralized office at the directorate level. Elements under the S&E Directorate included two associate directors, an Associate Director of Engineering and an Associate Director of Management; two offices, Reliability and Quality Assurance and Research and Technology; and eight laboratories. The laboratories were being restructured to emphasize system engineering and integration, analysis, design, and testing. The in-house capability to manufacture, inspect, and check out major hardware projects had been eliminated, with the associated supply, warehousing, and procurement capability.

5 March

Reorganization in MSFC's Administration and Program Support Directorate included consolidation of Center-wide automatic data-processing functions within A&PS. Offices under A&PS would include Technology Utilization, Manpower, Financial Management, Facilities, Procurement, Management, Services, Logistics, and Computer Services. (NASA Ann, 5 March 74; NASA Release 74-76; MSFC Release 74-31)

- Edward Z. Gray, NASA Assistant Administrator for Industry Affairs and Technology Utilization, and a number of users of NASA technology testified in FY 1975 NASA authorization hearings before the House Committee on Science and Astronautics' Subcommittee on Aeronautics and Space Technology: In the past year NASA had developed programs with the Environmental Protection Agency in instrument development and testing, with the Dept. of Interior's Bureau of Mines in mine safety, with the Dept. of Housing and Urban Development in fire and lead-paint detection, and with the Dept. of Transportation in highway and rail safety.

  Medical applications of NASA-developed technology included a stethoscope for diagnosing respiratory diseases, developed from technology used to analyze aircraft and rocket engine sounds; a low-friction replacement for arthritic human ball-and-socket joints; magnetometers to analyze orthopedic diseases, a method to measure heart action with the computer technology used to enhance Mariner spacecraft TV pictures of Mars; and an isolation garment to protect patients from infection. (Transcript)

*5-6 March:* NASA Spacelab Program Director Douglas R. Lord testified 5 March during FY 1975 NASA authorization hearings before the House Committee on Science and Astronautics' Subcommittee on Manned Space Flight that, with the European commitment to develop Spacelab, the program had become international. Two parallel structures, the NASA Spacelab Program Office headed by Lord and a European Space Research Organization Spacelab Program Office under Jean-Pierre Causse, directed the program on either side of the Atlantic. Activities were coordinated by the Joint Spacelab Working Group, which met monthly, cochaired by Lord and Causse.

Causse testified that the ESRO Spacelab employment level was at 60 and would reach 100 by December 1974. ESRO member countries, except Sweden, had committed $369.6 million to the Spacelab effort, in the following proportions: West Germany, 54.1%; Italy, 18%; France, 10%; United Kingdom, 6.3%; Belgium, 4.2%; Spain, 2.8%; The Netherlands, 2.1%; Denmark, 1.5%; and Switzerland, 1%. ESRO would attempt to spend in each country an amount proportional to that country's contribution. Phase B studies—to provide a data base for program commitment and the final go-ahead—had been completed and one of the two competing firms, Messerschmitt-Boelkow-Blohm and ERNO Raumfahrttechnik GmbH, would be selected as prime contractor in June.

M/G Robert H. Curtin (USAF, Ret.), Director of the Office of Facilities, testified 6 March that shuttle construction projects totaled $86 020 000 and included $71 950 000 for launch and landing facilities, $7 480 000 for ground-test facilities, and $6 590 000 for solid-fueled rocket booster production and test facilities. (Transcript)

- The House Committee on Science and Astronautics' Subcommittee on Space Science and Applications held hearings on the FY 1975 NASA authorization. Dr. William Nordberg, Chief of Goddard Space Flight Center's Laboratory for Meteorology and Earth Sciences, said data from

*Erts 1* Earth Resources Technology Satellite (launched 23 July 1972) had demonstrated savings in land use planning and regulation of 10 to 100 times conventional costs in both dollars and personnel. Every cloud-free photo taken by the satellite had been sold to the general public at least once. One photo of an Alaskan naval petroleum reserve showed a striking alignment of lakes indicating that the deposits might extend the known petroleum area.

Dr. Warren A. Hovis, GSFC scientist, testified that the Heat Capacity Mapping Mission (HCMM) satellite to be launched in 1977 would use a thermal mapper, with a resolution of 0.5 km in optimum orbit, to determine maximum daily temperature variations and thermal inertia of surface material, identify mineral resources, map natural and man-made thermal effluent, and investigate geothermal source location by remote sensing.

Data from the SEASAT–A satellite to be launched in 1978—or from one of its follow-on systems—would provide optimum, minimum-time ship routing around storms and away from adverse wind, waves, currents, and ice conditions, Dr. John R. Apel, National Oceanic and Atmospheric Administration meteorologist, testified 6 March. The savings from a one-day decrease in transatlantic crossing time could amount to millions of dollars each year for shipping interests. (Transcript)

*6 March: Explorer 51* Atmosphere Explorer, launched into orbit 15 Dec. 1973, was moved in a series of maneuvers down to 139.8 km above the earth, its lowest perigee to date and its first successful excursion into the lower thermosphere. Thirteen of the fourteen experiments aboard the spacecraft performed as expected and worthwhile data were acquired, but both filaments of the closed-source neutral mass spectrometer had failed and that instrument could no longer provide data. Evidence indicated that oxidation, compounded by thermal stress, had weakened the filaments in the oxygen-rich environment. A failure review committee would study the anomaly and also evidence of further filament degradation in two other spectrometers. (NASA MOR, 19 March 74)

- The Federal Aviation Administration issued an airworthiness directive ordering all DC–10 jumbo jet aircraft operators to make certain that cargo holds were locked and pressurized before the aircraft left the ground. Evidence had increased speculation that the 3 March crash of a DC–10 Turkish airliner had been caused by a faulty rear cargo door. In addition, operators were required to observe normal pressurization of aircraft during the initial flight period. If pressurization problems did occur, the aircraft was to be depressurized and brought down at the nearest suitable airport. (AP, B *Sun*, 7 March 74, A2)

- Wind-tunnel tests of almost all supersonic portions of a typical NASA space shuttle flight were completed at Arnold Engineering Development Center. The simulated flight profile included separation of the orbiter and its fuel tank from the two large solid-fueled rocket motors, orbiter flight with and without fuel tank, and orbiter reentry into the earth's atmosphere. The tests, begun September 1973, showed orbiter flight characteristics and the heat levels segments of the system could be expected to experience while traveling $4\frac{1}{2}$ to 20 times the speed of sound. (AFSC *Newsreview*, June 74, 8–9; AEDC proj off, interview, 27 March 75)

*6–7, 12–13 March:* NASA Acting Associate Administrator for Aeronautics and Space Technology Edwin C. Kilgore and other NASA officials testified on the FY 1975 NASA OAST budget in hearings before the House Committee on Science and Astronautics' Subcommittee on Aeronautics and Space Technology. Kilgore testified 6 March that in FY 1975 OAST would modify its "program again to meet the changing needs of the Nation." Energy conservation "is an inherent part of our aeronautics and space research and I believe we have the capability also to provide significant support to civil needs for energy technology. At the same time, we must continue our effective response to the Nation's need for a quiet, clean, economical air transportation system, for support to military aviation, and for a viable program to exploit and explore space." Much of the aeronautics program [see 26 Feb.] already was "contributing to aircraft fuel conservation." Advanced avionics systems to relieve terminal congestion, more highly maneuverable aircraft and operational procedures, more efficient engines, composite materials for weight reduction, and aerodynamic drag reduction techniques—"*all* contribute to fuel economy." And the program was being examined to increase emphasis in areas related to the national energy problem, both in conservation and in long-term technology.

For space exploration, propellants with more powerful liquified-gas oxidizers—storable in cold space conditions—offered up to 30% increase in payloads over those permitted by earth-storable propellants. Nuclear energy studies included investigation of low-temperature thermionic converters, gaseous uranium reactor concepts, and use of heat from high-temperature reactor sources for industry use. In stationary power, NASA was working with the Dept. of Interior's Office of Coal Research to increase power produced from coal 40%. [See also During March.]

George W. Cherry, Deputy Associate Administrator for Programs, OAST, said 12 March that research and development on shuttle systems had contributed to substantial improvements in both high- and low-temperature reusable insulations, coatings, bearings, seals, and hydraulic fluids, all applicable on the earth. The requirement that a reusable space tug return to the shuttle for return to the earth had doubled the velocity requirement of the vehicle, speeding the development of a high-performance propulsion system. (Transcript)

*7 March:* A proposal to tap energy from tiny black holes, superdense objects in space, was presented by Dr. Lowell Wood, member of a three-man team from the Lawrence Livermore Laboratories, at a conference of international physicists organized by the New York Academy of Sciences. Some black holes within the orbit of the moon might be so tiny that their effects would not be evident beyond a few hundred meters. A tiny power plant could be placed in orbit to accompany the moving hole but remain at a discreet distance. Fusion fuel would then be fired from the power plant to the hole, and the hole's extremely powerful gravity would squeeze the fuel toward the vanishing point. The fuel would fuse, releasing vast amounts of energy, which could be collected and beamed to the earth by microwave frequencies. (Sullivan, *NYT*, 9 March 74, 12)

- The House of Representatives passed the Federal Energy Administration bill, H.R. 11793, which would reorganize and consolidate certain functions of the Federal Government in a new Federal Energy Administration to promote more efficient management of energy. (NASA LAR, XIII/26)
- The first system to use solar energy to heat a U.S. school began operation in Timonium, Md. The $568 000 project, jointly sponsored by the National Science Foundation and AAI Corp. [see 2 Feb.], was rushed

into operation in six weeks to use the remainder of the winter's chilly days to gather experimental data for future solar heating systems. (Richards, *W Post*, 8 March 74, C2)

**7, 12–14 March:** Dr. John E. Naugle, Associate Administrator for Space Science, testified on NASA's space science program before the House Committee on Science and Astronautics' Subcommittee on Space Science and Applications during FY 1975 authorization hearings: A Solar Maximum Mission (SMM) satellite was proposed for FY 1975 design study and launch in the next period of maximum solar activity, 1978–1979. SMM would investigate solar flares and related phenomena and their effects on the solar-terrestrial system, making, for the first time, simultaneous measurements of flares across the electromagnetic spectrum—within a national program of satellite, sounding rocket, and ground-based observations of the sun during maximum activity. The spacecraft would be shuttle-retrievable for refurbishment and reflight with advanced solar physics experiments in the 1980s.

Requested funding for the Large Space Telescope (LST) for FY 1975 was $6.2 million, to proceed with design of telescope optics, detectors, and support systems. NASA had originally planned to spend about $10 million in FY 1975 on the LST—to be launched on the space shuttle in 1980—but reduced the request to reflect the four- to six-month delay in shuttle development.

NASA—in cooperation with the Naval Research Laboratory, Sandia Laboratories, and West Germany—was developing an Aries sounding rocket to support the physics and astronomy program. Aries, which used the 2nd stage of surplus Minuteman intercontinental ballistic missiles, would provide a greater lift and longer flight time than possible at the price of rockets now in use.

Cost overruns on the 1975 two-spacecraft Viking mission to Mars could total 10% of the original $838-million estimate for the program because of hardware development problems. Major problems with the biology instrument included packing the sophisticated instruments into a small volume, development of very small valves to inject sufficiently small amounts of nutrients into Martian samples, and thermal control. The instrument was running $21.2 million over the budgeted cost. Difficulties also had been encountered in producing enough 0.005-cm plated magnetic wire of uniform characteristics for the onboard computer. Dr. Naugle said that, although the overall complexity of the spacecraft had been underestimated, "we expect these problems to be solved and to launch both Viking spacecraft on schedule in 1975." (Transcript)

**8–19 March:** NASA launched the United Kingdom's *Miranda* (UK–X4) experimental satellite from Western Test Range at 7:22 pm PDT on a four-stage Scout D booster, after the launch had been twice delayed by a malfunctioning pitch rate gyro. The spacecraft carried five experiments into orbit with a 926.8-km apogee, 727.7-km perigee, 101.1-min period, and 97.8° inclination.

Experiments were designed to demonstrate an attitude control accuracy of better than three arc minutes using a gas jet system, measure the performance in orbit of operational infrared sensor components, demonstrate an inexpensive Canopus sensor of high reliability, measure the density of sun-reflecting particles near the spacecraft, determine interference generated by the propane gas control system, demonstrate a basic sensing element with application in digital sun and earth albedo sensors, and measure degradation of silicone solar cells in orbit.

NASA's objective of placing the satellite in an earth orbit that would permit achievement of the scientific objectives was met, and the mission was adjudged a success 19 March. Under a December 1972 Memorandum of Understanding between NASA and the U.K.'s Dept. of Trade and Industry, NASA would be reimbursed for launch vehicle and services.

Developed for the U.K. by Hawker Siddeley Dynamics Ltd., the 83.5- by 66.5-cm, 92-kg spacecraft was powered by a four-panel 250-cm-long solar array of 1800 solar cells. *Miranda* was deployed in a fully sunlit polar orbit with a yo-yo despin system and its attitude control system acquired the sun through three sun sensors.

The eighth NASA launch of a U.K. satellite and the fourth on a Scout vehicle—*Miranda* carried the names of 80 000 Boy Scouts from nine U.S. states on a microdot. (NASA MOR, 21 March 74; NASA Releases 74-36, 74-53; UN Reg; Newport News, Va, *Daily Press*, 22 Feb 74; *SBD*, 12 March 74, 59)

*8 March:* Preliminary reports for Skylab principal investigators indicated that large-scale zero-g production facilities in space could manufacture alloys, composites, and crystals of a homogeneity and perfection far greater than that obtainable on the earth, Marshall Space Flight Center announced. A sample from a Skylab brazing experiment demonstrated an almost perfect brazed joint, suggesting that it would be possible to assemble large facilities in space using welding and brazing techniques. Two ends of 2.5-cm tubes of nickel or steel joined with a sleeve by brazing with a silver-copper alloy showed a metallic structure that could not form under gravity conditions.

Crystals of germanium selenide, typically 2 to 3 mm on the earth, were grown in zero g to a length of 25 mm with straight edges and no surface structure defects. A mixture of metals that would not form alloys under gravity conditions showed at least some alloying under zero g. A strengthening fiber of silicon carbide whiskers introduced into composite materials was uniformly distributed, suggesting a potential for space production of materials of greater strength and lighter weight. (MSFC Release 74-33)

- The 1974 Dr. Robert H. Goddard Memorial Trophy was presented to Rep. Olin E. Teague (D-Tex.) at the National Space Club's 17th Annual Goddard Memorial Dinner in Washington, D.C. Rep. Teague was honored for his leadership as charter member of the House Committee on Science and Astronautics and Chairman of the Subcommittee on Manned Space Flight since its 1963 formation, leading to a "much higher level of congressional understanding in support of the National Space Program."

The Astronautics Engineer Award was presented to Skylab Program Director William C. Schneider, NASA Acting Associate Administrator for Manned Space Flight, for "his exceptional personal leadership of the Skylab Space Station Program." The Nelson P. Jackson Aerospace Award was presented to the NASA Ames-TRW Systems Group Pioneer 10 team for its "historic achievement in obtaining close-up photographs and measurements of the Planet Jupiter, after a 600 million mile [970 million km] journey from Earth, in the process of becoming the first spacecraft to fly beyond the orbit of Mars, the first to penetrate the asteriod belt, and, eventually, the first man-made object to escape the solar system."

Winners of the Dr. Robert H. Goddard Historical Essay Award were Dr. Norriss S. Hetherington of Kansas Univ. for "Winning the Initiative: NASA and the U.S. Space Science Program, 1958–60" and Dr. James A. Malloy, Jr., of American Univ. for "The Dryden-Blagonrarov Era of Space Cooperation, 1962–65." The National Space Club Press Award went to John N. Wilford of the *New York Times* for "his superlative chronicle of the Space Age from Houston to Zvezdny Gorodok; from the first Gemini rendezvous to the last Skylab Space Station."

The Dr. Hugh L. Dryden Memorial Fellowship was awarded to John W. Edwards of Flight Research Center for his ability to apply new techniques to practical flight research, and the Dr. Robert H. Goddard Scholarship went to Miss Rosemarie Szostak of Georgetown Univ. for her research in inorganic chemistry. (Program)

- Analyses of light flashes seen by crew members of the last Apollo missions, tentatively attributed to cosmic ray nuclei penetrating the head and eyes, were reported by Univ. of Houston and Johnson Space Center scientists in *Science* magazine. Three kinds of flashes—the spot or star-like flash, the streak, and the cloud—had occurred randomly, with no apparent preference for one eye over the other. Astronauts saw many more flashes during translunar coast experiment sessions than in transearth coast, although the cosmic ray flux appeared to be the same inside the spacecraft during both coasts. No physiological explanation was found for this difference in frequency of flashes. (Pinsky, *et al.*, *Science*, 8 March 74, 957–958)

- NASA was presented the Principality of Monaco's gold and diamond-studded Silver Jubilee plaque at the Annual Awards Luncheon of the American Astronautical Society in Washington, D.C., for its contribution to worldwide television. The award was presented to Dr. James C. Fletcher, NASA Administrator, by U.S. Information Agency Deputy Director Eugene P. Kopp on behalf of Prince Ranier III of Monaco, who had said that NASA's live telecasts from the moon deserved recognition as "one of the most dramatic television events of the past quarter century." (JSC *Roundup*, 29 March 74, 1)

*9–12 March:* Soviet space probes *Mars 6* and *7* launched 5 and 9 Aug. 1973, flew past the planet Mars but failed to achieve their objectives. *Mars 7* approached the planet 9 March. A descent module was separated from the spacecraft but, because of a malfunction in an onboard system, it passed by the planet a distance of 1300 km. *Mars 6* neared Mars on 12 March. Its descent module was separated at 48 000 km and *Mars 6* continued past Mars within 16 000 km of the surface and into heliocentric orbit. The descent module transmitted data to *Mars 6* for 148 sec as it descended, but radio contact was lost just before landing. The Western press later reported that high winds might have destroyed the spacecraft. Data received reportedly indicated several times more water vapor in the Martian atmosphere over some areas than had been thought, traces of ozone, broad oscillations in atmospheric pressure, identification of a hydrogen corona reaching to 19 300-km altitude, confirmation of erosion and formation of what appeared to be river beds, and a magnetic field 7 to 10 times greater in the planet's immediate environment than in planetary space.

*Mars 6* and *7*—continuing to transmit data on solar wind and radiation from heliocentric orbit—were part of a series of four automatic stations launched to explore the Martian atmosphere and surface. *Mars 4* and *5*, launched 22 and 25 July 1973, had approached the planet 10 and 12 Feb. *Mars 4* flew by Mars when its braking engine failed to fire. *Mars 5* went into a successful orbit and continued to transmit data to the earth. Earlier, *Mars 3* had softlanded a capsule on Mars 2 Dec. 1971 during a windstorm; the capsule had transmitted 20 sec before falling silent. (Tass, FBIS-Sov, 15 March 74, U1; 21 March 74, U1; Pravda, FBIS-Sov, 27 March 74, U1; *W Post*, 15 March 74, A29; GSFC *SSR*, 31 March 74; *Av Wk*, 25 March 74, 17; 1 April, 11)

*9 March:* The small tactical aerial mobility platform (STAMP)—a wheelless "flying jeep"—was undergoing tests at the Marine Corps Air Station in El Toro, Calif. The two-man 300-kg vehicle, produced under a $500 000 Marine contract to the Garrett Corp., could hover as well as fly 120 km per hr at any altitude between 2.5 cm and 1500 m for 30 min at a time. The marines were testing the vehicle for possible reconnaissance, target spotting, courier work, mine detection, and rescue. (AP, B *Sun*, 9 March 74, A3)

*11 March:* *Explorer 46* Meteoroid Technology Satellite, launched 13 Aug. 1972, was officially adjudged a success. The primary objective of measuring meteoroid penetration rates on a bumper-protected target in the near-earth environment was achieved despite the failure of half of the bumper panels to deploy fully. Because of the anomaly, the spacecraft batteries had overheated and data collection from the main telemetry system had been discontinued; since 22 Aug. 1972, data from the primary experiment had been transmitted through the backup system, which operated on solar cells only. Flight data showed that the bumper concept was six times more effective than a single-wall structure, twice the anticipated effectiveness.

Because the secondary experiments had been powered only by the spacecraft batteries, no data were received from these experiments after 22 Aug. 1972. One secondary experiment—to measure impact flux of small mass meteoroids—had recorded 1850 penetrations, however, enough to show conclusively that the microphone meteoroid sensors flown on early Explorer and Discoverer satellites had greatly overestimated the population of small meteoroids. (NASA MOR, 11 March 74)

- A General Accounting Office report disclosed by Sen. Thomas F. Eagleton (D-Mo.) had questioned Pentagon plans to begin production of 34 radar-carrying aircraft, part of the Airborne Warning and Control System, the *Washington Post* reported. The GAO report said the radar planes' mission, defense against bomber attack, had changed and the project should be delayed until it was certain the system would work in its new role. There had been no demonstration that the aircraft could work effectively in a massive air battle. In addition, GAO questioned whether the aircraft's large radar would be susceptible to enemy jamming from distances as great as 320 km and whether the planes would become high-priority targets. Sen. Eagleton therefore had requested the Senate Armed Services Committee to deny AWACS production in 1974. (Getler, *W Post*, 11 March 74, A2)

*11, 21 March:* Joint hearings on the Technology Resources Survey and Applications Act, S. 2495, were held by the Senate Committee on Aeronautical and Space Sciences and the Committee on Commerce. Dr. James

C. Fletcher, NASA Administrator, said NASA supported the broad objectives of the bill, which would amend the National Aeronautics and Space Act of 1958 to establish within NASA an Office of Technology Applications to support the proposed National Resources Council. However, NASA already had an effective technology utilization program and did not require additional authority to ensure that its technological capabilities were effectively used. Dr. Fletcher said he could give more support to the bill if the responsibility for a proposed survey of technological resources were given to the existing Domestic Council.

Dr. H. Guyford Stever, National Science Foundation Director, said technology coordination and planning analysis could be more effectively done by an augmented Federal Council for Science and Technology than by the proposed Council. In addition, although NASA had great technological resources, it was essential that mission agencies should be held to their originally defined responsibility.

Dr. Philip Handler, President of the National Academy of Sciences, said the bill unduly emphasized technological resources of NASA and the aerospace industry, when the real issue was how to use the entire Federal science and technology effort in combination with private industry resources within coherent national programs and plans. Instead of direct authorization to NASA, a similar amount should be made available to the President or a council, such as the Council of Economic Advisors or the Council on Environmental Quality, to allocate and coordinate.

Dr. Robert C. Seamans, Jr., President of the National Academy of Engineering and former NASA Deputy Administrator, favored a compact organizational structure with leadership appointed by the President and a research staff independent of all agencies and industry. NASA-developed technology could be used with that in other agencies and private industry. NASA laboratories should be used whenever appropriate. Employees and facilities of Lewis Research Center, which was preeminent in the fields of air-breathing and space-propulsion systems, could be applied effectively to energy research and development. (Transcript)

*13 March:* The Senate Committee on Aeronautical and Space Sciences reported out H.R. 11864, Solar Heating and Cooling Demonstration Act of 1974. The Committee amended the bill, which had been passed by the House 13 Feb., to delineate agency responsibility more specifically. NASA would have responsibility for the necessary system research, development, hardware procurement, and delivery. The Dept. of Housing and Urban Development would develop dwelling and system performance criteria, construct the dwellings, and install the solar heating and cooling systems. Transfer of responsibility to the Energy Research and Development Administration, should it be established, was authorized but not mandatory. (S Rpt 93-734)

- The Senate passed H.R. 11793, a bill to establish the Federal Energy Administration, after amendment to substitute the language of the Senate-passed S. 2776. The Senate appointed conferees to settle differences from the version passed by the House 7 March. (*CR*, 13 March 74, S3706)

- NASA Associate Administrator for Applications Charles W. Mathews testified on applications programs before the Senate Committee on Aeronautics and Space Sciences during NASA FY 1975 authorization hearings. In NASA's weather and climate program, satellites were providing increasingly valuable information for short-range forecasting: day and night cloud-cover images around the globe, temperature profiles (although still cloud-limited) and surface temperature maps, gross moisture profiles, and sea ice mapping. NASA had begun microwave sounding experiments to overcome the cloud problem. Still needed were improved sounding accuracy and coverage, refined moisture profiles, and simpler and less expensive random-access data-collection systems.

For one- to two-week prediction, space capabilities were contributing to progress but lacked complete coverage of critical wind and temperature areas around the globe. New measurement approaches and improved cloud tracking were planned. For severe storm and disaster warnings—which contained "the greatest potential payoff"—a long-term effort was studying advances in computer technology and display devices to assimilate and convey the mass of data produced. NASA had begun to develop primitive climate models for farmers that would have "high utility value to the agriculture section of our economy." Remote sensing from space and research in a laboratory carried on the space shuttle could give data on weather modification possibilities and results. (Transcript)

- Johnson Space Center and the Mexican National Commission for Outer Space had begun a remote-sensing test project to assist in eradicating the screwworm from Mexico, NASA announced. The experimental phase was under way and was expected to continue for a year. Sensor data provided by *Erts 1* Earth Resources Technology Satellite (launched 23 July 1972) and orbiting meteorological satellites including *Noaa 2* and *3* (launched 15 Oct. 1972 and 6 Nov. 1973) would be combined with data from a Mexican remote-sensing aircraft on soil temperature, moisture, and vegetation—all of which affected breeding patterns of the screwworm fly. Sterile flies could then be distributed to potential breeding grounds of the fly, which frequently caused $200 million annual losses to cattle, poultry, and wildlife. (NASA Release 74-60)

- Sen. Frank E. Moss (D–Utah) and Sen. Barry M. Goldwater (R–Ariz.) introduced S. 3175 to amend the National Aeronautics and Space Act of 1958 to authorize NASA to contract for tracking and data reply satellite services. (*CR*, 13 March 74, S3548; NASA *LAR*, VIII[XIII]/30)

*14 March:* Gerald M. Truszynski, NASA Associate Administrator for Tracking and Data Acquisition, testified on the proposed Tracking and Data Relay Satellite System (TDRSS) in FY 1975 authorization hearings before the House Committee on Science and Astronautics' Subcommittee on Aeronautics and Space Technology: TDRSS essentially would be two tracking network stations in synchronous orbit to support all low-earth-orbit satellites, giving real-time control and data with a significant increase in reliability. The system would provide close to six times the coverage given by ground networks because supported satellites would be visible to the orbital stations during 85% of the orbits rather than the 15% from ground stations. Close to continuous visibility would also provide greater flexibility in programming satellite use and would simplify operations. Number of ground tracking stations could be reduced from the 1974 total of 18, plus one ship and two aircraft, to 8 by 1981. (Transcript)

- The U.S.S.R. was developing two new supersonic bomber aircraft and another missile-firing submarine, the Baltimore *Sun* reported. A new heavy bomber, the Backfire, was already in production. The Backfire was believed to have a top speed of 2100 km per hr and, with inflight refueling, a range of 9700 km. The second bomber, a Sukhoi prototype, could take off at a weight of 136 000 kg and reach altitudes of 21 000 to

24 000 m traveling at 3400 km per hr. The new submarine was to carry 16 missiles with ranges of 6400 km and was comparable to a development planned by the U.S. Navy for 1979. (Corddry, B *Sun*, 14 March 74, A1)

*15 March:* NASA and the National Science Foundation announced an agreement to design, construct, and test an experimental 100-kw wind turbine generator at Lewis Research Center. The large windmill would be tested to determine performance and characteristics of systems for future commercial production of electric power [see 11 Jan]. (NASA Release 74-63)

- The revised estimated cost of the Rockwell International Corp. B-1 bomber program—up $1.3 billion, to $15 billion—was sent to Congress in a Dept. of Defense letter. New estimates of the cost of the program to produce 244 aircraft resulted from a decision to build more prototypes and allow for an annual inflation rate of 3.3% instead of the previous 2.57%. (Reuters, *NYT*, 19 March 74, 17)

- Aviation pioneer Henri Pequet had died at his home in Vichy, France, at age 86, friends reported. Pequet had flown in a free-floating balloon in 1905 and was licensed as France's 88th pilot in 1910 after teaching himself to fly. On 18 Feb. 1911 he piloted the world's first official mail flight, between Allahbad and Nani in India. (UPI, *W Post*, 17 March 74, B7)

*16 March:* The Air Force launched an unidentified satellite from Vandenberg Air Force Base by a Thor-Burner II booster into an orbit with an 878-km apogee, 781-km perigee, 101.4 min period, and 98.9° inclination. The press later identified the satellite as part of the Defense Meteorological Satellite Program (DMSP). (Pres Rpt 74; *SBD*, 26 March 74, 138; *Av Wk*, 25 March 74, 22)

- The velocity of NASA's *Mariner 10* interplanetary space probe, launched toward Venus and Mercury 3 Nov. 1973, was successfully changed by 17.8 m per sec along the sunline of the spacecraft by the third trajectory correction maneuver. A 51.1-sec burn of the onboard engine at 7:54 am EDT changed the trajectory from a sun-side flyby of 12 359 km to a dark-side flyby of 860 km on 29 March. *Mariner 10* had continued to perform satisfactorily since its 5 Feb. encounter with Venus despite anomalies [see 8-31 Jan.]. A down-link transmitted power problem apparently corrected itself and the received power levels at Deep Space Network sites were back to normal. The return of power would permit real-time TV pictures from Mercury at the 117-kilobit data rate. Spacecraft attitude-control gas, used at an increased rate during a 28 Jan. maneuver, was being conserved by rotating the solar panels to maintain roll control. (NASA MOR, 21 March 74)

- U.S. officials were increasingly concerned by strong pressure from developing countries, backed by the U.S.S.R., to restrict the use of satellites to broadcast TV programs from one country directly to the home sets of another, the *Washington Post* reported. Old cultures were afraid of being "irremediably debased by the 'hard sell' techniques so commercially lucrative in the 'advanced' societies." As the world's technological giant, the U.S. was the most suspect and the trend among the developing countries was "to tie up . . . the use of satellites by the 'space-faring' nations . . . before they gain too great a head start." (Welles, *W Post*, 16 March 74, A18)

*18 March:* The F-14 Tomcat, the Navy's newest and most versatile fighter, made its first operational flights from the deck of the nuclear-powered aircraft carrier *Enterprise*. The flights of the $14.8-million, variable-sweep-wing F-14 began integration of the two Tomcat squadrons into the *Enterprise's* air wing. Two more squadrons would be assigned to a carrier in the Mediterranean or Atlantic. (Middleton, *NYT,* 21 March 74, 29)

- NASA announced two Kennedy Space Center contract awards for space shuttle facilities. Morrison-Knudsen Co., Inc., had been given a $21 812 737 contract to construct a space shuttle runway with overruns, apron, taxiway, and access roads. The runway—to be completed within 850 days after the notice to proceed—was to be built northwest of the Vehicle Assembly Building.

  Reynolds, Smith, and Hills received a $1 463 000 fixed-price contract to design modifications to Launch Complex 39 and Mobile Launcher 3 for shuttle operations. Complex 39, Apollo and Skylab launch site, would next be used for a manned flight in 1975, for the U.S.-U.S.S.R. Apollo-Soyuz Test Project. (NASA Releases 74-66, 74-67)

- The British government announced that it was seeking new talks with France on the Concorde supersonic airliner after heavy losses were predicted for each airliner built. Official estimates had indicated that Britain would lose $450 million to $517 million in producing 16 aircraft and, even with a selling price of $40.25 million to $47.5 million, losses would rise until sales reached 70 or more. (Reuters, B *Sun,* 19 March 74, C9)

*18-22 March:* The Fifth Annual Lunar Science Conference, at Johnson Space Center, was attended by more than 550 lunar scientists from the U.S. and several foreign countries. Some 200 papers were presented on the lunar regolith, interior, crust, impact effects, mare basins, and the interchange of material and energy between the moon and its environment.

Dr. A. Albee, California Institute of Technology scientist, said that analysis of an *Apollo 17* lunar rock sample indicated that the sample was 4.6 billion yrs old and an example of a very early differentiate. The study suggested that the lunar crust formed very early in the evolution of the moon by extensive melting associated with differential, gravitational separation of various other rock types.

The conference showed the trend among lunar scientists had been shifting away from the initial dating and chemical analysis of rock to cross-discipline studies of possible implications of the lunar samples to understanding of solar system chronology. (JSC Release, 7 March 74; JSC *Roundup,* 29 March 74, 1)

*19 March:* Deputy Associate Administrator for Space Science Vincent L. Johnson testified during FY 1975 NASA authorization hearings before the House Committee on Science and Astronautics' Subcommittee on Space Science and Applications that the Thor-Delta launch vehicle improvement program begun in 1971 was nearing completion. The most modern Thor-Delta configuration, the 2914, was expected to fly for the first time during the April launch of Westar-A. Final decision for the launch was pending recommendation of the Failure Review Board set up at Goddard Space Flight Center to investigate the 2nd-stage failure of a Thor-Delta during the 18 Jan. *Skynet IIA* launch.

NASA believed that, with the 2914 Delta, the present family of vehicles could launch any mission, but RCA Corp. had determined that a 30% to 35% increase in the 2914 Thor-Delta capability would permit a 24-channel, rather than the usual 12-channel, satellite to be launched into synchronous orbit. RCA and McDonnell Douglas Corp. had agreed to share the cost of the uprating and McDonnell Douglas would reimburse NASA for technical direction. NASA had not agreed to make the new 3914 Thor-Delta a part of its launch family but probably would do so following successful proof flight. The first flight was scheduled for 1975. (Transcript)

- Instrumented aircraft overflights of Volusia County, Fla., to determine if soils could be classified by remote-sensing techniques were begun by Kennedy Space Center in cooperation with the Dept. of Agriculture. USDA was developing soil classification maps of Florida by traditional field sampling and laboratory analysis. The overflights, by KSC's twin Beechcraft and a C-46 under contract to Johnson Space Center, would use multispectral scanning to attempt to speed the procedure, which would otherwise take 40 yrs. (AP, LA Her-Exam, 19 March 74)
- Soviet scientists had offered a theory explaining why Comet Kohoutek had not been as bright as expected, an East Berlin newspaper was quoted as saying. Leningrad physicists had found in a laboratory test that Kohoutek was surrounded by a high-density fluid of cosmic particles; this "armor" had not burst and released light, as expected, when the sun had heated and expanded the comet's nucleus. (AP, LA Her-Exam, 19 March 74)

*19–20 March:* NASA launched a record series of 79 sounding rockets from eight sites as part of a program to determine daily temperature and wind variations in the upper atmosphere during the vernal equinox. The launches were made over a 24-hr period from North, Central, and South America, the Caribbean Sea, and the Atlantic Ocean. Single-stage Loki and Super Loki rockets carried meteorological instruments to about 70 km, where the payloads were ejected and returned to the earth on parachutes, telemetering data to ground receiving stations. (NASA Release 74–71)

*20 March:* Pioneer 11, launched 5 April 1973 toward Jupiter, completed the second passage of a spacecraft through the Asteroid Belt. The spacecraft had completed 70%, or 700 million km, of its 1-billion-km flight path to Jupiter and was traveling at 50 000 km per hr toward a December encounter. *Pioneer 11* and its predecessor *Pioneer 10*, which had successfully completed its crossing of the Asteroid Belt 15 Feb. 1973 and had gone on to a 3 Dec. 1973 encounter with Jupiter, found no danger from small high-speed particles in the belt. Particles in the center of the belt were found to orbit the sun at 61 200 km per hr—fast enough for a meteoroid with a mass of 0.001 g to penetrate 1 cm of aluminum. However, most particles detected by the two spacecraft were smaller and far fewer than predicted. Traveling out from the earth's orbit the smallest particles—0.001 mm—appeared to decline in number while larger particles—0.01 to 0.1 mm—were evenly distributed throughout the belt. The number of particles 0.1 to 1.0 mm was three times greater in the center of the belt than near the earth. Particles larger than 1.0 mm appeared to be very thinly spread. (ARC Release 74–9)

- President Nixon was briefed on the space shuttle and the Apollo-Soyuz Test Project at Johnson Space Center. Following a tour, the President and Dr. James C. Fletcher, NASA Administrator, presented the *Skylab 4* astronauts with NASA's Distinguished Service Medal. Nixon praised the efforts of the "great American scientific, mechanical, and clerical communities" for their contributions to the space program. "You have contributed to . . . the spirit of a great country, which means always look out toward the unknown, go there, take any risk, make any sacrifice, and never be discouraged because sometimes you may fail." Of the ASTP, President Nixon said that although systems of government were different, the U.S. wanted "to cooperate with the Russian people and with all the people on the earth in anything that will advance the cause of science, the cause of health, the cause of a better life for all our children, as well, of course, as the cause of peace." (JSC *Roundup*, 29 March 74, 1; *PD*, 25 March 74, 345–347)

- NASA officials testified on the proposal to lease a Tracking and Data Relay Satellite System and on the closing of Plum Brook Station, during FY 1975 authorization hearings before the House Committee on Science and Astronautics' Subcommittee on Aeronautics and Space Technology.

  Associate Deputy Administrator Willis H. Shapley said NASA recommended lease instead of Government ownership and operation for four reasons. First, contracting for services would be as cost-effective as a conventional network of ground stations. Second, expenditures would be deferred to years when resulting savings could be realized. Funding required to procure a NASA-owned TDRSS would reduce the resources that could be applied to important mainline programs. Third, industrial technology was ready to provide the needed services. Fourth, the proposed arrangement was consistent with long-standing policies and practice of the Government to rely on industry for communications services.

  Acting Associate Administrator Edwin C. Kilgore of NASA's Office of Aeronautics and Space Technology testified that Lewis Research Center's Plum Brook Station, which was to be closed by June 1974 because of FY 1974 Civil Service work force cuts, would keep its Space Power Facility on standby. The Air Force, Navy, National Oceanic and Atmospheric Administration, Atomic Energy Commission, and other agencies were considering using the facility—the world's largest space environment facility, which included a unique solar simulation system. Also, NASA's cooperative program with the National Science Foundation to study wind-driven energy devices was based at Plum Brook. LeRC Director Bruce T. Lundin reported that reduction-in-force notices had been sent to 177 Civil Service employees 11 March, following 180 in the past year, out of LeRC's 400 separations. Dr. Kilgore said 50 engineers and technicians would be retained for the Space Power Facility at Plum Brook, which once employed 600 workers. (Transcript)

- Dr. Malcolm R. Currie, Dept. of Defense Director of Defense Research and Engineering, testified on DOD space programs before the Senate Committee on Aeronautical and Space Sciences during FY 1975 NASA authorization hearings: DOD would provide an interim, upper, orbit-to-orbit stage (OOS) for initial use with NASA's space shuttle until NASA could develop a permanent space tug. The OOS, which would provide the high orbits necessary for about 60% of military payloads, would be a low-cost modification of an existing upper stage and would be phased to coincide with the first operation of the shuttle orbiter in 1980. About $3.0 million would be spent in FY 1975 to select stage configuration and initiate detailed specifications. The OOS would have no payload retrieval capability and might not be reusable but could be used by NASA or any other shuttle user during early shuttle operations. DOD would install facilities to provide initial shuttle capability at Vandenberg Air Force Base in December 1982. Shuttle hardware, software, operating procedures, and ground equipment would be common and interchangeable to the maximum extent possible between KSC and Vandenberg.

During FY 1974 DOD had initiated the development and validation plan for the NAVSTAR Global Positioning System, using satellites to provide worldwide precise positioning and navigation, accurate to tens of meters. The system, usable for civilian purposes as well as military, would halt the proliferation of specialized systems.

The Navy's Fleet Satellite Communications System (FLTSATCOM), to provide reliable beyond-the-horizon communications for command and control of mobile DOD forces, was "well into development." Terminal equipment for the fleet would be installed in FY 1975 and FY 1976 and would be operated through minimal leased transponders in Communication Satellite Corp.'s Maritime Communications Satellite System pending full operation of FLTSATCOM in 1976.

Three operational early warning satellites were in geostationary orbit. DOD would seek to improve coverage and mission duration and continue development of a simplified processing station. Two Lincoln Experimental Satellites, comsats LES 8 and 9 scheduled for 1975 launch, would use radioisotope thermoelectric generators instead of solar cell arrays.

The Space Detection and Tracking System (SPADATS), which could cover up to 5600 km, would be expanded. DOD was working on detectors, target discrimination, data processing, and other critical optical and radar components, to reach a near-real-time ground-based capability by FY 1978 to detect, track, and identify all objects up to 36 000-km altitude. (Transcript; *Av Wk*, 1 April 74, 20)

- The European Space Research Organization marked the 10th anniversary of the convention which established ESRO. The organization—which included member countries Belgium, Denmark, West Germany, France, Italy, The Netherlands, Spain, Sweden, Switzerland, and the United Kingdom—had successfully launched 7 satellites and 180 sounding rockets. In the 10 yrs of ESRO's existence, the budget had increased from $101.4 million for 1964–1966 to $405.6 million for 1972–1974. Employment had risen from 450 in 1964 to 1200 in 1974. The anniversary would be the last for ESRO, which was to be incorporated within the new European Space Organization.

Dr. James C. Fletcher, NASA Administrator, wrote Dr. Alexander Hocker, ESRO Director General: "We on the 'working level' of NASA who are now working very closely with this organization add our congratulations to those officially transmitted and look forward with pleasure to the continued association in bringing Spacelab into being." (ESRO newsletter, 74–3; ESRO Release).

*21 March:* The success of Skylab's Apollo Telescope Mount solar experiment was more than a milestone in space exploration, Dr. John Eddy, solar physicist of the Boulder, Colo., High Altitude Observatory, said in a *New Scientist* article. This battery of the world's most advanced telescopes lofted to the ideal observation site had clarified previous theoretical sketches of the vertical extent of the sun's chromospheric structures. Spicules, which outlined the giant circulation cells of the chromosphere and extended through the transition region, disappeared at temperatures of 1 million K in the low corona, suggesting that the magnetic field lines diverged in the corona.

21 March

First results from the ATM x-ray observations destroyed the common concept of a homogeneous background corona, which had served as a basis of modern coronal physics, and revealed that the corona was composed almost entirely of closed loop structures fitting magnetic field lines calculated from observed photospheric fields.

Reasons for the ATM success included the large-scale and highly sophisticated instruments, large data-storage capability, efficiency of instrument operation, operation of the six instruments in concert, extensive ground support, and astronaut interest and training. Program results would be shared by scientists in many countries and the program appeared a good investment, in scientific yield per dollar spent. (*New Scientist*, 21 March 74, 738–741)

- Dr. James C. Fletcher, NASA Administrator, testified in FY 1975 NASA authorization hearings before the House Committee on Science and Astronautics that NASA supercritical wing technology was "ready now for both military and commercial people to pick up and use."

  Dr. Rocco A. Petrone, Associate Administrator, said demonstration tests for solar heating and cooling under way at Marshall Space Flight Center [see 11 Jan.] would provide enough baseline data within one year to 18 mos to install the system in a significant number of homes. (Transcript)

- George Cunningham, paralyzed from the neck down in a high school football game, was the first quadriplegic to use a device developed for astronauts disabled in space, the Baltimore *Sun* reported. He could place a telephone call, control a radio or TV, open the curtains in his hospital room, or call the nurses' station by blowing into a tube. The device, Nu-Life, made up of two tiny video display screens and a miniature computer, was operated by inhaling and exhaling into a plastic tube or controlled by eye rotation, a turn of the head, or a touch of the tongue. (AP, B *Sun*, 21 March 74, A3)

*23 March–5 April:* NASA's *Mariner 10,* launched 3 Nov. 1973, became the first spacecraft to explore Mercury, during a 12-day flyby of the planet—returning some 1700 TV photos and much new information on its atmosphere, surface, mass, and shape.

The incoming encounter sequence began 23 March at 5 600 000 km from Mercury, transmitting the clearest photos yet taken of the tiny planet. Although most of the photos were recorded on tape before transmission, real-time photos with reduced resolution were sent to analyze the reflective properties of Mercury at different lighting angles. By 28 March photos were showing the planet to be heavily cratered like the moon.

During the incoming near-encounter, from 17 to 3½ hrs before encounter, 18 frames were transmitted. Mosaics were made for locating smaller areas photographed later at higher resolutions. Spacecraft instruments identified an extremely thin atmosphere surrounding the planet, containing helium at concentrations above lunar levels and small amounts of neon, argon, and possibly zenon.

*Mariner 10* passed within 704 km of Mercury at 4:47 pm EDT 29 March, within 56 km of the aim point. Data were transmitted without interruption over the 150 000 000-km distance until earth occultation began at 4:50 pm EDT. Radio signals were recovered following the 12-min occultation. Photos of the planet—described by scientists at Jet Propulsion Laboratory as "superb"—showed details less than 1.6 km in size and covered almost half the planet's surface, which was pocked with craters as deep as 32 km and between 8 and 160 km wide. Giant cracks in the surface suggested high heat beneath. Some ancient volcanic activity was seen in a few rille-like ravines that crossed the craters and dome-like hills. Absence of halo craters indicated young craters probably had been covered by the dust storms that constantly swept the planet. At a JPL press briefing following the encounter, Dr. James C. Fletcher said onboard instruments showed that Mercury had an unexpected magnetic field, estimated to be 1% that of the earth. The magnetic field was measured at 90 to 100 gammas at the spacecraft's closest approach, with an extrapolated value at the surface of 100 to 200 gammas.

As *Mariner 10* raced by Mercury on 30 March, photos of the morning side of the planet showed a much smoother region than the rugged area seen under the afternoon sun. The north pole region appeared much less pitted and flatter than the south pole. Instruments found that Mercury was two-thirds iron, making its density greater than previously believed and probably responsible for the unexpected magnetic field. Concentrations of uranium and thorium were comparable to those on the earth.

Preliminary results, presented at a 31 March press conference at JPL, included a Mercury mass measurement 100 times better than previous ones, a detached well-defined bow shock wave, a temperature varying from 90 K on the night side to 700 K on the day side ($-300°F$ to $800°F$) with no thermal anomalies detected, opaque surface soil with the top few centimeters 1.0 to 1.5 times as dense as water and 50% as porous, and high-energy electrons in the mev range observed from 700 to 5000 km above the surface. An ultraviolet scanner had detected what appeared to be an egg-shaped moon moving in orbit around Mercury, but the Mariner navigational team later identified the object as a dim but very hot star, known as 31 Crater in the constellation Corvus.

Despite a power anomaly, *Mariner 10* continued to transmit TV pictures in real-time at 117.6 kilobits per sec and tape-recorded frames at 22.05 kbps. TV imagery was discontinued 5 April when the resolution became equal to that of earth-based telescopes.

All mission events were accomplished as planned and the mission was adjudged successful 22 April. The spacecraft was expected to swing back by Mercury in September. (NASA MOR, 24 April 74; JPL Press Kit, 19 March 74; *NYT*, 24 March 74, 30 March–1 April.74; *W Post*, 29 March–1 April 74; *W Star-News*, 28 March 74; *LA Times*, 20 March–1 April 74)

*23 March:* Key appointments in the space shuttle program at Marshall Space Flight Center became effective. Robert E. Lindstrom became Manager of the Space Shuttle Office, replacing Roy E. Godfrey, who became Special Assistant to the MSFC Director. Within the Shuttle Office, James R. Thompson became Project Manager for the space shuttle main engine. John A. Chambers became Deputy Manager of the Shuttle Engine Office and Frank M. Stewart became technical assistant to the Manager of the Shuttle Projects Office. (MSFC Release 74–41)

*24 March:* Manned space spectaculars were gone, at least until the 21st century, Thomas O'Toole said in the *Washington Post*. The similarities between the air age when men had begun flying and the space age were striking. Aviators had been considered stuntmen until they began to carry the mail. The Apollo flights, with their "vaguely circus air," had struck many people the same way. The space shuttle promised to change all that. The U.S. planned to operate five shuttles, each one carrying more than 27 000 kg of satellites and space machinery into orbit 12 times a year, to "improve the quality of life on earth and to find out if life exists elsewhere in the solar system." By the end of the 1980s the U.S. would have visited eight of the solar system's nine planets; the only exception was Pluto.

While one part of the space agency looked at the other planets, another would be looking down to learn more about the earth's health. The next 20 yrs would see an unprecedented use of earth orbit to observe and communicate with the earth. O'Toole quoted Dr. James C. Fletcher, NASA Administrator, as saying that the earth orbit might become overcrowded. Not that things would bump into each other but "there will not be enough radio frequencies to communicate back to earth." (*W Post*, 24 March 74, C1)

- New U.S. proposals might help lead to a new agreement at the stalemated Strategic Arms Limitation Talks, Leslie H. Gelb said in the *Washington Post*. Dr. Henry A. Kissinger, Secretary of State, arriving in Moscow, would submit to the U.S.S.R. two formulas to establish equivalence between nuclear forces. Formula one would bring overall equality in total numbers of land-based missiles, sea-launched missiles, and long-range bombers. Within the total, each side would determine its own mix of forces. Formula two would develop equality in land-based missile throw weight, a better way to guard against a Soviet first strike on U.S. missiles than to seek control of multiple independently targetable vehicles. No MIRV controls were proposed. Working principles contained no numbers, leaving open the question of whether U.S. missiles were to be increased, Soviet missiles decreased, or both lowered. (*W Post*, 24 March 74, 3; FONF, 30 March 74)

*25 March:* Data from U.S.S.R. space probes *Mars 4, 6,* and *7* [see 9–12 March] had indicated that several "tens percent" of the Martian atmosphere was made up of an inert gas, probably argon, Tass quoted Prof. Vasily I. Moroz as reporting in *Izvestiya*. The data would substantiate Soviet scientists' belief that several million or even several hundred thousand years ago Mars had an atmosphere with a density close to that of the earth. In addition, data from *Mariner 9* (launched 30 May 1971) and *Mars 5* had shown that the planet had had surface water and might "acquire it again, for such changes take place during the planet's geological history . . . with a definite periodicity." New data was increasing hope of finding the simplest biosphere. "In any case, in the not so distant geological past Mars apparently had . . . the conditions necessary for the emergence of life. The reappearance of such periods in the future is not ruled out."

The Western press later quoted U.S. scientists as saying that the presence of large amounts of argon would have "dramatic implications" and would tend to support the theory that the Martian climate had been going through a series of cycles during which ice ages alternated with warmer and wetter conditions. But Dr. Carl E. Sagan of Cornell Univ. said during a telephone interview that the 90% content of carbon dioxide in the Martian atmosphere had been well established, so that "at most 10% of argon was barely possible." (FBIS–Sov, 26 March 74, U1; Shabad, *NYT*, 7 April 74, 34)

- Dr. Edward U. Condon, a leading U.S. atomic scientist, died at the age of 72 in Boulder, Colo., of complications from heart disease. In 1940 Dr.

Condon had become a member of the National Defense Research Council and, in 1941, a member of the Roosevelt Committee on Uranium Research. During the World War II he played an important role in the development of the atomic bomb. In 1945 Dr. Condon had been appointed Director of the National Bureau of Standards but resigned in 1951 because of House Un-American Activities Committee investigations as a "weak" security link. In 1952 he appeared before the Committee and denied charges under oath.

From 1945 to 1951 Dr. Condon had been a member of the National Advisory Committee on Aeronautics. In the 1950s he developed a missile nosecone—later used by U.S. astronauts—that could withstand the high temperatures of reentry into the earth's atmosphere. In 1954 he returned to teaching and, while on the Univ. of Colorado faculty, he undertook an Air Force study to investigate sightings of unidentified flying objects. (Weil, *W Post*, 27 March 74, C4; Slusser, *W Star-News*, 27 March 74, B5; *Who's Who*; House Un-Am Activities Com Special Subcom Rpt)

*26 March:* The U.S.S.R. launched *Cosmos 637* from Baykonur Cosmodrome near Tyuratam, into orbit with 35 604-km apogee, 35 595-km perigee, 23-hr 46.6-min period, and 0.2° inclination—the first Soviet satellite in geosynchronous orbit. The press later quoted American analysts as suggesting that *Cosmos 637*, which was stationed above the Indian Ocean, might be an engineering test for a full-fledged stationary communications satellite long planned by the Soviets to improve communications with the southern part of the U.S.S.R. and the Warsaw Pact countries. (GSFC *Wkly SSR*, 21–27 March 74; Shabad, *NYT*, 28 March 74, 20; *SBD*, 28 March 74, 153; *Av Wk*, 1 April 74)

- The Senate Committee on Aeronautical and Space Sciences' Subcommittee on Aviation held hearings to determine whether the Federal Aviation Administration and McDonnell Douglas Corp. had dealt adequately with cargo door problems on the DC–10 aircraft. Experts believed that the door had caused a 12 June 1972 near crash in Canada and the 3 March 1974 crash outside Paris that killed all 346 persons aboard.

  Documents had revealed that, following the Canadian incident, the FAA had drafted an airworthiness directive requiring three improvements in the suspect rear cargo door. McDonnell Douglas appealed to the FAA, which then agreed to let the company issue voluntary service bulletins. An inspection of all 134 in-service DC–10s after the Paris crash had showed that two aircraft, including the Turkish airliner, had not been modified despite McDonnell Douglas records to the contrary.

  Seven months after the Canadian incident the FAA, at the suggestion of the National Transportation Safety Board, had asked the company to consider a redesign of the vital systems, reinforcement of the passenger cabin floor, and more pressure vents. In a 25 Feb. 1974 letter to the FAA, McDonnell Douglas said that they did not "have the manpower available at this time to undertake the study, nor are we in a position to accept this burden alone." (Egan, *W Post*, 23 March 74, A4; 27 March 74, A2; Witkin, *NYT*, 21 March 74, 1; 26 March 74, 1)

- The Federal Energy Office directed oil refiners to provide airlines with sufficient fuel to meet allocations calling for 95% of 1972 supplies for trunk carriers and 100% for local service carriers, FEO also proposed that international airlines be authorized to draw on domestic fuel supplies if bonded fuel was not available at prices comparable to domestic costs. (FEO Release 74-116)

*29 March:* President Nixon wired congratulations to Dr. James C. Fletcher, NASA Administrator, on *Mariner 10*'s successful 5 Feb. flyby of the planet Mercury: "The successful flight . . . marks another historic milestone in America's continuing exploration of the solar system. With this mission we will begin to end centuries of speculation about our planetary neighbor closest to the Sun. On behalf of all Americans, I extend warmest congratulations to NASA and the *Mariner 10* team on their outstanding performance. The hard work, skill, and ingenuity that contributed to the success of *Mariner 10* is in keeping with that historical tradition which began when men landed at Plymouth Rock and has continued through the landing of men on the moon." (*PD*, 1 April 74, 374)

- President Nixon issued Executive Order 11775 abolishing the Energy Policy Office, which had been established 29 June 1973. The responsibilities of the Office would be taken over by the Federal Energy Office, established 4 Dec. 1973. (*PD*, 1 April 74, 356)
- A Black Brant sounding rocket launched from Peru by a team of U.S. and Peruvian scientists found evidence that auroral activity occurred not only in the polar regions but in the ionosphere near the earth's geomagnetic equator. The scientists—from Aerospace Corp., Univ. of Texas, U.S. Air Force, Peru's Geophysical Institute, and the Peruvian government—were studying recurring ionospheric disturbances, known as spread F, that had been discovered by Peruvian radar studies. (Miles, *LA Times*, 28 April 74)
- Immigration of scientists and engineers into the U.S. had dropped sharply, from the record 10 300–13 300 yearly between 1967 and 1972 to 6600 in 1973, the National Science Foundation reported. The decrease reflected the impact of the 4 Feb. 1971 restrictions in immigration regulations. (NSF *Highlights*, 29 Mar 74)

*31 March:* Sen. James G. Abourezk (D–S.D.) asked the Government Accounting Office to investigate the "evidence of solar energy feasibility" contained in a report by Atomic Energy Commission scientists. He said evidence that solar energy could be developed more quickly and cheaply than previously believed had been ignored when the AEC recommended, in a report to the President, that only $200 million out of a total of $10 billion for energy research should be spent on solar energy for a five-year period beginning in 1975. Sen. Abourezk wanted to know "why the A.E.C. has been sitting on scientific data pointing the way toward solar energy while major oil companies are quietly moving to take control of the means to produce solar power." (UPI, *NYT*, 1 April 74, 34)

*During March: Apollo 17: Preliminary Science Report* (NASA SP–330), dated 1973, was issued. The Taurus-Littrow landing site for the 7–19 Dec. 1972 final mission of the lunar landing program had been the most diverse in both geologic features and samples brought back. Seismic traverse experiments had indicated basaltic flows extended to a depth of 1.2 km. The traverse gravimeter experiment had provided limits to the density of underlying material and permitted development of models to help interpret mascons. Analysis of mare basalts from the *Apollo 11, 12, 15,* and *17* missions and from the U.S.S.R.'s *Luna 16* indicated that the filling of the mare basins took place mostly between 3.2 billion and 3.8 billion yrs ago. Formation of more than 90% of the cratering was believed to be 4 billion yrs earlier. Highland material collected on *Apollo*

*14, 16,* and *17* missions showed a widespread occurrence of breccias with an apparent age of 3.8 billion to 4.1 billion yrs, with strong evidence that rocks as old as 4.5 billion to 4.6 billion yrs existed on the moon. Dark mare regions were thought to be underlain by lava, but craters had been mostly caused by impacting projectiles. The moon was once active volcanically but was now quiet, with any activity since 3 billion yrs ago highly restricted or nonexistent. It had a crust more than 60 km thick and a negligible overall magnetic field.

Two theories of lunar evolution had come from mission data. The first was that the planet was chemically layered during its formation, with a low interior temperature gradually increasing, perhaps reaching the melting point in the last billion years, while the initial heat of the exterior gradually decreased. The second theory assumed that the moon, chemically homogeneous and molten during formation, underwent extensive chemical differentiation resulting in surface concentration of radioactivity very shortly after formation. None of the three theories of origin of the moon—separation from the earth, capture from a solar orbit, or formation from a dust cloud—could be absolutely eliminated by data collected to date, but chemical differences made it unlikely that the moon was torn from the earth. (Text)

- Experiments carried out during the U.S.S.R.'s eight-day *Soyuz 13* mission launched 18 Dec. 1973 were further evidence that the U.S.S.R. was continuing manned space station efforts, a *Spaceflight* article said. The flight had included the Oasis 2 experiment, part of a plan to demonstrate that biological closed-cycle life support systems could be used for large space stations and other long-duration manned missions. Observation of stars in the ultraviolet range with the Orion 2 telescope was also a continuing program related to space station activity. Cosmonaut Konstantin P. Feoktistov had said in a statement to the British Interplanetary Society that, in the years between the *Soyuz 11* tragedy in June 1971 and the *Soyuz 12* mission in September 1973, design engineers had been analyzing the performance of all spacecraft systems and making modification and improvements. He stressed that work on manned space stations was continuing because of the importance of these stations in the exploration of natural resources. *Soyuz 12* and *Soyuz 13* crews had made multispectral photos of the earth's surface. (*SF*, March 74, 110)

- NASA submitted its report *NASA Energy-Related Research and Development* to the Senate Committee on Aeronautical and Space Sciences, at the request of its chairman, Sen. Frank E. Moss (D–Utah). NASA's FY 1974 budget had included a $2-million authorization for energy R&D for projects in solar energy utilization; energy conversion, transmission, and storage; transportation and propulsion systems; and energy and environmental conservation. Long-range research included studies of satellite systems to collect solar energy, convert it to electricity, and transmit it to the earth by microwave beams. In one study the use of solar cells would be considered, while in another the possibility of a turbogenerator in space would be explored.

  Near-term projects included demonstration of residential solar heating and cooling at Marshall Space Flight Center, office building tests at Langley Research Center, and component studies at Lewis Research Center. LeRC was also sponsoring a study of crops that could be converted into clean fuels, and a NASA–National Science Foundation study was testing wind-electric generating devices. Projects at Jet Propulsion Laboratory and LeRC sought to design low-polluting but higher efficiency automobile engines, and Flight Research Center was testing configuration changes to reduce aerodynamic drag for trucks.

Energy conversion research included an LeRC study, in cooperation with the Dept. of Interior, of technical and economic advantages of a potassium topping cycle to permit higher peak cycle temperatures and produce 40% more electrical power for a given quantity of fuel than could conventional steam systems. (Com Print)

- U.S. trunkline air traffic rose to unexpected heights, but industry officials were hesitant to interpret gains during the first quarter of 1974 as a positive trend for the year. United Air Lines, Inc., revenue-passenger-miles rose 13.6% in March over March 1973; Delta Air Lines, Inc., 10.1%; and Eastern Airlines, Inc., 7.4%. Local service carriers also reported near-record traffic improvements for March, with many showing load factors above the 55% level, rather than the normal monthly 44%. High fuel costs, however, were forcing the load-factor level required for airlines to break even to all-time highs.

  International operations were an exception to the general rise: Pan American World Airways, Inc., reported a revenue-passenger-mile decline of 6.1% in system-wide scheduled operations for March; Trans World Airlines, Inc., a 10.9% dip. (*Av Wk*, 14 April 74, 29–30)

## April 1974

*1 April:* Construction of the $22-million runway for the space shuttle was begun at Kennedy Space Center after a brief ground-breaking ceremony at the construction site northwest of the Vehicle Assembly Building. Dr. Debus, KSC Director, said the shuttle would lead to a "new era of international transportation to and from space" and the "boom and bust of the 1960s of this community and the activities of the government will be changed to a more steady—more reliable—situation."

The 90-m wide, 4600-m long, 40-cm thick runway would accommodate a 75 000-kg spacecraft returning from space at a 15° to 20° angle and at a speed of nearly 295 km per hr. More than 8000 km of grooves in the surface would permit drainage and prevent hydroplaning while landing in the rain. To reduce the speed of the returning shuttle, astronauts would have the help of brakes, parachutes, and a net at the end of the runway. The runway was scheduled for July 1976 completion. (*Spaceport News,* 4 April 74, 1, 5; McElheny, *NYT,* 2 April 74, 22)

- Marshall Space Flight Center began a six-month study of the feasibility of adapting the backup Skylab Apollo Telescope Mount to fly on the space shuttle for solar observations. NASA anticipated that modified ATM instruments positioned in the shuttle payload bay could provide low-cost scientific returns on early shuttle flights. (MSFC Skylab Off, interview, Jan 75; Covault, *Av Wk,* 15 April 74, 24–5)

- A *New York Times* editorial noted that, within a matter of hours, instrument readings from *Mariner 10's* 29 March close flyby of Mercury had "made all previous scientific writing and theorizing about Mercury obsolete. Where nearly complete ignorance had existed earlier, there is now a large, diverse and important body of hard information. . . . Even if the crippled spacecraft is unable to continue to transmit data it will have earned an indelible place in the history of space exploration." (*NYT,* 1 April 74, 30)

- The U.S.S.R.'s three-week *Cosmos 605* mission—launched 31 Oct. 1973 with several dozen mammal, reptile, insect, fungal, and bacterial life forms aboard—had yielded major data on organism reaction to prolonged weightlessness, Soviet Academy of Sciences member Oleg G. Gazenko was quoted as reporting. Although the rats had remained active in space, their food intake had been lower than that of an earth-based control group. After the flight the space rats' oxygen requirements had declined 10% to 20%. Evidence also indicated that, although the level of salts in the organisms had remained unchanged, they had redistributed. The space animals were sluggish immediately following spacecraft recovery but their memory and coordination remained unimpaired. Capillary blood showed major changes in content, and bone marrow cells were markedly reduced in number. Organs taken from a group of rats that had undergone autopsy within 24 hrs of spacecraft recovery showed no major pathological changes except for a marked weight reduc-

1 April

tion in the lymphoid glands and a small increase in the adrenal glands. The changes were viewed as stress reaction to flight conditions. A degree of atrophy had been noted in the hind muscles. Most of the disorders disappeared in a second group of rats during a 25-day observation period before autopsy. Mushrooms grown aboard *Cosmos 605* were greatly deformed, with long twisted stems and large root systems. (*Av Wk*, 1 April 74, 40)

- Available geosynchronous orbit "parking space" for communications satellites to serve the continental U.S. was fast approaching saturation with authorized and applied-for comsat systems operating in the 4- to 6-gigacycle band, *Aviation Week & Space Technology* reported. This saturation was expected to provide added incentive for second-generation domestic comsats to shift to the new 11- to 14-gc band, permitting spacecraft to orbit closer to the existing comsats. (*Av Wk*, 1 April 74, 59)
- Transformation of the European Space Research Organization into the European Space Agency, originally scheduled for 1 April, had been postponed, *Aviation Week & Space Technology* reported. The delay would allow representatives of member governments additional time to resolve the remaining areas of disagreement and decide on a Director-General. The postponement was not expected to affect ESRO's Spacelab program. Legal actions taken by ESRO would be equally binding under the new agency's charter. (*Av Wk*, 1 April 74, 21; 22 April 74, 19)
- Analysis by Hale Observatories astronomers of light from the diffuse halo around the nearest quasar, known as BL Lacertae, indicated that it was an ancient galaxy whose core was undergoing a series of catastrophic explosions, the *New York Times* reported. The implication was that all quasars, the brightest and most distant visible objects in the universe, represented some explosive phase in the later history of a galaxy. Before the study, BL Lacertae had been thought to be a variable star in the constellation Lacerta, part of the Milky Way. Instead the study had showed it to be a massive object 1 billion light years from the earth with visible light whose brilliance fluctuated radically. (*NYT*, 1 April 74, 21)
- The Air Force Flight Dynamics Laboratory had awarded contracts to three companies to make configuration studies for an advanced fighter technology integration (AFTI) aircraft, *Aviation Week & Space Technology* reported: $750 000 to McDonnell Douglas Corp., $608 000 to Rockwell International Corp., and $597 000 to Fairchild Republic Co. The studies were the first step in designing a small, low-cost, high-technology demonstration aircraft in the initial phase of the USAF's AFTI program. (*Av Wk*, 1 April 74, 22)

*1–5 April:* Data from *Pioneer 10* taken during its 3 Dec. 1973 encounter with Jupiter had indicated that the upper atmosphere of the planet was 400 K (260°F), about 480 K (520° F) hotter than previously calculated. Dr. Arvydas J. Kliore, Jet Propulsion Laboratory scientist, said at a meeting of the American Astronomical Society. Temperatures deeper down in the atmosphere rose to as much as 700 K (800°F), the highest temperature found so far on any planet in the solar system. The findings suggested that the surface temperature could reach 5800 K (10 000°F), the surface temperature of the sun. The surprising heat might be explained by heavy dust layers in the upper levels of the atmosphere that absorbed solar radiation or absorbed heat radiating away from the surface of the planet.

Dr. Fred L. Whipple, retired Director of the Smithsonian Astrophysical Observatory, said that methyl cyanide and hydrogen cyanide, two compounds always identified with interstellar dust, had been found in abundance in Comet Kohoutek. The discovery marked the first time the chemicals had been identified with a comet and the first time they had been seen in space inside the solar system. "It . . . increases the suspicion that comets were formed . . . out at about 30 trillion km or halfway to the nearest star." (*W Post*, 5 April 74, A29; 7 April 74, A3)

- The second Meeting of Signatories of the International Telecommunications Satellite Organization (INTELSAT) was held in Acapulco, Mexico. The delegates determined that a 1.106226% investment share would entitle a signatory or group of signatories to representation on the Board of Governors. (INTELSAT Release 74–17; INTELSAT PAO, interview)

*2 April:* Dr. John E. Naugle, NASA Associate Administrator for Space Science, testified in FY 1975 authorization hearings before the Senate Committee on Aeronautical and Space Sciences that Orbiting Solar Observatory OSO-I, scheduled for 1975 launch into orbit, was expected to acquire ultraviolet solar spectra of the most detailed resolution ever obtained. Spacecraft and experiment assembly was to be complete by mid-1974. A cooperative program with the European Space Research Organization, called International Sun-Earth Explorers (ISEE), planned three satellites to make simultaneous measurements. Two spacecraft, "Mother" built by the U.S. and "Daughter" built by ESRO, would travel close together in the same highly elliptical orbit, measuring magnetic and electric fields, energetic particles, and plasma properties. The third spacecraft, called "Heliocentric" and built by the U.S., would be placed near one of the unique points in space where the earth's and sun's gravity balance each other. The spacecraft would travel around the sun and close to the earth to learn what radiation was impinging on the earth's magnetosphere. Data from the three satellites would help to untangle spatial from temporal changes in radiation fields around the earth. A cooperative effort with ESRO and the United Kingdom would be the International Ultraviolet Explorer (IUE) scheduled for launch into geosynchronous orbit in 1976 to observe high-resolution spectra in the UV with a telescope controllable from the ground in real time. Techniques gained would be applied later to NASA's Large Space Telescope.

Associate Administrator for Tracking and Data Acquisition Gerald M. Truszynski reported deactivation of NASA's transportable tracking station in Newfoundland and closing of Spaceflight Tracking and Data Network stations at Corpus Christi, Tex., and Carnarvon, Australia. The Newfoundland station, placed in caretaker status after the Skylab missions, would be reactivated for the 1975 Apollo-Soyuz Test Project mission and then permanently closed. The Deep Space Network portion of the Johannesburg station was being phased out beginning June 1974; the STDN facility would be phased out following the near-earth phase of the Viking mission. The closings would leave 15 stations, adequate to handle the projected workload for the next five years and consistent with long-range planning. (Transcript)

- The House Committee on Appropriations' Subcommittee on Transportation approved a Federal Aviation Administration budget request of $3 million to begin the Aeronautical Satellite (Aerosat) system. The two-satellite system, to be built in cooperation with the European Space Research Organization and Canada, would provide over-the-ocean communications for aircraft. In Subcommittee hearings FAA Administrator Alexander P. Butterfield said recent changes in the program had eliminated airline opposition. The changes would limit the size and cost of the program, preclude the imposition of user charges on aircraft operators for recovery of R&D costs, eliminate any commitment for a follow-on operational system, include a very-high-frequency capability for evaluation against L-band, and limit the number of earth terminals. Signing of a Memorandum of Understanding by the participants was pending Senate approval of the FAA budget request. (Subcom Staff Off, interview; *Av Wk*, 8 April 74, 21; *SBD*, 3 April 74, 190)

- NASA announced selection of General Electric Co., Grumman Aerospace Corp., and TRW Systems Group to negotiate three six-month, $600 000, fixed-price contracts to make parallel system definition studies for the Earth Observatory Satellite (EOS) mission. The studies were to design a low-cost modularized spacecraft with a space platform for testing sensors and collecting remotely sensed data for a variety of applications missions. EOS would be designed for launch by both conventional boosters and the space shuttle. (NASA Release 74-80)

- Federal Aviation Administrator Alexander P. Butterfield said in a written statement to the Senate Committee on Commerce's Subcommittee on Aviation that he had directed FAA offices to issue airworthiness directives to correct all safety-related aircraft design changes. Such directives carried the force of law and required FAA inspection. In addition, all safety-related functions of the FAA were being centralized under one official. The FAA was also studying ways to prevent rapid decompressurization on jumbo jets after accidents. The actions followed investigations of the 3 March Turkish Airlines, Inc., DC-10 crash outside Paris that had killed all 346 persons aboard. By 14 May, 43 airworthiness directives and 7 telegrams had been issued, nearly double the usual rate. (Egan, *W Post*, 3 April 74, A3; *Av Wk*, 20 May 74, 15)

- The U.S.S.R. had requested permission, during the February visit of a Soviet science delegation to Australia, to establish a jointly controlled satellite tracking station in Australia, the *New York Times* reported. The U.S. had expressed concern that the Soviet Union might use the scientific station to monitor U.S. messages to bases in Australia and reportedly believed the proposal would be rejected. (*NYT*, 2 April 74, 4)

*3 April:* The U.S.S.R. launched *Cosmos 638* from Baykonur Cosmodrome into orbit with a 30-km apogee, 189-km perigee, 89.5-min period, and 51.8° inclination. The spacecraft was identified by Maj. Gen. Vladimir A. Shatalov, chief of cosmonaut training for the U.S.S.R, at a 9 Sept. ceremony at Johnson Space Center as an unmanned test of docking systems and rendezvous for the Apollo-Soyuz Test Project mission. The spacecraft reentered 13 April. (GSFC *SSR*, 30 June 74; Transcript; *SF*, Sept 74, 394)

- Rep. Olin E. Teague (D-Tex.) introduced H.R. 13961 to amend the National Aeronautics and Space Act of 1958 to authorize NASA to contract for Tracking and Data Relay Satellite services. (*CR*, 3 April 74, H2584; NASA *LAR*, XIII/43)

- Indonesia would build a domestic communications satellite system including two satellites and 50 earth stations, the Indonesian Post and Telecommunications Directorate General announced to the press. Each satellite could serve as backup for the other and would have 12 transponders. Each transponder would be able to serve 12 telephone lines or 1 color TV channel.

The press later reported that Malaysia and the Philippines were planning to join the Indonesian system. The satellites would be put into orbits that would allow point-to-point telephone connections within each country and among the three countries. Construction of the system was expected to cost $90 million and take two years. (Jakarta Domestic Service and Hong Kong AFP, FBIS–Indonesia, 18 April 74, N1–2)

- Pan American World Airways, Inc., and Trans World Airlines, Inc., each petitioned the Civil Aeronautics Board for subsidies. Pan Am sought a $195-million-a-year mail subsidy because of a pending financial crisis following six years of losses, including an expected net loss of $65 million in 1974. In addition to a huge debt, Pan Am had been hit by foreign competition, sharp drops in traffic, overcapacity (half-empty Boeing 747 transports), and soaring costs, including a $200-million annual increase in fuel costs. TWA asked a temporary subsidy to offset fuel costs, estimating 1974 international losses would be $47 million—in contrast to a 1973 profit of $19.5 million.

  On 2 April the Dept. of Justice and Dept. of Transportation had discouraged Pan Am and TWA discussions of pooling transatlantic operations as an economy measure, telling CAB that discussions were premature. (O'Hanlon, *Fortune*, Oct 74, 123–7, 212–6; CAB Doc Section, Docs 26560 & 26563; *W Star-News*, 3 April 74, B9)

- Johnson Space Center had awarded a $6 598 000 cost-plus-fixed-fee contract to Charles Stark Draper Laboratory, Inc., for Apollo-Soyuz Test Project and space shuttle orbiter software development, JSC announced. (JSC Release 74–62)

*4 April:* The YF–17 prototype lightweight fighter aircraft was rolled out at the Northrop Corp. plant in Hawthorne, Calif. The YF–17 and the YF–16 being built by General Dynamics Corp. marked the first Air Force effort to place more emphasis on hardware performance and less on paper studies by using an advance prototyping concept. The USAF would evaluate the YF–16 and YF–17 in a 12-mo flight-test program before making a commitment to production. (DOD Release 131–74)

- Dr. Harrison H. Schmitt, *Apollo 17* astronaut and now Special Assistant to the NASA Administrator for Energy Research and Development, testified in Senate Committee on Aeronautical and Space Sciences hearings on FY 1975 authorizations: One of the most important contributions NASA could make to energy research and development was the application of managerial techniques developed in the space program, permitting programs, schedules, and budgets to be defined and then adhered to without major degradation of the original goals. "This mission-oriented approach to research and development and its broad integration of the great capabilities and imagination of industry will . . . be applied just as successfully to the critical problems that face the Nation and the world in energy." (Transcript)

- The Federal Aviation Administration announced a proposed new regulation requiring fuel tanks and fuel venting spaces on turbine-powered transports over 5670 kg to be equipped with explosion-prevention systems. Under the proposal, aircraft operators and manufacturers could use a system that maintained a continuous nonflammable atmosphere, such as nitrogen, or a system that arrested a fire or explosion once initiated. All turbine aircraft manufactured two years after the effective date of the final rule would be required to be fitted with this equipment within three years. (FAA Release 74-47)

*5 April:* Experiments had been approved for the first High Energy Astronomy Observatory (HEAO) mission, to make an x-ray survey of the sky from low earth orbit in 1977, NASA announced. Contracts totaling $23 349 266 had been let for a large-area x-ray survey experiment to map the celestial sphere for x-ray sources; a cosmic x-ray experiment to measure emissions and absorptions of diffuse x-rays and correlate results with radio and visible-light-ray emission; a scanning modulation collimator experiment to determine precisely the celestial position, size, and structure of selected cosmic x-ray sources; and a hard x-ray and low-energy gamma ray experiment to determine the intensity and other properties of certain x-rays and gamma rays. (NASA Release 74-79)

- The U.S.S.R.'s Aeroflot Soviet Air Lines made its inaugural flight from Washington's Dulles International Airport to Moscow's Sheremetyevo Airport, with a stopover in Paris, initiating regular weekly Washington-Moscow service. Under a U.S.-U.S.S.R. protocol signed 23 June 1973, Aeroflot also served the Soviet capital from New York with two flights weekly during peak periods and one flight per week at other times. (DOT Release 29-74)

- Large numbers of scientists and engineers trained in a variety of traditional and interdisciplinary fields would be required over the next several years by increased energy research and development efforts, a *Science* editorial said. Numbers of graduating scientists and engineers were dropping each year and some signs of a shortage were appearing already. Top-level representatives from R&D agencies—including NASA, the Federal Energy Office, Office of Management and Budget, National Science Foundation, and the Atomic Energy Commission—should examine manpower needs and set up or fund a program for continuing assessment over the coming decade. "Dollar budgets can be changed quickly. Manpower budgets require longer planning time if both dollars and manpower are to be used wisely." (Vetter, *Science*, 5 April 74, 11)

*6 April:* The backup Skylab Orbital Workshop was loaded for shipment by sea to Marshall Space Flight Center for long-term storage. (MSFC Release 74-56)

*8 April:* President Nixon transmitted to Congress the *Aeronautics and Space Report of the President: 1973 Activities*. In his transmission message the President said: "This year has been particularly significant in that many past efforts to apply the benefits of space technology . . . to problems on Earth are now coming to fruition." Data from Skylab and Earth Resources Technology Satellite *Erts 1* were being used for applications including resource discovery and management, environmental information, and land use planning. Communications satellites had become a principal method of international communication and important in meeting national defense needs. Skylab had provided new information on the energy characteristics of the sun that could aid understanding of thermonuclear processes and contribute to development of new energy sources. Skylab had proved that man could effectively live and work in space for extended periods of time. Advances in aeronautics had contributed to defense of the Nation and helped reduce aircraft congestion and provide quieter, safer, more economical and efficient aircraft.

The report said that in previous years NASA's emphasis in space had been to expand human knowledge of phenomena, develop and operate space vehicles, and preserve U.S. leadership in aeronautics and space technology. But 1973 had been the year of "cashing in" on the technology development and knowledge obtained from lunar and space exploration programs. (*CR*, 8 April 74, H2677; Text)

- NASA announced selection of McDonnell Douglas Corp. to negotiate a two-year, $13.2-million, cost-plus-award fee contract to support the space shuttle program in system analysis; avionics system engineering; mission planning, analysis, and software formulation; computer systems and software; and crew procedures and flight planning. (NASA Release 74-83)
- The U.S.S.R. was continuing plans to build an earth station in the Ukraine to operate with the International Telecommunications Satellite Organization's comsat system. The project—the second planned Soviet INTELSAT link—had been unofficially reported by Russians in discussions with U.S. Government and private representatives over the past few months. (The first station, near Moscow, had been initiated to provide a bilateral communications capability for the U.S.-U.S.S.R. Hot Line, via a Molniya and an Intelsat satellite). U.S.S.R. participation in the INTELSAT system as a commercial user would fill in a major link in the U.S. goal of a single global comsat system, as stated in the 1962 Communications Satellite Act, *Aviation Week & Space Technology* commented. (Off US Rep to INTELSAT, ComSatCorp, interview, 23 Jan 75; Johnsen, *Av Wk*, 8 April 74, 24)
- An omnidirectional wheel that could change a vehicle's direction of movement without changing the direction it was facing had been developed by Marshall Space Flight Center engineer Josef F. Blumrich, MSFC announced. The rim of the wheel consisted of several barrel-shaped segments that rotated independently when in contact with the ground. The wheel, which would be strong enough to negotiate rough terrain, could be used on a planetary explorer vehicle or on wheelchairs. (MSFC Release 74-53)
- The French aerospace industry was reported as recording $1.07 billion worth of exports delivered in 1973. About 15%—$170 million—was in cooperative European aircraft programs. Missiles and space deliveries amounted to $79 million. New missile and space orders totaled $169 million.

    British aerospace exports were later reported at almost $1.3 billion for 1973, a 25% increase over 1972 and nearly twice 1970 exports. The leading British market was the U.S., with $360 million worth of deliveries. (*Av Wk*, 8 April 74, 23; *Interavia*, April 74, 275)
- The Navy had selected International Business Machines Corp. Electronic Systems Center as prime contractor for the light airborne multipurpose system (LAMPS) helicopter, *Aviation Week & Space Technology* reported. Under a $13.8-million contract, IBM would develop and test a prototype antisubmarine warfare and avionics and weapon control system for installation and integration in a helicopter furnished by the Navy. (*Av Wk*, 8 April 74, 24)

*8 April–3 May:* Some 75 Soviet aerospace specialists—including U.S.S.R. Apollo-Soyuz Test Project Technical Director Konstantin D. Bushuyev, prime crewmen Aleksey A. Leonov and Valery N. Kubasov, Flight Director Aleksey S. Yeliseyev, and training officer for ASTP crews Valery F.

Bykovsky—joined U.S. ASTP astronauts and specialists at Johnson Space Center to continue technical discussions and planning for the July 1975 joint ASTP mission. Agreement was reached that specialists and flight crews would visit the U.S. launch site in February 1975 and the U.S.S.R. launch site in May 1975, for familiarization with the spacecraft and to test compatible equipment. NASA had defined no technical need for U.S. representatives at the Soviet launch site during actual liftoff; U.S. specialists would be in the control center.

Apollo docking module and Soyuz life-support-system tests were reviewed, in addition to safety assessment reports on control systems and spacecraft manufacturing test and checkout. Schedules for each of the five joint experiments—ultraviolet absorption, biological interaction, microbial exchange, multipurpose furnace, and artificial solar eclipse—were approved. Agreement was also reached on the basic approach and content of the joint flight readiness review, to be held in May 1975. The U.S.S.R. crew and U.S. ASTP Astronauts Thomas P. Stafford, Vance D. Brand, and Donald K. Slayton walked through hundreds of procedures for the first time as crews, to verify the integrity of the docking module that would connect the Soyuz and Apollo command and service modules.

At a 26 April press conference Stafford demonstrated his proficiency in the Russian language by answering newsmen's questions in Russian. Cosmonauts and astronauts agreed that "we've no problem in language." One cosmonaut said, "Our work is considerably better when the American crew speaks Russian and our crew speaks English."

Glynn S. Lunney, ASTP Technical Director for the U.S., said during a 3 May press briefing that the work in the U.S. and jointly was on schedule. The progress "has been good and . . . gratifying." Prof. Bushuyev said that "there can be no longer any doubt that we will be successful in preparing for and achieving our joint flight in July 1975." (JSC *Roundup*, 12 April 74, 1; ASTP Communiqué; Press briefing transcripts, 26 April 74, 3 May 74)

*8–12 April:* Mariner 10 closeup photos of Mercury taken 23 March–5 April showed that many large basinlike features on the planet appeared to be flooded by lava, Dr. Bruce C. Murray—California Institute of Technology scientist and head of the *Mariner 10* photo interpretation team—reported to the American Geophysical Union's annual meeting in Washington, D.C. The apparent volcanic outpourings indicated that Mercury like the earth had a massive iron-rich core, hidden beneath a battered, heavily cratered, moonlike crust, making Mercury "a unique planet in the solar system" and serving as important evidence in deciphering the primordial history of the solar system. The latest results from *Mariner 10*'s encounter with the planet showed the outer portions of Mercury to be made of silicate-rich rocks, suggesting the planet had undergone considerable melting in its history, with heavier materials sinking to the center of the planet and lighter silicates rising to the surface. Analysis also showed that craters might have existed from the time the planet was formed by a great influx of cosmic debris 4 billion to 4.5 billion years ago, with no great internal surface-shaping processes since then. (UPI, *NYT*, 13 April 74, 9; AP, *W Star-News*, 12 April 74, A5; NASA MOR, 24 April 74)

*9 April:* Analysis of *Pioneer 10* photographs taken during the spacecraft's 3 Dec. 1973 encounter with Jupiter indicated that the planet's great red spot was probably a towering mass of clouds rising from discrete thermal sources, Dr. Tom Gehrels, Univ. of Arizona scientist, said at a NASA

Jupiter science press briefing in Washington, D.C. The heat sources caused the atmospheric gases to rise locally to lower pressures. As the altitude increased the gases condensed, forming the aerosols that colored the clouds. The aerosols then sank and were again evaporated by the higher pressures and temperatures, creating an up-and-down circular motion within the red spot. Dr. Gehrels said that the spot "is not a unique feature." *Pioneer 10* photos showed other areas of rising clouds. "We see quite a few of these that occur in equatorial regions."

Dr. James A. Van Allen, Univ. of Iowa scientist, said that Jupiter had been identified for the first time as a source of energetic particle emissions. The earth had been known to be a feeble emitter but Jupiter emitted particles at a strength detectable from fractions of an astronomical unit up to four or five AU from Jupiter. The discovery had dramatically changed interplanetary science from its traditional hypothesis that energetic particles observed in space had come only from the sun or from local acceleration. (Transcript)

- NASA announced the Delta 100 Failure Review Committee had concluded that the failure of the Thor-Delta launch vehicle 2nd stage during the 18 Jan. *Skynet IIA* launch had been caused by an electronics package short circuit induced by a piece of conductive contaminant shaken loose during launch. Also, the insulation coating on the printed circuit boards had been ineffective at all sharp edges; tests had determined that the contaminant had probably welded across the exposed ends of component leads. The committee recommended proper insulation of exposed electrical points and an electronics package subjected to higher acceptance test vibration levels. (NASA Release 74–88)

- Award of a $64 620 000, fixed-price-incentive contract to McDonnell Douglas Corp. for Thor-Delta launch vehicle mission checkout, launchings, engineering support, and modifications through 30 Sept. 1975 was announced by NASA. Additional contract options totaled about $2 500 000. (NASA Release 74–84; Off Delta Mgr, interview)

- The Senate Committee on Government Operations' Subcommittee on Reorganization, Research, and International Organizations favorably reported S. 2744 to the full Committee. The bill would split the Atomic Energy Commission into two agencies, the Energy Research and Development Administration (ERDA) and the Nuclear Energy Commission (NEC). (House, Bill Status Off, interview, 24 Jan 75)

*9–10 April:* Passengers aboard two Delta Air Lines, Inc., flights on 5 and 6 April might have been exposed to radiation from an improperly shielded shipment of 32 curies of solid radioactive iridium 192 transported in the aircraft cargo compartments from Washington, D.C., to Baton Rouge, La., the Atomic Energy Commission announced. The highest exposure for any one passenger would have been about 8 roentgens on one flight and 5½ roentgens on the other, and these exposures would have occurred at only one seat location. The AEC also announced 10 April that it had ordered Value Engineering Co., the AEC licensee that had originated the shipment, to suspend further shipments of radioactive materials pending investigation. (AEC Releases T–161; T–162)

*10 April:* NASA FY 1975 authorization bill H.R. 13998, a clean bill introduced 4 April to replace H.R. 12689, was reported out of the House Committee on Science and Astronautics. The revised bill would authorize $3.253 billion in FY 1975 funding, an increase of $6 million over the original request. Authorizations for research and development increased $11 million to $2.357 billion, and construction of facilities decreased $5 million, to $146.5 million. Funding for research and program management remained the same.

Included in the R&D increases was a $20-million addition, to a total $112.3 million, for the space shuttle main engine, which had encountered technical problems during early development. The increase was proposed to add confidence in meeting schedules and hold total program costs at the current projection. A $4.3-million increase, to $170.7 million, was added for aeronautical research and technology. A $1.8-million increase brought the total for space and nuclear research and technology to $76.6 million. Manned space flight operations funding was decreased $15 million, to $308.3 million, including a $5-million decrease (to $109.6 million) for the Apollo-Soyuz Test Project and a $10-million decrease (to $165.2 million) in development, test, and mission operations.

Changes in facility construction authorizations included an increase of $4 million, to $10.04 million for construction of the infrared telescope facility in Mauna Kea, Hawaii.

The Committee also requested that NASA make a scientific and technological inventory of its capability to contribute to such critical national needs as energy, transportation, and environmental control. NASA was to report to the Committee by 1 Aug. (H Rpt 93-983)

*11 April:* Communications Satellite Corp. Laboratories and Telesat Canada had recently completed joint field tests of the transmission of digital data through Canada's *Anik 1* communications satellite (launched 9 Nov. 1972), ComSatCorp announced. Data were transmitted at 67.2 megabits per sec, believed to be the highest rate ever transmitted by satellite. The high rate allowed two digital TV communications system color channels to be carried through a single transponder. The tests confirmed previous system analyses and proved the ability of satellites to transmit the equivalent of 100 newspaper pages in one minute. (ComSatCorp Release 74-22)

• Communications Satellite Corp. filed proposed rates with the Federal Communications Commission for a new medium-speed digital-data satellite service called DIGISAT, to provide digital data channels between the U.S. mainland and Hawaii. The proposed monthly rate to international common carriers would be $2700 for 2.4 kilobits per sec, $3000 for 4.8 kbps, and $3900 for 9.6 kbps. Expansion to European points was expected later in the year. (ComSatCorp Release 74-21; ComSatCorp PAO, interview)

*12 April:* Two A-9 twin turbofan aircraft, Northrop Corp.'s unsuccessful contenders for the Air Force close air support aircraft, had been transferred to Flight Research Center, the FRC *X-Press* reported. One of the aircraft would be used in three research programs: as a support aircraft in the trailing vortices research program; as a mother ship to air-launch supersonic Firebee drones in the remotely piloted research vehicle program; and as a test-bed aircraft in the Center's nonpropulsive aircraft noise program. The second A-9 would be used for spare parts. (FRC *X-Press*, 12 April 74, 2; FRC proj off, interview, 21 Feb 75)

• The economic return from the U.S. space program amounted to 33% per year, exceeding the typical yield from investments in stocks, bonds, and real estate, Dr. Christopher C. Kraft, Jr., Johnson Space Center Director, said in a speech at Virginia Polytechnic Institute and State Univ. at Blacksburg, Va. Among returns not immediately visible to the public were some 30 000 inventions available by the end of the Apollo program. The price of *Erts 1* Earth Resources Technology Satellite—surveying the earth since its launch 23 July 1972—was less than 50 cents per American. "Given its potential for aiding us in the coming years, who would not support a program that costs less than a . . . half-pound [one fourth kilogram] of ground beef?" Dr. Kraft estimated that the space shuttle, at an annual cost of $5 for each American, was an investment that would "pay, perhaps, the greatest return to future generations." With such a vehicle we could not only perform a variety of scientific studies with immediate application to life on earth but build permanent space stations and launch planetary vehicles and eventually starships. The space shuttle program, with its international cooperation, would unite "many nations for the benefit of all." (Text)

- The U.S.S.R.'s Cosmonautics Day and 13th anniversary of the first manned space flight, made by Cosmonaut Yuri A. Gagarin in 1961, was marked by articles in the Soviet press.

   "Cosmonautics is one of the most important testing ranges of science in which many directions of science and technology are being synthesized into a single constantly perfecting system," Soviet Academician Valentin Glushko wrote in *Pravda*. The successes of cosmonautics in the preceding 15 to 20 yrs had given a powerful impetus to various branches of knowledge. The initial accumulation of facts by spacecraft had been completed. On the agenda now was the detailed study of dynamic processes, the interconnection of phenomena, and the verification of hypotheses and models.

   Maj. Gen. Vladimir A. Shatalov, chief of cosmonaut training, acknowledged in an interview with *Sovietskaya Rossiya* the importance of orbital stations for future space experiments. He also acknowledged that Soviet designers had not yet been able to develop a satisfactory space shuttle. He said the U.S.S.R. had no plans for a manned landing on the moon. His impassioned defense of the Soviet space program suggested that it, like the U.S. program, was under fire by domestic critics; "many millions of rubles" had been saved because of the use of meteorological satellites.

   In *Trud*, Academician Boris N. Petrov, head of the Council of International Cooperation in Space Exploration and Use under the Soviet Academy of Sciences, said that manned flights "must become more frequent and more extended" to produce "really creative work for specialists of many branches of science." (Tass, FBIS-Sov, 16 April 74, U1–2; Wren, *NYT*, 13 April 74)

*13–23 April:* NASA launched *Westar 1* (Westar-A), the first U.S. commercial domestic communications satellite, for Western Union Telegraph Co. from Eastern Test Range at 7:33 pm EDT on a Thor-Delta launch vehicle, the first with the TR–201 2nd-stage propulsion unit. One of the nine Castor motors strapped onto the booster failed to jettison until main engine cutoff, decreasing vehicle velocity. Propellant reserves in the 2nd stage provided sufficient additional velocity, however, to put the satellite in the planned transfer orbit with a 36 238-km apogee, 230-km perigee, and 24.7° inclination. Nonmission-related 2nd-stage maneuvering experiments, which were to use the propellant reserves, were not performed. NASA's mission objective—to place the satellite in a transfer orbit that would allow onboard propulsion to place it in a stationary synchronous orbit while retaining sufficient stationkeeping propulsion to meet mission lifetime requirements—was met and the mission was adjudged successful 23 April.

*Westar 1* was acquired by a mobile ground station at 9:38 pm EDT and control was turned over to Western Union. The satellite apogee motor was successfully fired at 5:21 pm EDT 16 April and *Westar 1* was placed in geosynchronous orbit with a 35 592-km apogee, 35 166-km perigee, 23-hr 35-min period, and 0.6° inclination. An onboard hydrazine reaction control system was moving the satellite toward its final position above the equator at 99° W longitude.

Each of 12 transponders aboard *Westar 1* could relay 1200 voice circuits, 1 color TV channel with program audio, or data at 50 megabits per sec. The Westar system—including one satellite to be launched in June, one satellite to be held in reserve, and five earth stations—would greatly increase Western Union's capacity for commercial and personal communications. Expected to be operational in August, the system would be the first to interconnect, under one company, satellites and an existing continent-wide, computer-switched, multipurpose terrestrial network. (NASA MORs, 19 March 74, 24 April 74; GSFC *Wkly SSR*, 11–17 April 74; NASA Release 74–77; McElheny, *NYT*, 15 April 74, 1; AP, *CSM*, 18 April 74; *SBD*, 26 April 74, 322; GSFC proj off, interview, 7 Nov 75)

*14 April:* Stratospheric contamination from supersonic aircraft engines such as those fitted to the Anglo-French Concorde and the Soviet Tu-144 might lower the earth's mean temperatures, with a catastrophic effect on agriculture, and increase the incidence of skin cancer from overexposure to ultraviolet radiation, according to a Dept. of Transportation study cited by the *Washington Post.* In a telephone interview with the *Post* Dr. Alan Grobecker, manager of DOT's climatic impact assessment program, confirmed the existence of the study, which was based on data collected by satellites, balloons, and instrumented aircraft. He emphasized that the pollution by the emission of nitrogen oxide and sulfur dioxide would have a worldwide impact because of the capacity of the stratosphere to retain the pollutants for nearly a year. A separate Agricultural Research Service study showed that stratospheric contamination could cut temperate climate crops by 60%. The DOT paper would be considered by committees of the National Academy of Sciences and the National Academy of Engineering at a special meeting in July. (Wilson, *W Post*, 14 April 74, A8)

- Appointment of Stephen E. Doyle as NASA's Deputy Assistant Administrator for International Affairs became effective. Before joining NASA Doyle had been Manager for International Organization Affairs, Office of Telecommunications Policy in the Executive Office of the President. (NASA Release 74–101)

*15 April:* Marshall Space Flight Center had begun a 20-mo investigation of the vibration characteristics of railroad freight cars as part of an extensive study of train dynamics, with the Federal Railroad Administration and the American Association of Railroads, NASA announced. Martin Marietta Corp.—under a $329 000 contract from MSFC—would make a mathematical model of truck-car dynamics for comparison with data from laboratory and operational tests of a freight car and components. (NASA Release 74–86)

- The European Space Research Organization was investigating systems for all-weather earth resources observations, *Aviation Week & Space Technology* reported. Thomson-CSF and British Aircraft Corp. had com-

pleted a nine-month study of a synthetic-aperture radar satellite with a 50-m-resolution imaging sensor. (*Av Wk*, 15 April 74, 11)
- "Skylab was an achievement that will ultimately benefit every inhabitant of this planet," Robert Hotz wrote in an *Aviation Week & Space Technology* editorial. Further, "in addition to their constant work as space mechanics, the Skylab crews demonstrated that man is a valuable addition to the loop of scientific experiments and activities that can be performed uniquely from space.... There is no doubt that man must be an integral part of any future space reconnaissance system....

    "But perhaps the real essence of Skylab's performance and its significance for future space operations were in its tremendous flexibility." Besides design, planning, and extra equipment, "it was a spirit developed between the ground crews with their considerable array of resources and the flight crews who were able to focus these resources on solving their problems aloft." One astronaut had said, "Skylab worked better broken than anybody had hoped for if it was perfect." (*Av Wk*, 15 April 74, 9)
- Radioactive material like that which had leaked from an improperly shielded container on two 5 and 6 April Delta Air Lines, Inc., flights, and germs, bacteria, and nerve gas were shipped routinely on commercial aircraft, the *Washington Post* reported. The *Post* quoted an Atomic Energy Commission investigation as saying that 1 million shipments of radioactive material had been made during 1973. Medical isotopes had accounted for 95% of the total. On 20 March the Air Line Pilots Association had asked the Federal Aviation Administration to ban the shipment of radioactive material by air temporarily. (Conlan, *W Post*, 15 April 74, A6)

*16 April:* "I am sure life will move out from the Earth, probably first to the Moon and then to Mars," Dr. John E. Naugle, NASA Deputy Associate Administrator, said in a speech reviewing NASA's planetary program before the National Space Club. Just what form of life would be transplanted, whether man or some simple organism, would be the subject of debate. Four planets—Jupiter, Venus, Mars, and Mercury—had already been visited. The next major event would be the Viking landings on Mars in 1976. Historically, two significant events in 1965 had determined the direction of the planetary program. The first was a study at Woods Hole, Mass., which had recommended that the space program, following Apollo, should search for life on the planets and give first priority to the study of Mars—while also maintaining a wide-ranging program of general planetary exploration—because Mars most closely resembled the earth and was the most likely to support life. The second event, the 14 July 1965 flight of *Mariner 4* past Mars, supported the recommendations when it found evidence of water in the Martian atmosphere.

"In retrospect, I think the decision ... has been sound." But a major decision that faced NASA this summer, one year before the launch of Viking, was whether there should be follow-on Viking missions. "Should Mars continue to have first priority in planetary exploration? ... Clearly, if we find evidence of life on Mars, ... Mars will have first priority." (Text)

*17 April:* Soviet scientists had determined that weightlessness might change heredity, Tass reported. Mustard seeds flown in space had reproduced brown instead of green seeds. Nikolay Dubinin, Director of the General Genetics Institute at the U.S.S.R. Academy of Sciences, was quoted as

saying that the more complex the organism, the greater the change. (UPI, *W Post*, 18 April 74)
- Marshall Space Flight Center had awarded a $3 850 000 contract modification to Rockwell International Corp. for architectural and engineering services and construction of test and industrial facilities for the space shuttle main engine, NASA announced. (NASA Release 74–97)
- President Nixon announced he would appoint John C. Sawhill Administrator of the Federal Energy Office. He would succeed William E. Simon, who would be nominated Secretary of the Treasury. Sawhill had been Deputy Administrator of EEO since 4 Dec. 1973. (*PD*, 22 April 74, 413)

*18 April:* American Telephone & Telegraph Co. and GTE Satellite Corp. filed a proposal with the Federal Communications Commission to operate a joint domestic satellite system. The system would serve the contiguous 48 states and Hawaii and could serve Alaska and Puerto Rico through earth stations owned by other companies. A joint system would improve network efficiency and reduce the number of earth stations to 7, from the 10 required for two separate systems. Three satellites operated by COMSAT General Corp. and provided to AT&T would serve the joint system and could handle 28 800 simultaneous telephone conversations. The FCC had approved separate domestic satellite systems for the two companies 12 Sept. 1973, directing them to work together to integrate the systems and negotiate agreements for handling communications traffic and equitable division of revenues from jointly rendered services. (Smith, *NYT*, 19 April 74, 49)

*18–19 April:* Dr. James C. Fletcher, NASA Administrator, presented the NASA Distinguished Service Medal to Dale D. Myers, President of Rockwell International Corp. North American Aircraft Operations Group and former NASA Associate Administrator for Manned Space Flight, for his contributions to the Skylab program in a Skylab Honor Awards Ceremony in Washington, D.C. Nine Headquarters employees and three from Goddard Space Flight Center received the Exceptional Service Medal.

Twenty-six individuals and four groups received awards in ceremonies at Johnson Space Center. Twenty-six individuals and five groups at Kennedy Space Center and thirty-four individuals and six groups at Marshall Space Flight Center received awards in ceremonies 19 April. (NASA Ann; NASA Release 74–90; Program)

*19 April: Pioneer 11*, launched 5 April 1973 toward a December 1974 encounter with Jupiter, was retargeted to pass within 42 000 km of Jupiter and then, with a gravity assist from the planet, fly on to Saturn in 1979. Ground controllers at Ames Research Center commanded the spacecraft's onboard thrusters to fire for 42 min 36 sec, increasing velocity 230 km per hr. On its new course, *Pioneer 11* would pass Jupiter nearer the south polar region, intersecting the planet's equatorial plane at 55°, and come three times closer than *Pioneer 10* had during its 3 Dec. 1973 encounter. (NASA MOR, 5 Nov 74; NASA Release 74–94; ARC Release 74–10)
- The Saturn IB 1st stage for the U.S.S.R.–U.S. Apollo-Soyuz Test Project launch in 1975 left Michoud Assembly Facility for Kennedy Space Center on the ocean-going barge *Orion*. The stage would be stored in the Vehicle Assembly Building until December, when the launch vehicle would be assembled on its mobile launcher. (KSC Release 70–74)

*20 April:* The U.S.S.R. launched *Molniya 1–27* communications satellite from Plesetsk into orbit with a 40 705-km apogee, 623-km perigee, 12-hr

17.6-min period, and 62.9° inclination. The satellite would help provide a system of long-range telephone and telegraph radio communications in the U.S.S.R. and would transmit Soviet central TV programs to the Orbita network. (GSFC *Wkly SSR*, 18–24 April 74; Tass, FBIS–Sov, 22 April 74, U1; *SF*, Sept 74, 394)

- Japan and the People's Republic of China signed a civil aviation agreement in Peking setting up commercial flights between the two countries. Japan Air Lines Co., Ltd., would be able to fly to Peking and Shanghai. Civil Aviation Administration of China flights to Tokyo would be the P.R.C.'s first service to a non-Communist country. The P.R.C. had purchased 10 Boeing 707s for its expanded international service, including possible future service to Canada, the U..S., and Latin America. In addition to ratification by Japan's Diet, technical details would have to be worked out before regular service could begin. Within hours after the agreement was announced the Foreign Minister of the Republic of China (Nationalist) announced termination of service by JAL and Taiwan's China Air Lines between Taiwan and Japan and a ban on JAL flights over Taiwan. (Mackenzie, *W Post*, 21 April 74, A1; Butterfield, *NYT*, 21 April 74, 1)

*21 April:* Design work on the primary rescue system for space shuttle passengers proceeded at Johnson Space Center. A prototype "beach ball" large enough for a person to crawl into, zip up, and pressurize had been demonstrated and given a go-ahead at NASA Hq. in December 1973. Assistant Chief for Shuttle Larry E. Bell in JSC's Crew Systems Div. had told a news interviewer the balls could be packaged into cylinders to accompany each passenger aboard the shuttle in the 1980s, "almost like a parachute." Simplicity, size, light weight, structural strength of the ball shape, and low cost had won approval over expensive and complicated spacesuits for nonastronauts on flights and eliminated the requirement for a rescue docking module.

In an emergency, passengers could enter the balls—each with its own window—an hour or two before rescue by a second shuttle. Oxygen containers connected to the shuttle supply system would be used until just before rescue; a device would remove carbon dioxide. Once disconnected, each container would last an hour; transfer through space to the rescue shuttle was expected to take 30 min. (Bell interview, 4 March 75; Hill, *H Chron*, 21 April 74)

- Aerospace companies that had succeeded in using the system approach to develop advanced technology for use in space were encountering difficulties in the field of urban mass transportation, a *New York Times* article said. Boeing Co. Aerospace Div. faced losses of up to $6 million on an automated "people mover" shuttle system in Morgantown, W. Va. Cost of the project, chosen four years ago as a pilot application of aerospace management techniques, had jumped from the original estimate of $13.4 million to $115 million. Rohr Industries, Inc., had lost more than $10 million on a project to build 450 commuter cars for the San Francisco Bay Area Rapid Transit System, which had threatened to sue Rohr over late deliveries and alleged unreliability of some cars. Observers had noted that aerospace technologists tended to "overengineer" transit systems, increasing costs, aggravating maintenance, and marring reliability. In addition there were problems in converting prototypes to hardware that could withstand the rigors of carrying throngs of people. But, despite the frustrations, more companies were entering the competition. A significant test of how the aerospace industry would fare in the next era of transit development would be the performance of Boeing's Vertol Div. in two projects to produce 230 streetcars for Boston and San Francisco and 100 cars for Chicago's rapid transit system. (Lindsey, *NYT*, 21 April 74, 3:1)

*21–25 April:* Observations of Comet Kohoutek in December and January and sudden changes in brightness of the more recent Comet Bradfield had led to the theory that comets were layered like onions, Dr. Edward P. Ney, Univ. of Minnesota scientist, reported at a joint meeting of the American Physical Society and the Optical Society of America. When some layers were exposed, they shone brilliantly; when a layer wore off, the comet might abruptly become dim. The onion model would explain why predictions of Kohoutek's brilliance were so inaccurate and why Bradfield had suddenly dimmed to 100th its previous brilliance.

Dr. Alastair G. W. Cameron, Harvard College Observatory scientist, said that an enormous spherical cloud of comets with a combined mass 20 times that of Jupiter might lie beyond the planets and form a far more massive part of the universe than the planets. This cloud, the Oort cloud, might be a repository for heavy elements formed since the infancy of the Milky Way.

Dr. R. W. Klebesadel of the Los Alamos Scientific Laboratory said that satellites, placed in a 100 000-km orbit as part of the Air Force Vela series developed to monitor the 1963 Nuclear Test Ban Treaty, has monitored 27 high-intensity explosions in space over four years. The explosions, which had produced unusually large amounts of gamma and x-ray energy, might be in the Orion arm of the galaxy, where the earth was. Dr. Philip Morrison of the Massachusetts Institute of Technology said nine current theories of the origin of the explosions included exploding stars, or supernovas; objects falling into neutron stars; and unusually large solar flares or sunstorms on distant stars. (Sullivan, *NYT*, 24 April 74, 13; Shurkin, *P Inq*, 26 April 74, 12)

*22 April:* An *Erts 1* photo map of the contiguous 48 states of the U.S., the first ever assembled from satellite images, had been completed for NASA by the Dept. of Agriculture's Soil Cartographic Division, NASA announced. The three- by five-meter map was composed from 595 cloud-free black and white images taken 25 July–31 Oct 1972 from an altitude of 912 km by the multispectral scanner on *Erts 1* (launched by NASA 23 July 1972). The mosaic would provide a base for regional compilation of investigation results, provide the first synoptic look at the U.S. for scientists examining subcontinental lineaments and faults, permit construction of a national surface water inventory, provide a base for a national-vegetation-cover and continuous-land-classification inventory, provide an accurate assessment of the national water drainage network, and document 1972 national land use.

The mosaic would be displayed publicly for the first time in ceremonies at NASA Hq. 26 April. (NASA Release 74–91)

- Reduction-in-force notices were delivered to 397 employees at Marshall Space Flight Center. Another 250 employees received notices that they would be reduced in grade. Earlier, MSFC officials had projectd a personnel reduction of about 500 employees, but retirements and attrition since then had reduced the number required. (MSFC Release 74–68)
- NASA had applied for a patent on a special absorptive coating developed by a Marshall Space Flight Center engineer. It would be applied to aluminum

panels used in the construction of the solar heating and cooling system being demonstrated at MSFC. The coating absorbed 93% of the total solar heat, which was transferred to water circulated through flow passages to a storage tank for use in heating or cooling; it reradiated only 6% of the infrared heat. (NASA Release 74-96)

- The Air Force was making additional efforts in the application of boron and graphite composites in 1974—to take advantage of weight savings, higher performance, and lower costs—the Air Force Systems Command announced. Tests would include a full-sized composite wing for the YF-16 prototype lightweight fighter, a composite speedbrake for the F-15 air superiority fighter, and a composite outer wing for the A-7 attack fighter aircraft. The composite maneuver augmentation program called for the design, manufacture, and evaluation of an advanced composite wing on a remotely piloted test vehicle. (AFSC Release OIP 029.74)

*23 April:* The U.S.S.R. launched eight Cosmos satellites on a single booster from Plesetsk. Orbital parameters were:

*Cosmos 641*—1483-km apogee, 1387-km perigee, 114.6-min period, and 74.0° inclination.
*Cosmos 642*—1481-km apogee, 1319-km perigee, 113.8-min period, and 74.0° inclination.
*Cosmos 643*—1482-km apogee, 1353-km perigee, 114.2-min period, and 74.0° inclination.
*Cosmos 644*—1483-km apogee, 1335-km perigee, 114.0-min period, and 74.0° inclination.
*Cosmos 645*—1483-km apogee, 1369-km perigee, 114.4-min period, and 74.0° inclination.
*Cosmos 646*—1486-km apogee, 1403-km perigee, 114.8-min period, and 74.0° inclination.
*Cosmos 647*—1484-km apogee, 1423-km perigee, 115.0-min period, and 74.0° inclination.
*Cosmos 648*—1480-km apogee, 1439-km perigee, 115.2-min period, and 74.0° inclination.

The press reported observers believed the satellites would operate in support of Soviet naval communications. (GSFC *Wkly SSR*, 18–24 April 74; *SBD*, 25 April 74, 313)

- *Skylab 4* Astronauts Gerald P. Carr, Dr. Edward G. Gibson, and William R. Pogue briefed members of the House Committee on Science and Astronautics on the Skylab program: Carr said that, in space, material samples such as paper, teflon, nylon, and neoprene-covered nylon were self-extinguishing when ignited, because of a lack of convection. Polyurethane, however, burned "like something soaked in gasoline. . . . I am very certain that we will never ever see polyurethane foam in a spacecraft." Dr. Gibson said that a leg muscle problem that had hindered previous Skylab crews in their ability to walk and run upon return to earth had been solved by a very crude treadmill device used to simulate walking and running. "We worked at it every day, and the condition we came back in was good testimony to the utility of that device."

Pogue narrated a dramatic presentation of slides taken from the Skylab Workshop—of mountainsides eroded by stripmining, forests denuded by the clearcutting of timber, cities obscured by smog, and rivers and lakes muddied by pollution and silt. He said that the astronauts had been constantly amazed how clearly such effects could be seen from their perch in space. "It has an emotional feedback." Carr said that it had become "very, very clear . . . that we don't have a whole lot of places to live comfortably on this Earth and we must take care of it."

Dr. Gibson said that the crew would like to see future missions make more use of man's capability to use his own judgment in data collection. "We would like to see instruments with high-data capability which could be pointed and selectively operated." Carr said that, "if you are going to use man in space, he is going to be happy if he can use his own judgment and . . . is productive. If he is nothing but a switch twiddler working against the clock, he is going to become very bored, and he is going to have psychological problems." (Transcript)

- Dr. Robert R. Gilruth, former Johnson Space Center Director and retired NASA Director of Key Personnel Development, was elected to the National Academy of Sciences during its 111th Annual Meeting. In addition to Dr. Gilruth, 95 other scientists and engineers, one posthumously, were elected in recognition of their achievements in original research. (NAS Release, 23 April 74)

*24 April:* The U.S.S.R. launched *Meteor 17* meteorological satellite from Plesetsk into orbit with a 892-km apogee, 863-km perigee, 102.5-min period, and 81.2° inclination. The satellite carried equipment to obtain cloud- and snow-cover images on the sunward and shadow sides of the globe as well as data on thermal energy reflected and emitted by the earth and the atmosphere. (GSFC *Wkly SSR*, 18–24 April 74; Tass, FBIS–Sov, 25 April 74, U1; *SF*, Sept 74, 395)

- NASA's Plum Brook Station had tested an experimental hypersonic ramjet engine in a hypersonic wind tunnel with simulated flight conditions of mach 5, 6, and 7, Lewis Research Center announced. The water-cooled ramjet engine, a cooperative effort of LeRC and Garrett Corp., was designed to operate efficiently from mach 4 to mach 8 by using hydrogen fuel, a translating spike inlet, and both subsonic and supersonic combustion modes. The tests had yielded valuable basic information for possible future aircraft development. (LeRC Release 74–14)

*25 April:* The House passed the NASA authorization bill, H.R. 13998, by a vote of 341 to 37. The total authorization was $3.259 billion, $6 million over the $3.253 billion suggested by the House Committee on Science and Astronautics 10 April and $12 million above the Administration request of $3.247 billion. On the floor, the House had approved an additional $3.9 million for research in coal extraction, coal energy conversion, and mine safety. An additional $2 million was authorized for research in hydrogen production and use. (*CR*, 25 April 74, H3201–31)

- Quasars—celestial objects thought to be among the oldest, most distant, and most radiant bodies observable from the earth—could also be manifestations of black holes, Dr. Elden C. Whipple, Jr., National Oceanic and Atmospheric Administration scientist, and Dr. Thomas E. Holzer of the National Center for Atmospheric Research reported in a paper at the joint meeting of the American Physical Society–Optical Society of America. Material falling into a massive black hole could produce the visible effects of a quasar without requiring the magnitudes of distance, velocity, age, and radiating power usually attributed to these objects. (NOAA Release 74–63)

- The Federal Aviation Administration proposed a new safety regulation that would require all shipments of radioactive material on passenger and cargo aircraft to be inspected for leakage and scanned with radiation monitoring instruments before flight. On passenger flights, radiation monitors would check the cabin floor above the cargo hold before departure and the cargo compartment after the aircraft had landed. The FAA proposal had been prompted by incidents of improper packaging that had exposed passengers and crew members to higher-than-normal radiation levels [see 9–10 April]. (FAA Release 74–59)

*26 April:* The U.S.S.R. launched *Molniya II–9* communications satellite from Plesetsk into orbit with a 40 700-km apogee, 599-km perigee, 12-hr 17.0-min period, and 62.9° inclination. The satellite would help provide a system of long-range telephone and telegraph radio communications in the U.S.S.R. and would transmit Soviet central TV programs to the Orbita network. (GSFC *Wkly SSR*, 25 April–1 May 74; Tass, FBIS–Sov, 30 April 74; U1; *SF*, Sept. 74, 395)

- Data from *Pioneer 10*, launched 2 March 1972 toward a 3 Dec. 1973 encounter with Jupiter, had indicated that sulfur was one possible element in the surface composition of Jupiter's moon Io, Marshall Space Flight Center announced. Scientists at MSFC had studied the reflective properties of Io and concluded that sulfur was one of the few materials that could have the high reflectivity and strong absorption features observed. The theory would be tested by measuring the changes in reflectivity of sulfur at different wavelengths and temperatures and observing the reflectivity change of Io in the fall when Jupiter would be in the evening sky. (MSFC Release 74–72)

- Wallops Station, Wallops Island, Va., was renamed Wallops Flight Center to describe its mission and operations more accurately. The Center had responsibility for development and launch of scientific payloads, using boosters ranging from the small Arcas to the four-stage, solid-propellant Scout. The installation had been established in 1945 as the Pilotless Aircraft Research Station of the National Advisory Committee for Aeronautics and had served as a base for launching small rockets. After NASA had been established in 1958, it became known as Wallops Station and assumed an expanded mission in support of space flight programs. (NASA Ann, 25 April 74)

- NASA had awarded the Planning Research Corp., Huntsville, Ala., a one-year $6-million cost-plus-fixed-fee contract, with an additional one-year $10-million option, to provide design engineering support services for development, test, and mission operations primarily in support of the space shuttle at Kennedy Space Center, KSC announced. (KSC Release 75–74)

- NASA had selected International Business Machines Corp. Federal Systems Div. for a $636 000 one-year contract, with options for renewal, to provide system engineering and software development support to the Design Engineering Directorate at Kennedy Space Center for space shuttle launch processing systems. IBM would help provide a flexible, reliable, and cost-effective means of system-testing, launch operations control, and status monitoring of the space shuttle, ground support equipment, and facilities during ground operations. (NASA Release 74–104)

- The U.S.S.R. was believed to have orbited between 8 and 10 radar-carrying satellites in the last four years that could monitor surface ship traffic around the globe, Thomas O'Toole said in a *Washington Post* article. Intelligence sources believed the satellites were still experimental and part of an effort to monitor submerged Polaris submarines. One radar satellite, *Cosmos* 626, launched 27 Dec. 1973 into an orbit 264-km high, separated into two parts 11 Feb. One part moved to a higher 382-km orbit and was thought to be the power supply for the orbital radar, using polonium 210, a "hot" source of radioactive heat that could generate as much as 2000 kw of electricity by thermoelectricity. (*W Post*, 26 April 74, A1)

**27 April–5 May:** Two full-scale models of the Spacelab manned orbital laboratory, to be developed by the European Space Research Organization for use with the shuttle, were shown to the international press and public for the first time at the 1974 German Aerospace Show in Hanover, West Germany. Visitors were able to see interior as well as the exterior equipment of the pallet and pressurized laboratory. A 1:50 scale model of the space shuttle was also on display. (ESRO Release, 22 April 74)

**29 April:** Bacteria found frozen in Antarctic ice and soil for a possible 1 million years had grown and reproduced in the laboratory when put into nutrient fluids, the National Science Foundation announced. If the estimate of the age proved correct, the bacteria would be the oldest examples of life ever found on the earth. Results of the project, sponsored by NSF and NASA's extraterrestrial life detection program, would add weight to the theory that some forms of life might be frozen into the permafrost of planets such as Mars. The bacteria had been found at a depth of about 130 m near the main U.S. base on Ross Island. Later samples also had been found in Taylor Valley 95 km away. A senior scientist in the project, Dr. Roy E. Cameron of Jet Propulsion Laboratory, said in a telephone interview with the *New York Times* that rod-shaped, club-shaped, and spherical specimens grew and reproduced, but one type, particularly interesting because it had been capable of spontaneous movement, could not be made to reproduce. (NSF Release 74–155; Schmeck, *NYT*, 30 April 74)

- Dr. James R. Schlesinger, Secretary of Defense, announced plans to consider full-scale development of an air combat fighter (ACF) as an alternative to costly tactical aircraft. The best option, he said, would be to follow the Air Force lightweight fighter prototype program—including the YF–16 and YF–17—with an ACF development program. Advocates later said the ACF could defend bases from fighter attack while larger aircraft such as the F–4 and F–15 carried the battle to an enemy hundreds of kilometers away. (DOD Release 171–74; AP, *W Post*, 30 April 74, A6)

- Total employment in the aerospace industry was expected to drop back to 948 000 positions in June 1974 after a temporary increase during the last quarter of 1973, according to the semiannual forecast released by the Aerospace Industries Association. September 1973 payrolls had been up 4000 positions over the June 1973 level of 949 000; December employment had added another 9000 positions; end of the year employment totaled 962 000 workers, the highest level since 1970. (AIAA Release, 29 April 74)

- The airline industry could not count on "quantum jumps" in technology to ease inflation woes, *Aviation Week & Space Technology* quoted former astronaut Frank Borman, Senior Vice President of Operations for Eastern Airlines, Inc., as saying in Washington, D.C. Eastern's furlough of more than 4000 employees over the past several months was a "sign of the austerity the industry must accept to offset inflation." He said service might have to be more spartan and less frequent. (*Av Wk*, 29 April 74, 15)

*30 April:* Johnson Space Center announced two personnel changes: Astronaut John W. Young, veteran of four space flights, had been named Acting Chief of the Astronaut Office in the Flight Operations Directorate. He was replacing former Astronaut Office Chief Alan B. Shepard, Jr., who would serve as senior adviser to Young.

Astronaut Russell L. Schweickart had been named Director of User Affairs at NASA Hq., effective 1 May, replacing Albert T. Christensen, who would return to private industry. The User Affairs Office maintained close ties with users of NASA's applications program and ensured a flow of information and that the programs were responsive to user needs. (JSC Releases 74-71; 74-72)

- The House of Representatives passed the special energy research and development appropriations bill, H.R. 14434, by a vote of 392 to 4. The bill appropriated funds to various agencies, including $8.9 million to NASA, for energy R&D in FY 1975. Of the NASA total, $4.5 million would be to implement the Solar Heating and Cooling Demonstration Act if enacted. An amendment to appropriate an additional $1 million to NASA was defeated by voice vote. (*CR*, 30 April 74, H3350-84)
- A portable, remote, patient-monitoring device called the Vitasign Attendant Monitor had ben developed by a NASA Biomedical Application Team at Southwest Research Institute, San Antonio, Tex., using monitoring techniques developed for manned space flight, NASA announced. The monitor, commercially available at a moderate cost, operated by three electrodes placed on the patient's chest. A sudden change in the electrocardiogram signal or the respiration rate automatically alerted medical attendants. (NASA Release 74-64)
- NASA had selected Computer Sciences Corp., Falls Church, Va., to negotiate a three-year $6.2-million cost-plus-award-fee contract for central computing support at Goddard Space Flight Center, NASA announced. CSC would provide software support for manned and unmanned missions in real-time mission programs, network scheduling, and acquisition-data generation and transmission and would monitor and control display programs. (NASA Release 74-113)

*During April:* Skylab Program Director William C. Schneider and the three Skylab crews had been selected as recipients of the 1973 Robert J. Collier Trophy for "proving beyond question the value of man in future explorations of space and the production of data of benefit to all the people on earth," the National Aeronautic Association announced. The Collier Committee—while recognizing contributions to the Skylab program by the more than 26 000 engineers, scientists, flight controllers, and technicians—also had unanimously agreed that the trophy should be presented to Schneider on behalf of the crews for his leadership in this "exceedingly complex enterprise." The trophy was presented by Vice President Gerald R. Ford in Washington, D.C., ceremonies 4 June. (NAA *News*, April 74)

- The report *For the Benefit of All Mankind: The Practical Returns from Space* was released by the House Committee on Science and Astronautics: America's space program could stand on its own feet as a "heroic manifestation of the evolutionary progress of humanity toward a higher and better life." During 1973 more than 2000 new examples of space-developed innovations and techniques had been reported under NASA's technology utilization program. Since the program's inception 11 yrs ago, more than 30 000 items had been reported. Hundreds of new

products and applications had become a part of daily life. Under liberalized licensing procedures, NASA had granted five exclusive patent licenses during the year, the largest since the inception of the program. In addition to publications announcing innovations, NASA's seven Regional Dissemination Centers worked directly with 2000 companies each year. In 1973 more than 57 000 industrial inquiries had been handled. (Com Print)
- Aircraft stalls and spins, greatest cause of fatal general-aviation accidents, had been intensively studied for the past 18 mos, NASA said in a report on its general-aviation technology program. More than 1000 spin tests had been made in Langley Research Center's spin tunnel; designs for test aircraft were nearly complete. Airport traffic flow and pilot approach-and-landing performance also were under study, and a program offering more precise flight path control and reduced pilot workload had significantly improved pilot landing ability. A five-year program was developing better training methodology to reduce pilot errors. (NASA Fact Sheet)

## May 1974

*1 May:* Kennedy Space Center had awarded a $1 228 000 contract to Seelye Stevenson Value and Knecht, Inc., for the design of an orbiter processing facility and modifications to the Vehicle Assembly Building to support assembly and integrated checkout of the space shuttle, KSC announced. VAB modifications would include redesigning high bays 3 and 4 to accommodate complete shuttles and widening the north door to permit orbiter entry. The design work was to be completed within 30 days of notice of the award. (KSC Release 77-74)

*2 May:* A full-scale simulated crash of an aircraft with a crew of four lifelike dummies was successfully completed at Langley Research Center as part of five years of tests in a program to study the crash-worthiness of light, general-aviation aircraft. The aircraft, a surplus twin-engine Piper Navajo, was suspended from the top of the test facility and swung by cables into the ground at 97 km per hr. An umbilical cable fed data on structural response back to a computer. The program was a cooperative effort with the Federal Aviation Administration and private industry to develop structural design techniques to improve the capability of light aircraft to withstand crash conditions and to increase the likelihood of passenger survival. (NASA Release 74-107; Newport News, Va., *Times Herald*, 9 May 74)

- NASA announced the appointment of Dr. David L. Winter as Director of Life Sciences succeeding Dr. Charles A. Berry, who had announced his retirement 21 Dec. 1973 but remained with NASA until April. Dr. Winter would manage programs in biomedical and bioscience research, medical aspects of manned space flight operations, man-machine integration, life science applications, aeronautical life sciences, extraterrestrial life research, and occupational medicine. He had been Deputy Director of Life Sciences at Ames Research Center. (NASA Release 74-118)

- Controversy over the parking position of Western Union Telegraph Co.'s second satellite—Westar-B, scheduled for June launch—was growing, the *Wall Street Journal* reported. Western Union had requested that the Federal Communications Commission assign the satellite a 119° west longitude position, but American Telephone & Telegraph Co. had petitioned the FCC to deny the request. AT&T would like to use that position for one of three comsats it planned to have in operation by early 1976. Positions were limited; communications satellites occupied orbits 35 500-35 900 km above the equator, and positions between 119° and 138° west longitude served the continental U.S., Hawaii, and Alaska best. In addition, comsats had to be spaced about 5° apart to avoid interference with each other. (Jacobs, *WSJ*, 2 May 74)

- Dr. Robert C. Seamans, Jr., former NASA Deputy Administrator and Secretary of the Air Force, had been reelected to a four-year term as President of the National Academy of Engineering, NAE announced. He was first elected to the office in May 1973 for a one-year term. (NAE Release, 2 May 74)

*3 May:* Kennedy Space Center had awarded a $382 000 contract to General Electric Co. for a multispectral-image-analysis system for data from the *Erts 1* Earth Resources Technology Satellite (in orbit since 23 July 1972), from Skylab's earth resources experiment package (carried on the Orbital Workshop in 1973), and from remote-sensing aircraft. (KSC Release 80–74)

*4–5 May:* The Lower Atmosphere Composition and Temperature Experiment (LACATE) balloon was launched at 12:00 midnight EDT by a joint Government-industry team from White Sands Missile Range in a Langley Research Center project. An onboard 10-band infrared radiometer, being tested for use on the planned Nimbus-G satellite, measured vertical profiles of radiance emitted by the earth's atmospheric limb. The profiles would provide basic data on temperature, ozone, nitrogen dioxide, water vapor, nitric acid, nitrous oxide, methane, and particles in the upper atmosphere and lower stratosphere, where high-flying jet aircraft operated. The 183-m-long, 137-m-wide, helium-filled balloon—the largest ever launched from WSMR—reached its maximum stable altitude of 50 km three hours after launch. Measurements were made before, during, and after sunrise so that both day and night atmospheric effects could be studied. At flight termination the gondola and its parachute were released from the balloon and returned to the earth. (*Langley Researcher*, 26 April 74, 1; LaRC PIO, interview, 20 Jan 75)

*6 May:* NASA FY 1975 authorization bill H.R. 13998 was reported to the Senate by the Senate Committee on Aeronautical and Space Sciences. The Committee authorization of $3.267 billion was $8.145 million higher than that passed by the House 25 April. It was equal to 1% of the total FY 1975 Federal budget and was $0.5 billion below the constant-level NASA budget endorsed by Congress two years earlier. The Committee had added $16 million to NASA's 4 Feb. budget request, to fund procurement and launch of a third Earth Resources Technology Satellite (ERTS–C) for continuity of remote-sensing data. The Committee also had added $6 million to support activities under way and the initiation and application of new ideas in energy and environmental areas. It refused to concur in $20 million added by the House to the space shuttle program. (S Rpt 93–818)

*6–10 May:* The possibility of self-sufficient colonies of men in space was discussed at a conference of physicists, astronauts, and space flight technologists at Princeton Univ. Dr. Gerald K. O'Neill, Princeton professor of physics, suggested that a colony for 2000 persons could be constructed on a space station at the point in the moon's orbit where the gravitational fields of the earth and the moon balanced. The cylindrical station could be built within 15 to 20 yrs using the space shuttle to transport the 9000 metric tons of necessary building materials from the earth. The abundant materials available on the moon could be transferred by a cargo-launching system. Dr. O'Neill pointed out that "virtually unlimited" resources existed in space, including a continuous source of energy from the sun and great chunks of almost pure nickel-iron from the Asteroid Belt. Increasingly large and complex stations could ensure that most "dirty" industry could be operating off the earth by the middle of the 21st century. Space colonies could also ensure the continuity of the human race. (Sullivan, *NYT*, 13 May 74, 1; OMSF, interview, Feb 75)

*6-21 May:* Fifteen nations, including the U.S., signed an international convention for the protection of copyrighted signals transmitted by satellite, following a week-long international conference in Brussels, Belgium. Each contracting country undertook to prevent any earth station in its territory from picking up and redistributing satellite signals without authority. Representatives of 60 nations had established the text of the convention at the conference sponsored by the United Nations Educational, Scientific, and Cultural Organization (UNESCO) and the World Intellectual Property Organization. The convention would remain open for signatures until March 1975. It would come into force three months after ratification by five signatory states. (Rpt to House Com on Sci & Astro, Com Print, July 74)

*7 May:* President Nixon signed H.R. 11793, the Federal Energy Administration Act of 1974, into Public Law 19-275, creating the Federal Energy Administration to replace the Federal Energy Office. In signing the bill President Nixon said that the FEA would provide a more firmly based organization to carry out the responsibilities of the FEO through 30 June 1976, including fuel allocation and pricing regulation, energy data collection and analysis, and broad energy planning, with particular emphasis on conservation and expansion of energy supplies. (*PD*, 13 May 74, 498-499)

• New synthetic materials developed from advanced aircraft research at Ames Research Center could be applied to civil aircraft tires and automobile brakes to improve wear and performance, NASA announced. A new polymer, resistant to heat buildup and the consequent frictional deterioration, promised up to 10 times the normal wear for braking systems. And another new polymer, with unique molecular properties, might provide aircraft tires that could make as many as 200 landings, instead of the customary 100, before retreading was necessary. The new materials would be tested in fleets of hard-use Government trucks and in commercial airline service using Boeing 727 aircraft. (NASA Release 74-117; ARC Chem Research Proj Off, interview, 12 March 74)

• Former astronaut John H. Glenn, Jr., defeated Sen. Howard M. Metzenbaum in the Ohio Democratic primary in his third try for nomination to the U.S. Senate. Glenn's first try for a Senate seat in 1964 had ended when injuries from a fall forced him out of the primary. In 1970 Metzenbaum defeated Glenn in the primary. Glenn became the first U.S. astronaut to orbit the earth 20 Feb. 1962. (Farrell, *NYT*, 9 May 74, 2: *A&A 1964*; Hope, *W Star*, 6 May 70, A1)

*8 May:* Siegbert B. Poritzky, Special Assistant for Systems and Research for the Air Transport Association, presented U.S. scheduled airlines' views on NASA's role and responsibility to the aircraft industry, before a House Committee on Appropriations' Subcommittee on HUD-Space-Science-Veterans hearing on FY 1975 appropriations. Poritzky said it was NASA's responsibility to build a technology base for aeronautics through innovative independent research in addition to undertaking directed research in response to specific problems. Also vitally important was NASA's responsiveness to the needs of other agencies. Airlines recommended a $24.5-million increase in funding for NASA's aeronautical programs—for propulsion environment-impact minimization, aircraft operations, alternative fuels, wake-vortex minimization, and noise research.

James J. Hartford, Executive Secretary of the American Institute of Aeronautics and Astronautics, testified that some cuts in the NASA budget were "penny wise and pound foolish." Additional but "relatively modest" funding would prevent the High Energy Astronomical Observatory (HEAO) program—which was to investigate the scientifically exciting black holes in space—from "limping along." Scientists had been skeptical of the usefulness of the space shuttle to science, but the Skylab Apollo Telescope Mount discoveries had so far exceeded expectations that astronomers "could hardly wait" for the Large Space Telescope, planned for launch on the shuttle. "That project unified the space science community" as "nothing ever did in the past."

Other "logical follow-ons" to the approved program were urged: Mariner Jupiter-Uranus 1979 mission, Jupiter orbiter, Venus atmospheric probe, and a follow-on to the Viking Mars 1975 with a rover capability.

Hartford also suggested that NASA and other Government agencies were not doing enough work in space applications. Cuts in NASA's communications satellite research and development program had been based on the incorrect assumption that private industry could take on full comsat R&D with no further Government stimulus. Early commercial exploitation of communications R&D had been rapid only because Government research had reduced risks to an acceptable level. Competition was now coming from Canadian, Japanese, and European organizations with "a lot of government backing" who had developed a technical capability approaching that of the U.S. and were willing to take high risks. (Transcript)

- Skyrocketing fuel prices had added almost $1.5 billion to the cost of producing aluminum, titanium, and magnesium, threatening serious inflation for military aircraft programs and NASA's space shuttle, the *Washington Post* reported. The price of aluminum had jumped from 6.5 cents for a half kilogram to 31.5 cents in the preceding six months. One aerospace company was quoted as telling the Pentagon and NASA that the rising aluminum prices would increase its program costs 16%. Grumman Aircraft Corp. had said it was paying 25% more for titanium and aluminum than one year ago. Grumman manufactured six models of aircraft, including the Navy's F-14, which used more titanium than any other aircraft. Rockwell International Corp. had raised the estimated delivery cost of each of its B-1 bombers from $45 million to $61.5 million, at least partly because of soaring aluminum and titanium prices. Each B-1 required 12 000 kg of titanium and 24 000 kg of aluminum. (O'Toole, *W Post*, 8 May 74)

- Methods developed to analyze images of the earth taken by the *Erts 1* Earth Resources Technology Satellite (launched 23 July 1972) had been adapted to enhance x-rays for medical use, NASA announced. The new technique, developed by a NASA cooperative student at Kennedy Space Center, included a microdensitometer that scanned the film and separated the grey shades into various densities. The untrained eye could detect 8 to 10 shades and the trained eye 32, but the microdensitometer could pick out 256 shades. Breaking down the image into more grey levels showed detail not perceived by the human eye. (KSC Release 81-74)

- Pan American World Airways, Inc., ordered a cockpit ground-proximity warning system installed on its entire 140-aircraft fleet. The device, designed by Sundstrand Data Control, Inc., would sound a loud "Whoop! Whoop! Pull up!" and flash a red light reading "terrain" if an aircraft was heading for a mountain slope or was too low on takeoff or landing. The decision to install the equipment had been hastened by recent aircraft accidents, including the 22 April crash of a Pan Am aircraft into a Bali, Indonesia, mountaintop, which had killed 107 persons. (Witkin, *NYT*, 9 May 74)

- *9 May:* NASA announced the appointment of John F. Yardley as Associate Administrator for Manned Space Flight. Yardley—who had been Vice President and General Manager of McDonnell Douglas Astronautics Co. Eastern Div. since 1972—would replace Dale D. Myers, who had returned to private industry. At McDonnell Douglas, Yardley had served as project engineer for Mercury spacecraft design 1958–1960, Launch Operations Manager for Mercury and Gemini Spacecraft 1960–1964, Gemini Technical Director 1964–1967, and as Corporate-wide General Manager for the Skylab project. (NASA Ann, 9 May 74)
- Federal Aviation Administrator Alexander P. Butterfield signed in London a Memorandum of Understanding for a joint program to test, evaluate, and demonstrate the Aeronautical Satellite (Aerosat) system, for improved communications and air traffic services over the North Atlantic. Other participants included Canada and European Space Research Organization members Belgium, Denmark, France, West Germany, Great Britain, Italy, The Netherlands, Spain, Sweden, and Switzerland. The space segment would include two synchronous-orbit satellites to be launched in late 1977 or early 1978 and would be jointly owned by ESRO, Canada, and a U.S. private sector coowner to be named later. FAA would lease use of the satellites from the U.S. coowner. (FAA Release 74–70)
- The Senate passed by voice vote the $3.267-billion FY 1975 NASA authorization bill H.R. 13998, as reported out by the Senate Committee on Aeronautical and Space Sciences 6 May. The Senate agreed on 22 May to a conference with the House on their differences. (*CR*, 9 May 74, S7461; NASA *LAR*, XIII/66)

*9–10 May:* Two *Mariner 10* trajectory correction maneuvers were made by an onboard propulsion system to take the spacecraft within 48 000 km of Mercury, for a second encounter in September. The correction would allow *Mariner 10*'s TV cameras to photograph the full planet from the sun side at a resolution of about 1 km. (NASA MOR, 3 Oct 74; NASA Release 74–121)

*10 May:* NASA's newest space communications facility was dedicated in a ceremony at the station site at Robledo de Chavela, Spain. The facility, equipped with a 64-m antenna—one of the world's largest and most sensitive—joined two other Deep Space Network facilities of the Madrid Space Communications Complex. Construction of the installation had begun in 1969 and became operational in the fall of 1973. The Madrid stations were operated for NASA and Jet Propulsion Laboratory by the Instituto Nacional de Técnica Aeroespacial under continuing agreements between Spain and the U.S. (NASA Release 74–119; NASA OTDA, interview, 17 March 75)

*11–12 May:* More than 500 weather balloons were launched from 55 stations in Marshall Space Flight Center's Atmospheric Variability Experiment (AVE–2) to provide data for comparison with data collected by NASA and Air Force weather satellites. The balloons—released by NASA, National Weather Service, National Severe Storms Laboratory in Oklahoma, and Eglin Air Force Base—flew as high as 30 km while measuring atmospheric pressure, temperature, humidity, and wind speed and direction. (MSFC Release 74–76; MSFC proj mgr, interview, March 75)

*12–17 May:* Scientists from the U.S. and U.S.S.R., participating in a Leningrad symposium on final results of the 17 Feb.–8 March 1973 joint Bering Sea experiment, concluded that satellite data could be used for accurate assessment of sea surface and related atmospheric conditions. Two research groups had independently surveyed the sea surface under identical atmospheric conditions, using microwave radiometry equipment aboard U.S. and U.S.S.R. research aircraft. Results had been compared with data obtained by ships sailing below. Findings offered convincing proof that scientists could use microwave measurements from satellites to assess sea ice distribution, motion, and stress and that multispectral observations could determine the content of liquid water and water vapor in the atmosphere above the sea surface. (NASA Release 74–139)

*13 May:* Dr. Harrison H. Schmitt became NASA Assistant Administrator for Energy Programs. He headed NASA's new Office of Energy Programs, supporting Federal agencies responsible for energy research and development. The office consolidated functions previously performed by other Headquarters offices. Dr. Schmitt, science pilot on the *Apollo 17* lunar landing mission, had been Chief of the Science and Applications Directorate in the Astronaut Office at Johnson Space Center. He had been on temporary assignment at Headquarters as Special Assistant to the Administrator for Energy Research and Development since 29 Jan. (NASA Ann, 16 May 74)

- An experimental scanning laser doppler system, originally developed at Marshall Space Flight Center to measure atmospheric winds and turbulence, was being tested at MSFC to determine the system's reliability in detecting and monitoring aircraft wake vortices. Flyby tests had been conducted by MSFC and the Federal Aviation Administration at Redstone Arsenal using a Boeing Co. 720 aircraft and two laser doppler units. Early results indicated that, in addition to detecting the vortices, the laser doppler system provided data on the air speeds within them. Two experimental systems would be used in field tests at Kennedy International Airport in New York after checkout of the units was completed. (NASA Release 74–127)

- Johnson Space Center had awarded a $2 104 900 cost-plus-fixed-fee contract to Westinghouse Electric Corp. to refurbish and modify 10 color TV cameras from the Skylab program, for use in the Apollo-Soyuz Test Project. Westinghouse would also provide two new cameras and equipment. (JSC Release 74–101)

- NASA announced the appointment of Judith A. Cole as Deputy Assistant Administrator for Legislative Affairs effective 19 May. She succeeded Gerald J. Mossinghoff, who had been appointed Assistant General Counsel 1 Jan. Miss Cole had been Staff Assistant in the Office of the Director at the U.S. Arms Control and Disarmament Agency. (NASA Ann, 13 May 74)

- NASA had begun a national aerospace fellowship program to encourage women and members of minority groups to seek careers in engineering and certain scientific fields. NASA had completed arrangements with seven colleges and universities to award 20, $2500 fellowships to students in the top third of the junior or senior class who had demonstrated a potential and interest in these fields of study. Recipients would also be

invited to work at NASA following graduation if appropriate job vacancies existed. (NASA Release 74–112)
- A New Orleans *Times-Picayune* editorial advocated international cooperation in space: The U.S. was committed to a reusable space shuttle for possible use in the future with a permanent manned space station. The U.S.S.R. would probably embark on its own space shuttle and space station program. "The cost of establishing a space station ... is such that it would be foolish for each nation to assume the full burden when a common station could provide everything a separate one could—even a 'secure section' for each side for passive military uses." (*T–Picayune*, 13 May 74, 23)

*14 May:* The U.S. Postal Service issued a stamp commemorating the Skylab program on the first anniversary of the launching of the *Skylab 1* Orbital Workshop. The yellow, red, blue, and black stamp depicted the Workshop and Astronauts Charles Conrad, Jr., and Dr. Joseph P. Kerwin as they worked in space to repair the spacecraft, which had been damaged during launch. (Tower, *NYT*, 28 April 1974, 30)
- President Nixon announced his intention to nominate Gen. George S. Brown, Air Force Chief of Staff, as Chairman of the Joint Chiefs of Staff succeeding Adm. Thomas H. Moorer, who was retiring. The President would nominate Gen. David C. Jones to replace Gen. Brown as Air Force Chief of Staff. The nominations were submitted to the Senate 15 May. (*PD*, 20 May 74, 527, 529)

*15 May:* Marshall Space Flight Center had requested proposals for two space tug studies, MSFC announced. The first study was to develop schedules, controls, and planning data for the space tug program from initial flight readiness to long-term mission operations. The second study was to recommend a space tug avionics system, with emphasis on rendezvous and docking avionics and data-management-subsystem requirements and configuration. (MSFC Release 74–79)
- The papers of Dr. Hugh L. Dryden, NASA Deputy Administrator from 1958 until his death in 1965, were opened to the public in ceremonies at Johns Hopkins Univ., Baltimore, Md. The papers would be housed in the university's Milton S. Eisenhower Library. (NASA Release 74–120; NASA Hist Off)
- Dr. William O. Davis, Chief of the Upper Atmosphere and Space Services Office of the National Oceanic and Atmospheric Administration and a leading rocket and space expert, died in Washington, D.C., at the age of 54. Dr. Davis was a member of the interagency team that planned and coordinated NASA's Earth Resources Satellite program. As Assistant Director of Laboratories at Wright-Patterson Air Force Base 1957–1958, he had directed Project Far Side, the earliest attempt to obtain radiation and magnetic data from outer space by launching a multistage rocket from a high-altitude balloon. (*W Post*, 15 May 74)

*16 May:* A new method using tiny electronic body sensors, developed by Ames Research Center for remotely recording physical responses during space flight training and research programs, was being used to measure the precise walking patterns of children with cerebral palsy, NASA announced. Patients could be free of cumbersome equipment that had hindered obtaining accurate gait measurements. (NASA Release 74–131)

*17–24 May: Sms 1* (SMS–A Synchronous Meteorological Satellite)—first meteorological satellite in synchronous orbit—was launched by NASA for the National Oceanic and Atmospheric Administration from Eastern Test Range at 5:31 am EDT. The 1st-stage propulsion of the three-stage Thor-Delta launch vehicle was less than normal. The 2nd stage burned to near propellant completion, compensating for the deficiency but leaving insufficient propellant to complete the planned second burn. The spacecraft entered a lower-than-planned transfer orbit with a 32 895-km apogee, 180-km perigee, 9-hr 37-min period, and 24.5° inclination. The apogee boost motor boosted the spacecraft into a near-geostationary orbit with a 32 828.3-km apogee, 30 747-km perigee, 20-hr 39-min period, and 2.0° inclination, with an eastward drift of approximately 56° per day.

The auxiliary propulsion system was fired at 3:00 am EDT 19 May to place the spacecraft in operational attitude. A planned series of APS firings between 19 and 24 May increased the orbit to geosynchronous altitude, eliminated orbital drift, and stopped the satellite at 45° west longitude. An early post-launch review of flight data revealed no spacecraft anomalies, and the first photos were transmitted 24 May.

NASA objectives—to launch the spacecraft into a geosynchronous orbit of sufficient accuracy to accomplish operational mission requirements and evaluate and check out the spacecraft—were met and the mission was adjudged successful 17 Dec. Operational control was turned over to NOAA 10 Dec. after checkout.

The 272-kg satellite (628 kg at launch) would, for the first time, provide continuous day and night images of cloud cover over the U.S. and Atlantic Ocean. Its data-collection systems would be key elements in the Global Atmospheric Research Program's Atlantic Tropical Experiment to be conducted during the summer, collecting meteorological, hydrological, oceanographic, seismic, and tsunami data. In addition, *Sms 1* would monitor solar flare activity for future manned spacecraft and supersonic aircraft flights.

*Sms 1* was the first in a series which would include two operational prototypes and one operational spacecraft (SMS-C, to become *Goes 1* on launch), developed and funded by NASA to meet the requirements of the Dept. of Commerce for its Geostationary Operational Environmental Satellite (GOES) system. The SMS project was managed by Goddard Space Flight Center under the direction of the Office of Applications. GSFC also had responsibility for the Thor-Delta launch vehicle. (NASA MORs, 10 May, 20 May, 18 Dec. 74; NASA, Mgr Small Launch Vehicles & Intl Prog, interview, 31 Jan. 75; GSFC, SMS Proj Mgr, interview, 4 Feb 75)

*17 May:* The U.S.S.R., with the cooperation of East Germany and Czechoslovakia, launched *Intercosmos 11* from Baykonur Cosmodrome near Tyuratam, U.S.S.R., into orbit with a 511-km apogee, 483-km perigee, 94.6-min period, and 50.6° inclination. The spacecraft, launched on the 250th anniversary of the U.S.S.R. Academy of Sciences, would study shortwave ultraviolet and x-ray radiation of the sun and the upper atmosphere. (GSFC *Wkly SSR*, 16–22 May 74; Tass, FBIS–Sov, 21 May 74, U2)

- A prototype direct-readout ground station, designed by Goddard Space Flight Center to receive high-resolution weather photos from satellites, left GSFC for Dakar, Senegal. It was to play a key role in the Global Atmospheric Research Program's Atlantic Tropical Experiment (GATE) to study tropical weather patterns. Using the ground station, planners could quickly advise scientists aboard participating ships to focus attention on specific weather conditions and could route aircraft into regions where cloud development was detected. The prototype unit, which produced real-time images as large negative or positive transparencies with a resolution four to eight times greater than units already in use, consisted of four vans weighing a total of 12 000 kg. Future models would weigh 270 kg and could be built at modest cost by other countries. (NASA Release 74-129)

- NASA announced selection of International Business Machines Corp. for an $11-million, two-year, cost-plus-fixed-fee contract to design, develop, and implement a ground-based computing and data-processing system for the space shuttle and other programs. (NASA Release 74-132)
- Science and technology might be stepping into a more important role in the conduct of foreign policy, a *Science* article said. During an April speech before the United Nations Dr. Henry A. Kissinger, Secretary of State, had urged that we "now apply science to the problems which science has helped to create." He proposed international action in four areas: agricultural technology, birth control, energy, and weather. The *Science* article noted that access to U.S. technology had been a major inducement toward détente for the U.S.S.R. and the People's Republic of China. Although the speech had made no specific commitments, the proposals, announced by the Secretary of State, became U.S. Government policy. By calling on science to help solve the world's problems, "Kissinger clearly has the monster in his sights . . . but whether a coherent science policy will emerge is not yet evident." (Wade, *Science*, 17 May 74, 780–781)

*18 May:* A new low-cost method of making solar cells could make solar energy available to individual homes and businesses within a few years, the *Washington Post* reported. The *Post* quoted Dr. A. I. Mlavsky, Director of Research at Tyco Laboratories, Inc., and R. Bruce Chalmers, professor of metallurgy at Harvard Univ., as saying the new process—developed by a university-industry team funded by NASA and the National Science Foundation—could cut solar cell production costs by 95%. With the new cells, a system that could produce 1 kw of electricity would cost $350; a system using conventional silicon-crystal cells cost between $20 000 and $100 000. In the new process, molten silicon material was pulled into long thin ribbons and cut into various lengths. (Leary, *W Post*, 18 May 74)

- India set off its first nuclear blast, an underground explosion equivalent to 10 000–15 000 tons of TNT, to become the world's sixth nuclear power. At a press briefing, the Chairman of India's Atomic Energy Commission, H. N. Sethna, said the blast was an experiment to determine the cratering effect of the earth and the cracking effect of rocks under the soil. (AP, *W Star-News*, 19 May 74, A1)
- The Soviet Union's first 1-million-kw nuclear reactor had begun operation near Leningrad and had generated its first billion kilowatts of electricity, *Pravda* reported. (Shabad, *NYT*, 30 May 74, 51)

*20 May:* Dr. James C. Fletcher, NASA Administrator, and Dr. Alexander Hocker, European Space Research Organization Director General, held a Spacelab program review in Paris to assess the proposed ESRO design, appraise an ESRO presentation of contractor evaluation and findings, and discuss follow-on production. The review indicated that size, volume, subsystems, transportability, and operational turnaround met or exceeded design requirements. The design payload weight had been reduced from phase B results. An ESRO team with NASA support would

further assess the weight, and the common payload support equipment and instrument pointing system would be studied for nine months. NASA would give a letter of commitment to buy one flight unit with follow-on production commitment scheduled after critical design review. (NASA Gen Mgmt Rev Rpt, 17 June 74; NASA *Spacelab Newsletter* 74–5; 3 July 74)
- Edwin C. Kilgore had been appointed Deputy Associate Administrator for Center Operations, NASA announced. He had assisted Dr. George M. Low, Deputy Administrator and Acting Associate Administrator for Center Operations, in organizing and managing the new office since its establishment 15 March. Before March, Kilgore had served as Acting Associate Administrator for Aeronautics and Space Technology. (NASA Ann, 20 May 74)
- NASA announced a 45-day $500 000 continuation of a 13 Feb. contract with Thiokol Chemical Corp. for the space shuttle solid-fueled rocket motor design study. In addition to original contract requirements, the continuation contract required that Thiokol assist NASA in idenifying materials, independent of the specific final motor design, that had long-lead-time supply times. (NASA Release 74–130)

*21 May:* The Senate by voice vote passed H.R. 11864, the Solar Heating and Cooling Demonstration Act of 1974, as amended by the Committee on Banking, Housing, and Urban Affairs and the Committee on Labor and Public Welfare 14 May. The Senate bill would authorize $5 million to NASA and $5 million to the Department of Housing and Urban Development in FY 1975 for development of new or existing solar technology and $40 million to HUD over the next four years to carry out the demonstration. The development and demonstration were to be carried out within five years. NASA functions and funds could be transferred to the Energy Research and Development Administration if ERDA was established. (*CR*, 21 May 74, S8761–77)

*21–23 May:* Some 120 scientists and engineers from industry, Government, and universities attended the Outer Planet Probe Technology Workshop at Ames Research Center to study effective and economical methods of probing the outer planets and their moons. Participants considered spacecraft to fly past a planet while sending a probe vehicle into its atmosphere to report on composition, clouds, pressure variation, and other characteristics; the possibility of visiting such planet-sized moons as Saturn's Titan; and the cost of combined outer planet missions, such as the addition of an atmosphere probe to a Uranus flyby, the use of a similar design for a Saturn atmosphere probe, and the combination of a Jupiter orbiter and probe mission. (ARC Release 72–21; ARC *Astrogram*, 23 May 74; ARC, *Proceedings*, Aug. 74)

*22 May:* ATS–F Applications Technology Satellite, scheduled for a 30 May launch, was "a very unusual NASA satellite," Dr. James C. Fletcher, NASA Administrator, said at an ATS–F press briefing in Washington, D.C. It was an advanced satellite with a huge transmitting antenna and "the first step towards . . . a satellite that can broadcast TV directly to your homes."

Secretary of Health, Education, and Welfare Caspar W. Weinberger said that HEW would be using the satellite for career counseling, teacher training, education of medical students, health education, and transmission of medical records to remote areas of Alaska, Appalachia, and the Rocky Mountains. The practical civilian application "justifies fully a continuance of . . . both the space program in its inner-space aspects as well as . . . bringing it down to earth and using it in our own daily lives." President Henry Loomis of the Corp. for Public Broadcasting said that, for the "first time, we will have current TV programming available in Alaska." TV videotapes now were sent to Alaska by mail and reaching the most remote 10% of the people cost as much as reaching the remaining 90%. ATS-F would greatly reduce that cost. NASA Deputy Associate Administrator for Applications Leonard Jaffe said that the ATS-F was "a major step forward" in the amount of radio frequency power that a satellite could generate to earth and focus very precisely on a small area. Because of this capability, very small, simple receiving stations would be adequate. (Transcript)

*23 May:* The first phase of the Pioneer direct-mode command system, which would permit Ames Research Center to control its spacecraft directly, became operational. The Pioneer command (PCOM) system would permit direct control without depending on Jet Propulsion Laboratory computer systems. Committed operational use of PCOM would begin on *Pioneer 10* (launched 2 March 1972) on 1 Sept., followed by *Pioneer 11* (launched 5 April 1973) on 15 Jan. 1975. (ARC *Astrogram,* 6 June 74, 1)

- The European-built A-300B wide-body transport began commercial service on Air France routes between Paris and London. The airbus was the world's fourth wide-bodied aircraft—the first for use exclusively on the internal European route network and the first powered by only two engines and built outside the U.S. The current configuration of the A-300B carried 251 passengers but an alternate one would accommodate 331. The aircraft was built by Airbus Industries, a Paris-based consortium formed to build the A-300B with participation by West Germany, Great Britain, The Netherlands, Spain, and France. Noteworthy achievements included aircraft delivery within days of the schedule originally set in May 1969 and development costs that had remained within expectations. (*Av Wk,* 20 May 74, 30; *WSJ,* 29 April 74, 6; Air France, NYC PIO, interview, 7 March 75)

- A *New York Times* editorial commented on the Atomic Energy Commission report *The Nation's Energy Future,* submitted December 1973 to President Nixon: The report, which was the basis for Administration policy, would allocate more than half the proposed energy research and development budget to nuclear projects. More than 25% of nuclear funds would be spent on the fast-breeder reactor. Three critical questions were in need of answers: risk of accident, security against theft, and the relative merit of alternative forms of energy. An AEC environmental impact statement issued in response to a court order in a 1971 suit by a group of scientists failed, the *Times* said, to weigh objectively the possible alternatives that might be environmentally less damaging, particularly solar energy. "But in making the choice between nuclear fission and solar energy, Congress and the public have to weigh the alternatives carefully. The A.E.C. can contribute to that decision by offering the nation a more careful and disinterested impact statement." (*NYT,* 23 May 74, 38)

*24 May:* Johnson Space Center had requested proposals from 21 firms to design, develop, fabricate, install, and check out one orbiter aeroflight simulator to support crew training during Orbiter 1 development flights in the space shuttle program, JSC announced. (JSC Release 84[74]-150)

- President Nixon announced his intention to nominate John C. Sawhill to be Administrator of the Federal Energy Administration. He had served as Deputy Administrator of the Federal Energy Office from 4 Dec. 1973 to 8 May 1974, when he became Acting Administrator. The nomination was sent to the Senate 28 May and confirmed 18 June. (PD, 27 May 74, 541–542; 3 June 74, 570; NASA LAR, XIII/82)
- The Navy announced that Texas Instruments Inc. had been awarded a $1 400 000, incrementally funded cost-reimbursement contract for preliminary design of the high-speed antiradiation missile (HARM). The contract contained an option for engineering development including weapon system integration, prototype, and pilot production at a total estimated cost of $40 000 000. (DOD Release 221–74)

*25 May:* Dr. Paul E. Hemke, a pioneer in aeronautical engineering, died at the age of 84. In 1910 he had assisted in designing one of the Nation's first aircraft, which flew to a height of 0.6 m. Dr. Hemke, an analyst at the Langley Memorial Aeronautical Laboratory 1924–1927, was responsible for establishing aeronautical engineering departments at Case Institute of Applied Science in 1931 and at Rensselaer Polytechnic Institute in 1936, where he headed the department until 1949. During World War II Dr. Hemke served on a Government team of scientists sent to Europe to persuade European scientists, especially rocket experts, to emigrate to the U.S. (*W Star-News*, 29 May 74)

*27 May:* The U.S.S.R. launched *Cosmos 656* from Baykonur Cosmodrome into orbit with a 322-km apogee, 185-km perigee, 89.6-min period, and 51.8° inclination. The press reported that *Cosmos 656* was probably a test flight of a Soyuz spacecraft. The spacecraft reentered 29 May. (GSFC SSR, 30 June 74; Tass, FBIS-Sov, 29 May 74, U1; O'Toole, *W Post*, 30 May 74, A18)

*29 May–19 June:* The U.S.S.R. launched *Luna 22* unmanned lunar probe from Baykonur Cosmodrome near Tyuratam at 1:57 pm local time (4:57 am EDT). The automatic station was separated from a platform rocket system in earth orbit and put into a close-to-planned trajectory for the moon. A flight-path correction maneuver 30 May aimed the spacecraft toward a selected area of near-moon space.

*Luna 22* went into circular orbit of the moon 2 June, with a 220-km apogee, 220-km perigee, 130.0-min period, and 19.6° inclination. The Soviet press reported 3 June that all instruments aboard the spacecraft were functioning normally and scientific investigation of the moon had begun. *Luna 22* was transferred into an elliptical orbit with a 244-km apogee and 25-km perigee 9 June to take detailed TV photos of the lunar surface and to study surface relief using an onboard altimeter. Chemical composition of rocks was obtained by determining the gamma radiation. After completion of the photography program the spacecraft was moved to an orbit with a 299-km apogee and 181-km perigee. By 19 June *Luna 22* had completed 187 orbits of the moon; all equipment continued to function normally. (GSFC SSR, 30 June 74; Tass, FBIS-Sov, 30 May 74, U1; 4 June 74, U1; 20 June 74, U1; SF, Nov 74, 434)

*30 May–28 June:* NASA launched *Ats 6* Applications Technology Satellite from Eastern Test Range at 9:00 am EDT on a Titan IIIC launch vehicle. The booster ejected the transtage and spacecraft into an elliptical parking orbit with a 626-km apogee and 152-km perigee. At the second equatorial crossing the first burn of the transtage placed the spacecraft in a transfer orbit. Launch, parking orbit, and transfer orbit were close to the program plan. At 3:29 pm EDT, *Ats 6* was injected into geosynchronous orbit with a 35 845.6-km apogee, 35 774.1-km perigee, 23-hr 57-min period, and 1.8° inclination. The near-perfect orbit eliminated need for corrections, leaving nine kilograms of fuel for operational contingencies. The spacecraft separated from the transtage at 3:33 pm EDT.

An orbital correction 7 June changed the 0.27°-per-day westward drift to a 0.05° eastward drift. Checkout of spacecraft systems and onboard experiments was completed 28 June with no significant anomalies. The mission was adjudged a success 26 Sept.

Primary objectives of the mission were to inject the spacecraft into near-geostationary orbit, erect and evaluate a nine-meter antenna to provide a good-quality signal to small inexpensive ground receivers, stabilize the spacecraft using a three-axis control system, and support user-oriented applications experiments. Secondary objectives were to demonstrate new technology in aircraft and maritime traffic control and infrared earth observations; acquire system data for communications applications in space, including spacecraft-to-spacecraft relay; test spacecraft control; and measure particles and radiation. The 20 experiments included a health-education telecommunications experiment to broadcast programs in several languages simultaneously to citizens in remote areas of the country. The satellite also would aid in remote medical telediagnosis and teleconsultation, transmitting patients' medical records to faraway hospital centers for emergency assistance.

One year after launch *Ats 6* would be moved over Lake Victoria in Central Africa. India would then use the satellite to broadcast programs teaching agricultural techniques, family planning and hygiene, school courses and teacher education, and occupational skills to 5000 villages. The $600 ground receiving terminals each consisted of a receiver, converter, and three-meter chicken-wire antenna.

*Ats 6* was the sixth in a series of ATS satellites developed to advance technology in design, propulsion, and stabilization systems; meteorological concepts; applications; space environmental studies and measurement systems; and communications. *Ats 1* (launched 6 Dec. 1966), *Ats 3* (5 Nov. 1967), and *Ats 5* (12 Aug. 1969) were still transmitting useful data. The ATS program was managed by Goddard Space Flight Center under the direction of the Office of Applications. (NASA MORS, 24 May, 3 June, 8 July, 2 Oct. 74; NASA prog off, interview, 5 Feb. 74; NASA Release 74–111)

*30 May:* The Senate approved the House-Senate Conference report on H.R. 13998 recommending a $3.267-billion FY 1975 NASA authorization. The compromise between the 9 May Senate bill and the 25 April House bill would authorize $805 million for the space shuttle, a $5-million increase over the President's requested funding instead of the $20-million increase voted by the House. An $18.8-million increase in space applications, instead of the Senate's $23-million increase, dropped specific designation of $13 million for the ERTS–C spacecraft, but the Conference report recommended that the project be promptly initiated and it retained the Senate's $3-million addition for a Thor-Delta to launch ERTS–C. The Conference bill would decrease manned space flight operations funding by $10 million, instead of the Senate's $5-million cut and the House bill's $15-million cut. It would retain the Senate's $5.1-million addition for aeronautical research and add 4.9 million for energy research, as well as the House-and-Senate-voted $2 million for research in short-term weather phenomena and the House-added $1 million for ground propulsion research. (*CR*, 30 May 74, S9349–52)

*31 May:* Dr. John F. Clark, Goddard Space Flight Center Director, announced appointment of a committee to review all aspects of the Thor-Delta launch vehicle program, including all anomalies in recent Thor-Delta launches. Pending completion of the review, the 10 June launch of Westar-B—the second of Western Union's domestic communications satellites—had been postponed. (NASA Release 74-147)

- The U.S. Army Electronics Command was awarding a $1 689 116 cost-plus-fixed-fee contract to Cincinnati Electronics, Inc., to develop, fabricate, and test ultrahigh-frequency satellite-line-of-sight manpack transceiver models. (DOD Release 231–74)

*During May:* Preliminary discussions aimed at an eventual global system of second-generation meteorological satellites were held in Geneva, sponsored by the 139-member World Meteorological Organization. Delegates considered establishment of a worldwide forum for presentation of data requirements by users, creation of a mechanism for filing launch plans and distributing advance information on capabilities to potential users of planned satellites, and prospects for training users to interpret and apply data from advanced satellites. Talks were held concurrently on technical standardization of systems, including preliminary standards for ground stations, coordination of transmission times and frequencies, and standardization of automatic-picture-transmission installations. Agreements made at the informal sessions would be taken back to the technical agencies of the various countries and implemented, without the necessity of official review. One result was agreement on L-band frequencies, with U.S. satellites to transmit on 1690 mc.

Soviet sources revealed that the U.S.S.R.'s first synchronous meteorological satellite would be launched in 1977 or 1978 and placed over the Indian Ocean at 70° E. The press reported that it was expected to be technically compatible with U.S., European Space Research Organization, and Japanese satellites in its class. (*Av Wk*, 20 May 74, 21)

## June 1974

*3–20 May:* NASA launched *Hawkeye 1* (*Explorer 52*) scientific satellite from Western Test Range at 4:09 pm PDT on a five-stage Scout E launch vehicle. The 4th and 5th stages separated in parking orbit and the 5th-stage burn placed *Hawkeye 1* in polar orbit with a 124 477-km apogee, 469-km perigee, 49-hr 59-min period, and 89.8° inclination. The launch was the first on the five-stage Scout E configuration, incorporating an Alcyone 1A motor with a 5th-stage transition section.

Primary mission objective was to investigate the interaction between the solar wind and the earth's magnetic field. Secondary objectives included locating magnetic neutral points or lines during both quiet and disturbed solar wind conditions and studying magnetic field topology at large radial distances over the polar caps.

By 20 June all booms and electric antennas had been extended and the spacecraft had despun to the planned five to six revolutions per minute. The transmitters and all experiments were operating satisfactorily.

*Hawkeye 1* was the sixth of the Univ. of Iowa's Injun series, which had begun with the 29 June 1961 launch of *Injun 1* and provided a comprehensive study of charged particles trapped in the earth's magnetosphere. The Hawkeye program was managed by Langley Research Center under the direction of the NASA Office of Space Science. LaRC also had responsibility for the Scout launch vehicle. (MORs, 13 May 74, 20 June 74; NASA Release 74–138K)

*3 June:* Italy would build a ground station to receive data from NASA's Earth Resources Technology Satellites, NASA announced. Under a Memorandum of Understanding between NASA and Italy's communications satellite organization Societa per Azioni per le Comunicazioni Spaziali (Telespazio), the station, at Fucino, would complement existing stations in Fairbanks, Alaska; Goldstone, Calif.; Greenbelt, Md.; Prince Albert, Canada; and Cuiaba, Brazil. NASA's *Erts 1* had been transmitting data since 23 July 1972 launch; ERTS–B was scheduled for spring 1975 launch. (NASA Release 74–142)

- Operation of a demonstration solar heating and cooling unit was begun at Marshall Space Flight Center, using three house trailers to simulate a residence. Continuing studies would combine NASA and industry capabilities to improve efficiency and minimize cost. Efforts included development of a computer-simulation procedure for technical and economic evaluation of the systems, studies to explore mass-production effects on solar-collector costs, definition of a commercially practical absorption-cycle air conditioner that could be operated with water at 353 K (176°F) to 357 K (183°F), verification of the absorptive coating stability and application of coatings to lower-cost materials, and evaluation of other thermal storage techniques, such as melting and freezing waxes, as used in spacecraft. (MSFC Release 74–77)

- Directors of Lockheed Aircraft Corp. and Textron Inc. announced tentative agreement on a management reorganization to aid the financially troubled Lockheed. Textron would invest $85 million in Lockheed shares and stock. Textron's Chairman and Chief Executive Officer G. William Miller would also become Chairman and Chief Executive Officer for Lockheed; Lockheed's Chairman Daniel J. Haughton would become Vice Chairman. Lockheed was to remain a separate corporation. The arrangement provided for the release of the Federal Government from its 1971 guarantee of up to $250 million in bank loans to Lockheed. The agreement was subject to several conditions, including requirements for Lockheed to bring firm production commitments of its L–1011 TriStar aircraft to a total of 180 and to change accounting procedures. (Textron Release, 3 June 74; Witkin, *NYT*, 4 June 74, 1)

*4 June:* Dr. James C. Fletcher, NASA Administrator, presented Dr. Richard T. Whitcomb, Head of the Transonic Aerodynamics Branch at Langley Research Center, a $25 000 cash award for his invention of the supercritical wing. The cash award, made in a NASA Hq. ceremony, was the largest ever made by NASA to an individual. An improved airfoil design, the supercritical wing would permit significant increases in the speed and range of supersonic aircraft without increased fuel consumption. (NASA Release 74–148; NASA photo 74–H–422)

- Boeing Co. and the U.S.S.R. State Committee on Science and Technology signed an agreement on aviation cooperation, including exchange of information and technology and eventual commercial exchange. (FBIS–Sov, 11 June 74, B8)
- Dr. H. Guyford Stever, Director of the National Science Foundation and Chairman of the President's Advisory Committee on Energy Research and Development, predicted that solar heating and cooling systems would be commercially available in five years. Following a speech at the Edison Electric Institute's annual convention in New York, Dr. Stever said the technology would be ready before "society is ready to solve the economic issues surrounding it." A choice would have to be made between "building a lower priced house with added cost for a solar heating-cooling system, or spending more on higher fuel prices in a conventional house." (Smith, *WSJ*, 5 June 74, 65)

*4–18 June:* A team of 15 scientists led by Dr. Thomas A. Mutch of Brown Univ. tested the Viking lander camera in an area of Colorado with geologic conditions similar to those expected on Mars. With photo-diodes in the focal plane instead of film, the camera used a nodding mirror to reflect the image through the lenses onto the diode one line at a time. The entire camera was then rotated to scan the next line. Because the image was sequentially acquired, at five lines per second, several minutes were needed to obtain a complete photograph. The lander's two-camera system would provide color, black-and-white, infrared, and stereoscopic views of the Martian surface and photometric information to help determine sizes and composition of Martian soil samples, monitor the opacity of the Martian atmosphere, and record the position of the sun and brighter planets for precise location of the lander on Mars. (NASA Release 74–257; LaRC proj off, interview, 9 June 75)

*5 June:* The space shuttle would introduce a new mode of space research in the 1980s, Dr. John E. Naugle, NASA Deputy Associate Administrator, said at a space transportation system press briefing in Washington, D.C. Sounding rockets gave the scientist an immediate return of data but only for a very short-duration flight. Satellites provided longer mission times but it might be five to seven years from the initiation of the project until data were received. Experiments flown on the shuttle could remain in orbit from 7 to 30 days and then be returned immediately for evaluation. This capability would be a powerful tool in the development of new technology.

Dr. Myron S. Malkin, Space Shuttle Program Director, said NASA's commitment to Congress called for a total development cost of $5.15 billion in 1971 dollars for two operational shuttle vehicles plus $10.5 million for each of 725 missions projected for 1980–1990. Mission costs would be shared by participating experimenters, and estimates were now running at $9.05 million per flight because of reduced estimates for external tank development. Cost for a shuttle flight, which could take 29 500 kg into orbit, contrasted with $18 million for a 4500-kg payload launched on an expendable Atlas booster and up to $58 million to orbit 14 000–18 000 kg on a Saturn IB.

Shuttle development was on schedule; the first approach and landing tests were to be in the second quarter of 1977 and the first manned orbital flight in 1979. Orbiter manufacture was under way. The main engine oxygen burner had been tested 14 April and the first big engine firing was scheduled for the end of 1975. NASA would need a fleet of seven orbiters at $250 million each if the projected missions were flown. Elwood W. Land, Jr., of the Space Shuttle Systems Office said a 25-astronaut flight corps probably would be maintained to fly missions averaging more than one a week.

LeRoy E. Day, Space Shuttle Program Deputy Director, reported the main engine was being designed to make 55 starts before refurbishment was necessary. Engine lifetime would be much longer. The solid boosters were being designed for 20 flights. Phillip E. Culbertson, Director of Mission and Payload Integration, said that all but the smallest expendable launch vehicles would be phased out about three years after the shuttle became operational. (Transcript)

- The European Space Research Organization decided to award a $226-million, six-year contract to VFW-Fokker/ERNO Raumfahrttechnik GmbH as prime contractor heading a European team to design and develop Spacelab, the reusable, manned laboratory to be carried into orbit on NASA's space shuttle in the 1980s. The unanimous decision, made at a Paris meeting of ESRO's Administrative and Finance Committee, followed six weeks' evaluation of two industrial proposals. The second proposal had been submitted by Messerschmitt-Boelkow-Blohm. The contract provided for delivery of one Spacelab flight unit, fully qualified and ready for installation of experiments, by April 1979. Two engineering models and three sets of ground support equipment and initial spares would also be delivered.

  Spacelab, to be reusable for 50 missions of 7 to 30 days each, would permit scientists and engineers to work in earth orbit in shirtsleeve comfort and without extensive astronaut training. The first launch was scheduled for 1980. (ESRO Release, 5 June 74)

- The Federal Communications Commission authorized Western Union Telegraph Co. to launch its second domestic communications satellite. Western Union had requested that Westar-B be assigned a 119° west longitude position [see 2 May], a position the FCC reserved for systems authorized to serve Alaska and Hawaii. However, the FCC took into account petitions filed by Western Union's competitors, also requesting the location, and assigned the satellite a temporary station at 90° west longitude. The satellite was to be moved to 91° within 30 days after NASA's *Ats 6* was moved from 94°. If Western Union was later authorized to serve Alaska and Hawaii, the FCC could authorize a move to a more appropriate location. (FCC Memorandum Opinion, Order and Authorization, FCC 74-584)

- Satellite observations had been used by National Oceanic and Atmospheric Administration scientists to map the fading—or "recovery phase"—of the mammoth, layered ring of hot charged particles that encircled the earth during magnetic storms and temporarily canceled the earth's magnetic field, NOAA announced. *Explorer 45*, launched 15 Nov. 1971 into an elliptical orbit with a good view of the ring current, had measured the distribution of protons before, during, and after a major magnetic storm 17 Dec. 1971. The NOAA study found that interactions of hot and cold plasmas in the ring changed the pitch angle of the protons in relation to the magnetic lines of force, permitting some to escape. By mapping distributions of proton pitch angles, researchers could predict what would happen during the recovery phase of a magnetic storm. (NOAA Release 74-86)

- Air traffic activity during 1973 had increased 6% over 1972 activity at Federal Aviation Administration airport control towers, air route traffic control centers, and flight service stations, the FAA reported. The increase was the largest recorded by towers since 1968 and followed three years of declining activity. Takeoffs and landings at the 386 FAA-operated control towers during 1973 totaled 56 533 953, up from 53 255 919 in 1972. (FAA Release 74-86)

*5-6 June:* Results of five years of aerodynamic loads research were presented at a YF-12 Flight Loads Symposium at Flight Research Center. Wind-tunnel tests and flight measurements of loads and deflections for the 3200-km-per-hr research aircraft were shown to be in excellent agreement, and deflection due to loads and thermal effects could be predicted by ground calibrations. Recommendations included development of more accurate aerodynamic theories in the transonic range and thermal stress analysis and prediction methods for supersonic cruise aircraft with much larger transient temperature gradients from the skin to the primary structure than those experienced on the YF-12. (NASA Gen Mgt Rev Rpt, 17 June 74, 52)

*6 June:* A solar energy recorder to evaluate the solar energy available in any area of the world with an ambient temperature range of 233 K ($-40°$ F) to 325 K (125° F) was being developed by International Business Machines Corp. Federal Systems Div. under contract to Marshall Space Flight Center, MSFC announced. The recorder would measure and store sunfall data while simultaneously measuring direct and total (direct plus diffuse) solar radiation. MSFC engineers said the measurements were vital to efficient development of solar energy converters before any large-scale construction. (MSFC Release 74-98)

- COMSAT General Corp. announced the award of a contract to Scientific-Atlanta, Inc., for 100 shipboard antennas and associated terminal equipment as part of the maritime communications satellite (Marisat) system to provide improved communications to commercial ships at sea. The contract was expected to exceed $2 million and included options to purchase up to 300 additional terminals. (COMSAT Gen Corp Release 74-31)

*6, 11 June:* Hearings on converting solar energy into electricity by photovoltaic devices were held by the House Committee on Science and Astronautics' Subcommittee on Energy as part of a continuing series of hearings on solar energy. John V. Goldsmith, Jet Propulsion Laboratory scientist, said the total annual U.S. production of photovoltaic solar cells was 50 to 75 kw, mostly for spacecraft systems. In space, solar-cell systems were without parallel but large-scale applications on the earth could not be achieved until systems could be produced in adequate volume at commercially competitive prices. Goldsmith recommended a large-scale photovoltaic technology development program and said that a terrestrial photovoltaic system, with advantages of an inexhaustible energy source and minimal environmental impact, could be reduced in cost from its current $30 per watt to an economically feasible $0.50 per watt by 1985 but would require Federal funding of more than $250 million.

Dr. Harrison H. Schmitt, NASA Assistant Administrator for Energy Programs, said NASA was continuing investigations of orbiting satellite power stations. A microwave energy project was studying problems of processing large amounts of power. However, an orbiting power station would weigh 11 million kg and require a second-generation shuttle in addition to a nuclear-powered shuttle to put it into synchronous orbit. (Transcript)

*7 June:* Images sensed 1 Jan. 1973 by NASA's *Erts 1* Earth Resources Technology Satellite, launched 23 July 1972, had dramatically recorded a Santa Ana wind condition in Southern California, Univ. of California scientists reported in *Science*. The Santa Ana, rarely recorded by satellite, increased the danger of brush fires, crop and structural damage, and wind erosion in the open desert. The *Erts 1* images provided scientists their first view of the entire dynamics in progress at one time. Monitoring such environmental conditions might aid resource management decisions. (Bowder *et al.*, *Science*, 7 June 74, 1077-8)

- Evidence revealed during U.S.–U.S.S.R. planning for the joint 1975 Apollo-Soyuz Test Project had indicated the Soviet Union was never in the race to the moon, the Associated Press reported. U.S. officials had learned the Soyuz capsule was designed for relatively brief earth orbital flights and for ferrying cosmonauts to and from a space station, but it could not make a journey to the moon. AP quoted Astronaut Eugene A. Cernan, a member of the ASTP planning team, as saying, "Our Apollo spacecraft is far, far more sophisticated and has got a greater capability than their Soyuz. That doesn't say the Soyuz is no good. It was just built for a different job." In addition, Soyuz was designed to be controlled from the ground or by preautomated devices. "Their spacecraft . . . is designed around a philosophy that it doesn't need a man to fly." Cernan said that the U.S.S.R. "had no major over-all goal like landing on the moon. Their goals were all to be first. Get the first iron ball in orbit. Get the first man there . . . the first woman . . . the first multiman spacecraft." By proceeding in this way, politically they "gained some significant firsts." (Benedict, *H Chron*, 7 June 74)

- The Nation's scheduled airlines, despite the energy crisis, had set service records in 1973 by carrying more than 200 million passengers and taking in more than $1 billion in freight revenues, the Air Transport Association announced. The airlines, with more than 12 000 daily flights, had accounted for almost 80% of the intercity public passenger-kilometers and 95% of all passenger travel between the U.S. and foreign countries. The industry had earned $223 million in 1973 on revenues of $12.4 billion, up from the $215 million earned in 1972 on revenues of $11.2

billion. Revenue passenger-kilometers were up 6.3% from 1972, while freight-ton-kilometers had risen 12.3%. (ATA Release, 7 June 74)

- U.S. District Court Judge Joseph C. Waddy denied a request by the American Federation of Government Employees for a preliminary injunction that would have blocked NASA plans to lay off or reduce in grade 657 employees at Marshall Space Flight Center. The denial followed a 30 May temporary restraining order issued by Judge Waddy in response to an amendment to a 1967 suit filed by the AFGE against NASA and the Civil Service Commission. The suit challenged NASA's alleged hiring of private contractors for vacated civil service jobs. In a November 1973 decision, the Federal District Court judge had ordered CSC to review contract jobs at MSFC to determine whether they met standards set forth in the Pellerzi rule, which guided CSC in evaluating contracts for personal services. (NASA Off Gen Counsel, interview, 23 Jan 76; Casebolt, *Huntsville Times*, 31 May 74; Cramer, *W Star-News*, 1 June 74)

*8 June:* An *Economist* article, "No Voice in Space," commented on Europe's role in the future of educational satellites: The U.S., with its 30 May launch of the *Ats 6* Applications Technology Satellite, was fast gaining the technology to transmit uncensored TV broadcasts directly into homes in foreign countries, "commanding the eyes of the uncommitted, undeveloped world." Europe could develop comparable satellite equipment but had given up attempts to develop a launch vehicle. "The Americans can hardly be asked to launch Europe's propaganda for it"; perhaps European launch vehicle development should be continued. "The ATS satellite may not look like much of a menace now, but before a decade is out the earth is liable to be girdled with its offspring. . . . To say that Britain should not be up there claiming . . . attention is equivalent to saying . . . Europe has nothing to contribute. It would be a more rewarding venture than some of the follies European countries waste their resources on now." (*Economist*, 8 June 1974)

*9 June:* Northrop Corp.'s YF–17 lightweight tactical fighter prototype aircraft made its first flight, from Edwards Air Force Base. The mach 2 twin-engine jet remained aloft for 1 hr 5 min, reaching an altitude of 5500 m and a maximum speed of 740 km per hr. All flight objectives were met. The YF–17 joined the General Dynamics Corp. YF–16, which had made its first flight 2 Feb., in the lightweight fighter flight-test program. The two designs were being evaluated by the Air Force for possible full-scale development as the new air combat fighter. (AFSC *Newsreview*, July 74, 4)

- Development of a new generation of communications satellites—such as *Westar 1* (launched 13 April), *Ats 6* (Applications Technology Satellite launched 30 May), and three others scheduled for launch in 1974 and 1975—would bring the fifth revolution in human communications, Dr. Robert Jastrow, Director of NASA's Goddard Institute for Space Studies, said in a *New York Times* article. The satellites would provide a "quantum jump in the capacity of Americans to talk to and see one another." The most immediate effect would be to reduce cost of long-distance communications and increase number of words and pictures exchanged. Overseas calls originating in the U.S. had jumped from about 10 million a year in 1965 to 60 million in 1973. Dr. Jastrow predicted that by 1982 the U.S. would originate 1 billion overseas calls and a call halfway around the world would cost the same as a call next door.

Other consequences might be a world language and the ability of people to live where they pleased. With an inexpensive satellite-transmitted dial-up conversation by color TV, people would not have to live in cities; "every house can be transformed into an office, a theater, or a classroom by pressing a few buttons." Long-term effects might knit mankind into a global society, enriched by freer exchange of ideas and reduced fear between nations. A unified global society, however, would be an organism of great collective power in which the individual would be restricted to one role. "The prospect is not pleasant. But the transformation ... will take place slowly ... and most of our descendants will be conditioned to accept their more restricted options." (*NYT*, 9 June 74, 4:6)

*11 June:* A giant nylon parachute was demonstrated at Kennedy Space Center as a possible design for the recovery of the space shuttle solid-fueled rocket booster casings. The parachute—built by Goodyear Aerospace Corp. to provide knowledge of handling heavy parachutes—had a base diameter of 40 m, base canopy circumference of 120 m, and total weight of 940 kg. Plans called for the use of three or four chutes of this kind in addition to a drogue chute to land the shuttle's two 73 000-kg booster casings in the Atlantic, where boosters and parachutes could be retrieved by ships. (*Spaceport News*, 27 June 74, 5)

• Kennedy Space Center awarded two contract extensions. A one-year $13 257 787 cost-plus-award fee extension was awarded to Federal Electric Corp. for support services, including launch instrumentation and communications support, for the Apollo-Soyuz Test Project. The new award brought the total contract amount to $144 962 628. Bendix Corp. Launch Support Div. was awarded a $20 640 874, 16-mo extension to continue launch support services at KSC, including operation and maintenance of Launch Complex 39. The extension brought the total contract amount to $261 498 038. (KSC Releases 93–74; 94–74)

*11, 12, 13, 18 June:* Hearings on H.R. 10392, a bill authorizing NASA to conduct research on ground propulsion systems, were held by the House Committee on Science and Astronautics' Subcommittee on Space Science and Applications. Rep. Charles A. Vanik (D–Ohio) said that he was becoming impatient waiting "for an energy efficient American automobile which can meet our pollution standards." It was "time we truly involved NASA—on a massive scale—in solving our earth-bound problems." Carl E. Nash of the Public Interest Research Group suggested that serious consideration be given the Dept. of Transportation or the Environmental Protection Agency, rather than NASA, when giving authority for ground propulsion research and development, because these agencies had had more experience in the field.

Sydney L. Terry, Vice President for Public Responsibility and Consumer Affairs at Chrysler Corp., said many had questioned whether the Federal Government had a legitimate role in automotive propulsion technology, but Chrysler would "welcome NASA's interest in the automobile with enthusiasm and hope." NASA's skill and experience in long-range basic research and in the fields of thermodynamics, heat transfer, fluid mechanics, and aerodynamics could contribute substantially to the development of efficient transportation. The way to proceed was by a "thoughtful division of research and development work between Government and industry." (Transcript)

*12 June:* Effects of the shapes of trucks on air drag and fuel consumption were being studied by NASA and the Department of Transportation, NASA announced. Trucks traveling 80 km per hr used more than half their engine horsepower to overcome drag. Preliminary tests had shown that a rectangular vehicle with rounded corners had 30% less air drag at 90 to 95 km per hr and used 15% less fuel. Other configurations would be tested at Flight Research Center and DOT's Transportation Systems Center using technology and instrumentation developed in FRC's aeronautics flight-test program. (NASA Release 74-152)

- The House approved by voice vote the House-Senate conference report on H.R. 13998, recommending a $3.267-billion FY 1975 NASA authorization. The action cleared the bill for the President's signature. (CR, 12 June 74, H5057)
- Marshall Space Flight Center announced the award of a $5 986 930 contract modification to Chrysler Corp.'s Space Div. for stage and system engineering for the 1st stage of the Apollo-Soyuz Test Project Saturn IB launch vehicle. The modification, effective 1 May 1974 through 31 July 1975, brought the existing cost-plus-award-fee contract total to $64 455 696. (MSFC Release 74-104)
- No one was arguing against using the sun as an unlimited energy source but few saw it as an energy panacea, a *Christian Science Monitor* article said. With a firm commitment from the Government the sun could provide as much as one fifth of U.S. energy needs before the end of the century. But solar energy faced high capital costs for development; inefficient solar-collecting devices; unsolved corrosion and freezing problems for collectors; interfering building codes; lack of trained maintenance personnel; scarcity of solar technology designers, architects, and engineers; and insufficient mortgage money available. Solar energy also competed for funding with a Government "bias" toward nuclear energy. Government research dollars ran "25 to 1 in favor of nuclear energy." (Hoyt, *CSM*, 12 June 74)

*13 June:* The Anglo-French Concorde supersonic transport set a new speed record for a transatlantic flight of a commercial aircraft, flying from Paris to Boston in 3 hrs 9 min. The press quoted pilot André Turcat as saying the Concorde, which had been invited for the dedication of a new terminal at Boston's Logan International Airport, had cruised from Paris at an altitude between 15 000 and 17 000 m. (AP, B *Sun*, 14 June 74, A11)

- Glynn S. Lunney, Apollo Spacecraft Program Manager, received the Arthur S. Flemming Award—given each year to 10 outstanding young Federal employees by the Washington, D.C., Jaycees—during a ceremony in Washington. Lunney's selection was based on his work as U.S. Technical Director of the Apollo-Soyuz Test Project. The award cited his leadership and technical ability as important elements in the cooperation and progress achieved in the project by the two participating countries. (JSC Release 74-110; JSC *Roundup*, 21 June 74, 1)

*13-14 June:* A complex molecule called methyl cyanide had been discovered in Comet Kohoutek in December 1973 by scientists at the National Radio Astronomical Observatory, supporting the theory that comets and possibly planets were born out in space between the stars, a NASA spokesman said at a Comet Kohoutek Workshop at Marshall Space Flight Center. Twenty-two papers were presented by U.S. and European scientists on early results from observational programs including Skylab, ground observatories, rockets, and aircraft. Dr. Charles R. O'Dell, project scientist for NASA's Large Space Telescope program, said Kohoutek had been a landmark in the study of the solar system. "We found large numbers of heavy molecules that we had never seen before. We always thought they were there but now we have proof of it." (NASA Release 74–109; Michelini, Birmingham *Post-Herald*, 14 June 74)

*14 June:* Dr. James C. Fletcher, NASA Administrator, announced that NASA's Mississippi Test Facility had been renamed the National Space Technology Laboratories (NSTL), effective immediately, and would become a permanent field installation reporting directly to NASA Hq. MTF, whose activities now included developmental testing of the main engine for the space shuttle and an earth resources laboratory, had been established in 1961 as part of Marshall Space Flight Center, to perform static tests of Saturn V launch vehicle stages used in the Apollo program. (NASA Release 74–159)

- Kennedy Space Center's Earth Resources Office was studying the use of radar imagery for land use planning, water resources management, and detection of pools of water under canopies of dense vegetation, KSC announced. Three Brevard County, Fla., sites had been surveyed by a C–46 Curtis Commando aircraft equipped with a two-wavelength, multiplex, synthetic aperture, sidelooking radar. Results indicated that radar was promising for rural and urban land use applications and water resources management, but water detection experiments were less encouraging. The study was being conducted because photographic and TV sensors on spacecraft were limited to the visual or near-infrared wavelengths. They were unable to photograph in darkness or through clouds. (KSC Release 97–74)

- Overall U.S. aerospace exports during the first three months of 1974 had totaled $1713 million, up from $1245 for the same period in 1973, the Aerospace Industries Association of America announced. (AIA Release 74–6)

- President Nixon established the Committee on Energy. The Cabinet-level advisory body would coordinate development of energy policy within the Executive Branch. (*PD*, 17 June 74, 607–8)

*15 June–23 September:* The Global Atmospheric Research Program's Atlantic Tropical Experiment (GATE) was conducted over a 51-million-sq-km area of tropical land and sea to collect data on the behavior of the tropical atmosphere and its ultimate effects on global weather. Sponsored by the United Nations' World Meteorological Organization, more than 4000 scientists, technicians, and support personnel from 72 countries—including the U.S., Brazil, France, East and West Germany, Mexico, The Netherlands, United Kingdom, and U.S.S.R.—used 38 ships, 65 instrumented buoys, 13 aircraft, 6 kinds of satellites, and nearly 1000 land stations to observe and record phenomena from the top of the atmosphere to 1500 m below the sea surface. Objectives of GATE were to extend the time range and scope of weather predictions, develop means to assess pollution, determine feasibility of large-scale weather modification, and establish new bonds of international cooperation.

The National Oceanic and Atmospheric Administration coordinated U.S. participation, which included NASA, Dept. of Defense, Dept. of State, Dept. of Transportation, National Academy of Sciences, National Science Foundation, and National Center for Atmospheric Research, as well as the private sector of the scientific community.

Instrumented aircraft, including NASA's Convair 990, flew at altitudes between 90 and 9150 m to probe the cores of tropical cloud clusters and investigate the intertropical convergence zone, ocean waves and surface temperatures, radiation, tropical disturbances and dust layers moving off the African continent, and day-night weather changes in the coastal areas. Ships from 10 countries, including NASA's *Vanguard* tracking ship and eight other U.S. vessels, investigated tropical cumulus cloud clusters, uppermost layers of ocean affected by local atmospheric conditions, and the broad systems of ocean currents near the equator. Satellites—including *Sms 1, Ats 3, Noaa 2* and *3, Nimbus 5,* and DMSP Defense Meteorological Satellites from the U.S. and polar orbiting Meteor-class satellites from the U.S.S.R.—furnished day and night information on cloudtop heights and temperatures, cloud liquid-water content, wind speed and direction, temperature and moisture in the atmosphere, and sea surface temperatures. Land stations provided surface and upper-air data; instrumented buoys collected data from the sea surface to 10-m altitude.

Each nation contributing vehicles would process and validate the collected data. By March 1976 all data would be sent to two World Data Centers, operated by the U.S. and U.S.S.R., where they would become available to scientists. (NOAA, GATE News Press Kit, 9 May 74; GATE Release NI 74–45; FBIS–Poland, 25 June 74, G6; NOAA Release 74–154)

*16 June:* The Air Force Air Weather Service began operational testing of the Airborne Weather Reconnaissance System (AWRS) aircraft over the Gulf and East Coasts. The modified WC–130 aircraft—using the most advanced meteorological and atmospheric sensors to provide continuous accurate measurements of wind speed and direction, temperature, pressure, and dewpoint-humidity at flight levels—would study the structure and internal forces of hurricanes, storms, and other atmospheric phenomena. The aircraft also carried the first airborne, integrated Omega-inertial navigation system for positional accuracies of about 2 km on a 6000-km overwater mission. (AFSC Release OIP 116.74)

*17 June:* A Boeing 747 aircraft had been chosen to transport the space shuttle orbiter and related hardware from the West Coast to Kennedy Space Center, NASA announced. The 747 would also be used in approach and landing tests of the reusable orbiter. Bought from American Airlines, a used 747 would be modified and equipped with fittings to permit quick installation of the orbiter or hardware on top of the aircraft. Flight profile tests would begin immediately; modifications were to begin after November. Ground and flight tests would begin in late 1976. Takeoff weight with the orbiter was estimated as 35 000 kg. (NASA Release 74–160)

• Dr. William R. Lucas assumed duties as Director of Marshall Space Flight Center succeeding Dr. Rocco A. Petrone, who had been transferred to NASA Hq. as Associate Administrator [see 5 March]. Dr. Lucas, who had been with MSFC and predecessor organizations for 22 yrs in scientific and management positions, had been part of the group transferred from the Army in 1960 to form the nucleus of MSFC. He had served as Deputy Director since 1971. (MSFC Release 74–109; *Marshall Star,* 17 Feb 71, 1)

*17–18 June:* The international Mission Definition Group, sponsored by NASA and the European Space Research Organization at Ames Research Center, concluded that backup hardware from the *Pioneer 10* and *11* missions (launched 2 March 1972 and 5 April 1973) could be used for an orbital mission to Jupiter in the 1980s. The U.S. and European scientists brought together to discuss the possible mission found the Pioneer-class spacecraft best suited for a study of Jupiter's unusual magnetic field. The group recommended that 70% of a 40-kg payload be designed to study energetic particles and magnetic fields, with the remainder for visual, ultraviolet, and infrared images. The spacecraft might also carry a probe to explore Jupiter's atmosphere.

The possible mission to Jupiter was one of eight potential areas of space cooperation identified by NASA and ESRO during a joint review of space science program planning in February. After studying the Group's report, NASA and ESRO would decide whether further consideration should be given a future international mission to Jupiter. (NASA Release 74–168; NASA proj mgr, interview, 5 May 75)

*17–28 June:* A nine-man NASA delegation participated in joint tests with Soviet representatives in Star City, near Moscow, to evaluate lighting and facilities for TV and photography in the Soyuz spacecraft. Tests were made in a mockup of the Soyuz spacecraft. The U.S.–U.S.S.R. Apollo-Soyuz Test Project mission, scheduled for July 1975, would include TV coverage from both spacecraft transmitted to the U.S. and U.S.S.R. simultaneously. (JSC Release 74–114; NASA prog off, interview, 30 April 75)

*18 June:* Marshall Space Flight Center announced the award of a $69 316 six-month contract to International Telephone & Telegraph Corp. ITT Gilfillan Div. to analyze a two-way optical data link, using laser beams and operating between the ground and payloads in low earth orbit. The study, which could lead to a laser communications system for the space shuttle, was expected to provide information on the most effective lasers and frequencies and the effects of orbital dynamics. (MSFC Release 74–105)

*21 June:* The Air Force awarded a $42.8-million fixed-price-incentive firm-award-fee contract to Rockwell International Corp. to design, develop, fabricate, launch, and operate three prototype satellites in the NAVSTAR Global Positioning System. Included in the contract were associated hardware, qualification model, and computer programming and services. In addition, two navigation subsystems would be delivered for use aboard a NAVSTAR satellite to be developed by the Naval Research Laboratory. When completed, the 24 satellites in the NAVSTAR system would provide worldwide precise positioning and navigation, day or night, and under all weather conditions. (DOD Release 268–74)

*22 June:* President Nixon signed NASA's authorization bill, H.R. 13998, into Public Law 93–316. The $3.267-billion act authorized $2.37 billion for research and development, $144 million for construction of facilities, and $750 million for research and program management. (*PD*, 1 July 74, 726)

*23 June:* NASA announced selection of 39 scientists to provide experiments for the 1978 Pioneer twin missions to Venus. The selection, made from 162 scientists who had submitted proposals to NASA, included one scientist each from Germany and France. The scientists would participate in a detailed study of the planet, including composition, structure, and radiation field of its atmosphere; nature of the clouds; and magnetic and radiation environment of the planet. (NASA Release 74–155)

*24 June–12 July:* Eight U.S. astronauts—Thomas P. Stafford, Vance D. Brand, Donald K. Slayton, Alan L. Bean, Ronald E. Evans, Jack R. Lousma, Robert F. Overmyer, and Karol J. Bobko—took part in training in Star City, near Moscow, for the Apollo-Soyuz Test Project. A ninth, Eugene A. Cernan, took part in the first two weeks of training, representing the Johnson Space Center Apollo Spacecraft Program Office. In addition to studying Soyuz spacecraft systems, the astronauts trained with the Soviet flight crews in simulators and mockups. They also continued planning procedures and checklists for the joint mission. (JSC Release 74–121; JSC *Roundup*, 30 Aug 74)

*25 June–19 July:* The *Salyut 3* research station and—nine days later—the *Soyuz 14* spacecraft carrying a two-man crew were launched into near-earth orbit to establish a U.S.S.R. manned orbital workshop.

*Salyut 3* was launched at 4:55 am local time 25 June (7:55 pm EDT 24 June) from Baykonur Cosmodrome near Tyuratam into orbit with 252-km apogee, 213-km perigee, 89.2-min period, and 51.6° inclination. Tass announced the purpose of the mission was to improve the station's design, onboard systems, and equipment and to perform scientific research during space flight. By 30 June *Salyut 3*, which consisted of two cylinders more than 21 m in length and weighing some 18 500 kg, had been raised to a near-circular orbit with a 268-km apogee and 265-km perigee.

*Soyuz 14* was launched 3 July at 11:51 pm local time (2:51 pm EDT) from Baykonur Cosmodrome carrying Cosmonauts Pavel R. Popovich, commander and veteran of the 12 Aug. 1962 *Vostok 4* mission, and Yuri P. Artyukhin, flight engineer, to rendezvous with *Salyut 3*. The launch was seen by Soviet citizens by delayed broadcast on Moscow TV. The spacecraft entered orbit with a 338-km apogee, 201-km perigee, 89.9-min period, and 51.5° inclination. Maj. Gen. Vladimir A. Shatalov, chief of cosmonaut training, said during a 4 July Tass interview that the main objectives of the mission were "testing methods and systems of controlling the craft, improvement in the life-sustaining complex, [and] exploration of new possibilities of using both separate craft and groups of space objects to accomplish scientific and applied tasks in a near-earth orbit." Tass also reported that a 4 July trajectory correction placed the spacecraft in an orbit with a 277-km apogee, 255-km perigee, 89.7-min period, and 51.6° inclination. All spacecraft systems were successfully checked out and the crew was reported to be in good health.

The *Soyuz 14* cosmonauts rendezvoused and docked with *Salyut 3* on 5 July during their 17th orbit, successfully demonstrating techniques the Soviets would use during the July 1975 U.S.–U.S.S.R. Apollo-Soyuz Test Project maneuvers. After docking, Popovich and Artyukhin checked the station's onboard systems and microclimate and entered the 99-cu-m three-room laboratory.

During their 15-day mission the cosmonauts studied geological-morphological objects on the earth's surface, atmospheric formations and phenomena, physical characteristics of outer space, and manufacturing techniques, as well as testing the Salyut space station's improved design and solar-panel energy system. The crew also studied the influences of space on the human organism and exercised to prevent loss of muscle tone, using an onboard comprehensive trainer to simulate running, jumping, and weight lifting.

Tass reported 15 July that some Soviet scientists considered terminating the mission because unexpected, intense solar flares 4–8 July dramatically increased the radiation exposure. However, because the flares subsided and the space station's orbit prevented excessive exposure, the crew's work was not disrupted.

While the cosmonauts worked in space, two Soviet researchers lived in a duplicate station on the ground, simultaneously performing the same work. All commands relayed to the orbiting *Salyut 3* were first checked on the ground and alternate versions of decisions for unforeseen situations were calculated. On 8 July the U.S. Apollo-Soyuz Test Project crew, in training at Star City for the joint U.S.-U.S.S.R. mission in 1975, telegraphed congratulations to the *Salyut 3* cosmonauts on their "great success." They said they looked forward to meeting the two after their "happy return."

By 18 July the crew completed most experiments and began transferring equipment to *Soyuz 14*. After checking out spacecraft systems, the cosmonauts undocked 19 July. Retrorockets were fired, parachutes deployed, and *Soyuz 14* softlanded at 5:21 pm Baykonur time (8:21 am EDT) 140 km southeast of Dzhezkazgan 2000 m from the planned target after 15 days 17 hours 30 min in space. The crew left the spacecraft unaided, and on-the-spot medical checks showed the cosmonauts' health was "good." All mission tests were reported completed. For the first time, a U.S.S.R. orbital mission had been in touch with mission control continuously because of maritime tracking.

*Salyut 3* was the third space station orbited by the U.S.S.R. *Salyut 1* had been launched 19 April 1971. A three-man crew on *Soyuz 10* docked with the station 24 April but returned to earth without entering. A second crew, launched 6 June 1971 aboard *Soyuz 11*, entered the orbiting station and conducted scientific experiments for 24 days. The three died during reentry when an open valve evacuated air from the spacecraft. *Salyut 2* was launched into orbit 3 April 1973 but apparently an explosion or wildly firing thrusters sent the station out of control, tearing off the solar panels and making the station uninhabitable. (GSFC *Wkly SSR*, 20–26 June, 3–10 July 74; GSFC *SSR*, 30 June, 31 Aug 74; Tass, FBIS-Sov, 25 June–23 July 74; UPI, *W Star-News*, 14 July 74; Browne, *NYT*, 9, 20 July 74; *SBD*, 9, 27 July 74; *A&A* 1973)

**25 June:** A new flight research program, the NASA terminal-configured vehicle (TCV) program studying problems in aircraft operations near airports, was under way at Langley Research Center. A modified Boeing 737 flying laboratory with its advanced digital electronics equipment would be used to study present and future terminal-area operating environments, precision four-dimensional [longitude, latitude, altitude, time] flight paths, aircraft performance requirements, automatic systems, displays, and pilot workload. Particular emphasis would be given to the human factors and the interaction of crew members. Research tasks would include close-in, minimum noise, curved approaches to touchdown on a runway with zero visibility. Rollout after touchdown and high-speed turnoffs would also be investigated. (NASA Release 74–163)

**25 June–3 July:** President Nixon traveled to Brussels (25–26 June) to meet with the government heads of most of the North Atlantic Treaty Organization countries and then to Moscow (27 June–3 July) for a summit meeting with General Secretary Leonid L. Brezhnev of the Soviet Communist Party's Central Committee. During the Moscow summit meeting, agreements were signed 28–29 June outlining U.S.-U.S.S.R. cooperation—including exchange of information and scientists—in the fields of energy, housing and other construction, heart research, economics, industry, and technology. In addition, two treaties were signed 3 July limiting antiballistic missile systems and underground nuclear weapon testing. Under the treaties, both countries agreed to one antiballistic missile site each rather than the two originally agreed on in the U.S.-U.S.S.R. treaty signed 26 May 1972, to prohibition of underground nuclear explosions above 150 kt, and to discussions on the dangers of environmental warfare. The treaties were sent to the Senate for ratification 19 Sept.

During a 2 July televised address to the Soviet people President Nixon said the joint Apollo-Soyuz Test Project mission, scheduled for July 1975, was "symbolic of the new relationship we are building between our two nations" for several reasons. First, the rocket technology built for war was being used for peace. Second, during the mission the astronauts and cosmonauts would "join with one another—just as we are doing and must continue to do." Third, the newly standardized docking technology, making international space rescues possible, would make space travel safer "just as our new relationship can make life on earth safer for the people of both our countries." And, finally, because the process of "working and building together . . . is the way that together we can build a . . . peace that will last." (*PD*, 8 July 74, 727–763; 23 Sept 74, 1170)

*26 June:* The House passed, by a roll call vote of 407 to 7, H.R. 15572, the Dept. of Housing and Urban Development-Space-Science-Veterans FY 1975 appropriations bill. The bill—passed as recommended to the House by its Committee on Appropriations 21 June—appropriated $3.203 billion to NASA, $44.1 million below the FY 1975 budget request and $63.9 million below the FY 1975 authorization.

Research and development funds totaled $2.327 billion, construction of facilities $135.67 million, and research and program management $740 million. An additional $4.435 million had been provided in the Special Energy Bill.

The House cut $6.2 million authorized as planning funds for the Large Space Telescope, noting that the LST was not among the top four priority telescope projects selected by the National Academy of Sciences. Also the House deferred the $8.0 million requested to initiate SEASAT and recommended that the satellite be configured to fly on an early mission of the space shuttle. The Committee urged NASA to reprogram funds to begin work on ERTS-C as soon as possible but directed the agency not to reprogram any funds for cost overruns on the Viking project without requesting additional funds from Congress. Other cuts from the authorization included $15.8 million for the construction of facilities and $9.6 million for research and program management. (*CR*, 26 June 74, H5746–73; Com Rpt 93–1139)

- Dr. Noel W. Hinners' appointment as NASA Associate Administrator for Space Science became effective. He succeeded Dr. John E. Naugle, who had become Deputy Associate Administrator. Dr. Hinners, Director of Lunar Programs in the Office of Space Science before his appointment, had joined NASA in 1972 as Deputy Director and Chief Scientist of Apollo Lunar Exploration in the Office of Manned Space Flight. (NASA Ann, 26 June 74)
- A National Academy of Sciences panel report to President Nixon recommended that the White House create a Council for Science and Technology. The Council would recommend how science and technology could

be most effectively incorporated into the policy-making process. Three full-time members—drawn from the sciences, engineering, and related fields—would be appointed by the President with the advice and consent of the Senate. A Chairman, who would also serve as a member of the Domestic Council, would report to the President, and a staff of 25–30 experts would attack special problems. NAS also recommended that the Council actively participate in the work of the National Security Council and play a role in areas of foreign policy affected by science and technology. (Dr. H. Guyford Stever, National Science Foundation Director, had been assigned the role of Science Adviser to the President following the abolishment of the Office of Science and Technology by President Nixon in mid-1973.) (Sullivan, *NYT*, 27 June 74)

*27 June:* Dr. James C. Fletcher, NASA Administrator, announced his decision to award Thiokol Corp. immediately a 5.5-million, 180-day letter contract to develop the space shuttle solid-fueled rocket motors. The definitive contract would be awarded later. Lockheed Propulsion Co. had protested to the General Accounting Office the original, 20 Nov. 1973 contract award to Thiokol. GAO validated NASA's selection procedures 24 June 1974. GAO stated, however, that the NASA cost analysis had contained an overstatement of the cost differential favoring Thiokol's proposal over Lockheed's and recommended the NASA Administrator determine whether the award should be reconsidered. Dr. Fletcher advised the Comptroller General that the decision to proceed with Thiokol was based on the conclusion that the rationale used for the initial selection remained valid. (NASA Release 74–178; 74–180; *A&A 1973*)

- NASA announced plans for a joint mission with Great Britain in June 1975 to launch an x-ray telescope on a British Skylark sounding rocket to study the Puppis A supernova remnant. NASA would design, fabricate, and assemble the flight telescope; the British would integrate the payload into the sounding rocket and provide the high-resolution position-sensitive detector, launch facilities, and operations. The combination of the Wolter Type I glancing incidence telescope and the detector would permit the structural details of the regions responsible for the supernova's soft x-ray emission to be studied with unprecedented resolution. (NASA Release 74–179)

- Dr. Vannevar Bush, who directed the development of the atomic bomb and mobilized U.S. scientific and technical resources during World War II, died in Belmont, Mass., at the age of 84 following a cerebral hemorrhage. Dr. Bush was Dean of Massachusetts Institute of Technology during the 1930s, when he developed the Bush differential analyzer, which led to development of the modern computer. He had been President of Carnegie Institution in Washington 1938–1955 and while there had served as Science Adviser to President Roosevelt. In 1938 he was elected Vice Chairman of the National Advisory Committee for Aeronautics and was NACA Chairman 1939–1941. President Roosevelt appointed him Chairman of the National Defense Research Committee in 1940 and Director of the Office of Scientific Research and Development in 1941. In that capacity, Dr. Bush organized an unprecedented team of engineers, scientists, industrialists, and military men and oversaw the development of the atomic bomb, radar, proximity fuse, rocketry, and problems of technical warfare. (Washington, *W Post*, 30 June 74, B8; NAE Release, 19 April 66; *24th, 25th, 27th Ann Rpt of the NACA*, 1938, 1939, 1941)

*28 June–1 July:* NASA and the Air Force Cambridge Research Laboratory, in participation with more than 100 U.S. and foreign experimenters, launched 54 sounding rockets from Wallops Flight Center as part of the Atmospheric Layering and Density Distribution of Ions and Neutrals (Project ALADDIN, '74) program. In addition to the 54 rockets launched during a 24-hr period beginning at 1:00 pm EDT 29 June, a pre-ALADDIN chemical experiment was launched aboard a Nike-Iroquois 28 June and four post-ALADDIN meteorological rockets were launched 1 July. Chemical releases made yellow to yellowish green fluorescent aurora-like vapor trails visible over the East Coast.

The program was studying dynamic and structural conditions of the atmosphere 48–137 km above the earth. Using small meteorological probes, chemical releases, falling spheres, and more sophisticated spectrometer payloads, the rockets tracked and measured winds, wind shears, temperature and diffusion parameters, upper-atmosphere patterns and dynamics, and structure and composition. Data received from the rockets would be compared with data obtained by *Explorer 51* (launched 15 Dec. 1973). The data would be used for predicting communications problems and satellite orbits as well as for weather forecasting and pollution studies. (WFC Releases 74-4, 74-5; AFSC Release OIP 065.74; AP, *W Post*, 30 June 74, B3)

*30 June:* Permanent NASA employment had decreased 4.2% during FY 1974, dropping from 25 955 to 24 854 employees, NASA's Office of Personnel reported. Despite a numerical decrease of 470, employment in professional occupations had risen to 61.4% of the NASA total. The minority component increased to 6%. The increase in average age slowed to 0.2 year, reaching 42.7 years, while the average GS grade remained the same at 11.0. The 2358 separations were almost twice as many as the 1246 accessions, with most of the reduction-in-force separations occurring in technical support, wage grade, and clerical occupations. (NASA, "The In-house Work Force," FY 1974)

*During June:* NASA began "Outlook for Space," a comprehensive planning study to identify potential roles for space exploration and exploitation 1980–2000. The year-long study, headed by Deputy Director Donald P. Hearth of Goddard Space Flight Center, would seek to relate goals of potential civilian space activities to national interests, to develop a list of desirable and practical U.S. civilian space activities, to group activities with specific sets of goals, to define research and development tasks required for potential commercial and operational uses of space, and to identify social and economic challenges that could benefit from the use of space. NASA would consult representatives from universities, other Government agencies, and industry. The study would supplement "The 1973 NASA Payload Model" planning effort, which had evaluated potential uses of the space shuttle, space tug, and Spacelab and which had been completed in October 1973. The new study would continue through June 1975. (*NASA Activities*, 15 Sept 74, 2–6; GSFC proj off, interview, 17 July 74)

• The National Academy of Sciences issued *Scientific Uses of the Space Shuttle*, report of a July 1973 conference in Woods Hole, Mass., sponsored by the NAS Space Science Board. The conference, attended by 61 U.S. and European scientists from seven scientific disciplines, concluded the shuttle could be an important asset to scientific research. Increased size, weight, and simplicity permitted for payloads would be of chief importance. But many potential advantages depended on innovative management and clear-cut, efficient, and flexible procedures for flying multipurpose missions. Small experiments could be flown cost-effectively on small standardized pallets with integrated experiments, put on standby and flown when space was available. The projected 28-day sortie mission would be valuable for most disciplines but major programs requiring longer observation times should be carried out on free-flying automated spacecraft launched from a pallet in the shuttle's payload bay.

Cost-effectiveness of recovering and refurbishing payloads in orbit required additional study. A sophisticated Large Space Telescope—proposed for launch on the shuttle—might be worth the cost to revisit and service or return to earth. But visits to less expensive spacecraft in inconvenient orbits might prove too costly. Limits on recoverable payload weight might also restrict recovery.

Scientists expressed concern that weight requirements for men in addition to a crew might limit scientific payloads. A manned Spacelab module was considered essential for biomedical research but other disciplines could often use a smaller module housing a payload specialist and his console. Because of weight penalties, real-time control and evaluation might be best supplied by a ground-based control using a shuttle-to-ground communications system, such as the proposed Tracking and Data Relay Satellite System (TDRSS), giving continuous global coverage with a high rate of data interchange. (Text)

## July 1974

*1 July:* The Assembly of Engineering was established as part of the National Research Council, to advance efforts of the National Academy of Sciences and the National Academy of Engineering in applying engineering resources toward solving national problems. Initially Dr. Robert C. Seamans, Jr., NAE President, would serve as Chairman of the Assembly and Dr. J. H. Mulligan, Jr., NAE Secretary and Executive Officer, would be Executive Director. First study areas would include alternative aircraft fuels, uses and limits of technology assessment, operational safety in marine mining, and outer continental shelf oil and gas safety. (NAE, *Bridge*, July 74)

- A *Washington Post* editorial urged the importance of scientific advice to the White House: Although drought, famine, continuing energy crisis, overpopulation, and meteorological change had been byproducts of the sudden advancement of science and technology in the past century, it had become increasingly clear that only more progress could provide the knowledge needed to keep this planet reasonably habitable. "We also need a great deal more political wisdom to apply this knowledge effectively and cooperatively. . . . The President needs to be informed and forewarned to help avert . . . potential disasters by the wise and humane use of technology." To this end, a committee sponsored by the National Academy of Sciences had recommended establishment of a Council for Science and Technology as a staff agency in the White House [see 26 June]. Although President Nixon might remain hostile to the idea and a given President might choose some other method, "some orderly way of assuring science's service to government cannot be long delayed. We cannot arrest rapidly advancing scientific and technological developments. The question is whether these developments carry with them side effects . . . that are destructive or benign. No President can know this without the best continuing analysis and advice, close at hand." (*W Post*, 1 July 74)

*2 July:* *Ats 6* Applications Technology Satellite, launched 30 May, transmitted the first instructional TV course broadcast directly by satellite. The course, on career education for elementary school children, originated at the Univ. of Kentucky and traveled by land wires to NASA's Rosman, N.C., antenna for transmittal to the satellite. The spacecraft rebroadcast the program to more than 300 graduate education students throughout eight Appalachian states. Signals came through strongly in all the receiving areas. Students would receive university credit for successful completion of the course, which included taped lessons, programmed instruction, laboratory sessions, and live seminars. Seminars would use *Ats 3*, launched 5 Nov. 1967. The course was part of the Dept. of Health, Education, and Welfare's Health Education Telecommunications (HET) experiment to determine the effectiveness of satellite-transmitted educational TV. (NASA Release 74-172; McElheny, *NYT*, 3 July 74, 37)

- A *Mariner 10* trajectory correction maneuver was made to ensure that the spacecraft would pass within 48 300 km of Mercury for its second encounter 21 Sept. The spacecraft would pass at a greater distance from the planet than it had during the 29 March encounter, to photo-map the entire sunlit side of the planet as well as the south polar area. (NASA MOR, 3 Oct 74; Pasadena *Star-News*, 3 July 74)
- *Erts 1* Earth Resources Technology Satellite, launched 23 July 1972, stopped transmitting useful stored global data, following continued degradation of its tape recorder. After two years of operation, *Erts 1* was providing only real-time imagery, making world-wide coverage no longer possible. Good continuous U.S. coverage was being maintained by ground stations in Canada, Alaska, California, Maryland, and Brazil. The satellite had been designed for one-year operational lifetime. (GSFC proj off, interview, 30 July 75; *Av Wk*, 19 Aug 74, 17; NASA Release 73–62; NASA, ERTS-A Fact Sheet, 12 July 72)
- During the third quarter of the Federal energy conservation program, energy savings in the nondefense agencies had been 19% of anticipated energy use, and savings by the Dept. of Defense had been 31%, President Nixon said in a statement following receipt of a report from Federal Energy Administrator John C. Sawhill. Total savings for the first nine months of the Federal program had equaled 75 million barrels of oil, or $600 million in reduced costs. Secretary of Commerce Frederick B. Dent had reported that industrial energy consumption had been reduced by 5% per unit of output October 1973–January 1974. This rate, which was expected to be maintained or exceeded in 1974 by all of commerce and industry, would bring annual savings equal to 425 million barrels of oil. (*PD*, 8 July 74, 788)

*3 July:* The U.S.S.R. launched *Soyuz 14* from Baykonur Cosmodrome near Tyuratam, carrying Cosmonauts Pavel R. Popovich and Yuri P. Artyukhin to work in the orbiting *Salyut* 3 research station launched 25 June [see 25 June–19 July]. (GSFC *Wkly SSR*, 3–10 July 74; Tass, FBIS-Sov, 5 July 74, U1)
- International Business Machines Corp. and Communications Satellite Corp. announced plans to enter the domestic satellite communications field. Subject to Federal Communicatitons Commission approval, IBM and ComSatCorp subsidiary COMSAT General Corp. would acquire two-thirds interest in CML Satellite Corp. from Lockheed Aircraft Corp. and MCI Communications Corp. for $5 million. COMSAT General already owned the remaining one third. IBM would own 55% of the joint organization and COMSAT General 45%; shares would be offered to the public. CML—developing plans for a domestic satellite system to relay specialized voice, image, and data communications—was expected to operate its first satellite system in the late 1970s. (IBM & ComSatCorp Release, 3 July 74)
- NASA announced the appointment of J. Lloyd Jones as Deputy Associate Administrator for Aeronautics Technology in the Office of Aeronautics and Space Technology. Jones had been Director of the Aerodynamics and Vehicle Systems Div. in OAST since 1972, when he had come to NASA Hq. from Ames Research Center. He had been at ARC since 1954.

In an OAST reorganization providing for two deputies instead of one, Deputy Associate Administrator for Technology Robert E. Smylie became Deputy Associate Administrator for Space Technology. (NASA Release 74–187)

*5 July:* Photographs from the Skylab missions, *Erts 1* Earth Resources Technology Satellite, and a NASA U-2 aircraft had helped end a legal confrontation between land developers and the State of Florida, Kennedy Space Center reported. Proposed heavy development of a section of the Green Swamp had threatened to endanger a subsurface rock formation, the Florida Acquifer, which provided 90% of the state's water. In conflict were the rights of owners to develop their property and the public's right to protect vital water resources. The NASA photos demonstrated the effects of land development and led to a settlement beneficial to both parties, permitting development in large tracts with carefully planned drainage and without chemical control of aquatic weeds. (KSC Release 90-74)

- Three Johnson Space Center nutrition scientists—Dr. Malcolm C. Smith, Jr., Chief of the Food and Nutrition Branch; Dr. Norman D. Heidelbaugh, Chief of Food Science; and Dr. Paul C. Rambaut, Chief of Nutrition—would receive the 1974 Underwood-Prescott Memorial Award for "their contributions to the field of food science through their research in space feeding and nutrition . . . for space missions to the Moon and in Skylab." The scientists were also responsible for food systems on the space shuttle and the Apollo-Soyuz Test Project. The award, presented annually by the Massachusetts Institute of Technology, would be made at an MIT luncheon 1 Oct. (JSC *Roundup*, 5 July 74, 1)

*7 July:* Aleksander Y. Bereznyak, Soviet pioneer in jet aircraft development, died in Moscow at the age of 61. Bereznyak, one of a group of design engineers who worked on rocket and jet engines during World War II, had helped design a rocket-powered aircraft flight-tested in May 1942. (*NYT*, 14 Aug 74, 26)

*8 July:* An optical memory system that would increase computer memory storage capability by 30% had been developed by scientists under NASA contract, NASA announced. The new technique used organic compounds whose chemical reactions could be manipulated by low-power lasers to produce phase holograms (optical image-storing units) that could hold 100 million pieces of information per square centimeter. The new system also would be more economical, more compact, and more reliable and would require less energy and fewer moving parts. Advanced computer memory capability, required for future NASA missions, also could be useful for applications in business, industry, government, and education. (NASA Release 74-185)

- Dr. Kurt H. Debus, Kennedy Space Center Director, and Univ. of Florida officials signed an agreement to permit the University to operate a coastal and oceanographic engineering laboratory at KSC. Telemetric monitoring of the Atlantic Ocean environment would be augmented by visual observation of ocean climate and beach erosion. (KSC Release 113-74)

- The space shuttle had no imaginable use which could not be met more cheaply and expeditiously by rockets already in existence, columnist Nicholas von Hoffman charged in the *Washington Post*. Of the justifications for the space shuttle program, one of "the most persuasive is the large amounts of money that will be pumped" into an ailing economy. However, "it would be difficult to think up a more inflationary way to provide jobs for people." (*W Post*, 8 July 74)

*9 July:* The U.S.S.R. launched *Meteor 18* meteorological satellite from Plesetsk into orbit with an 891-km apogee, 863-km perigee, 102.5-min period, and 81.2° inclination. The satellite would photograph clouds and their underlying surface and study the distribution of precipitation and ice zones and the influence of corpuscular flows on upper layers of the atmosphere for rapid weather data supply. (GSFC *Wkly SSR*, 3–10 July 74; Tass, FBIS-Sov, 11 July 74, U1; *SF*, 1 Jan 75, 35)

*10 July:* Australian and Japanese astronomers had identified the chemical signature of methylamine in the dense interstellar cloud Sagittarius B, which was 30 000 light years from the earth near the center of the Milky Way, the *Washington Post* reported. Methylamine was an organic chemical that mixed freely with formic acid to create glycine, the simplest of the amino acid building blocks of life. Identification of methylamine brought to 29 the number of chemical footprints identified in interstellar space and brought scientists one step closer to understanding the processes creating life. (O'Toole, *W Post*, 10 July 74, A3)

*11 July:* NASA announced the award of a $483 000 000, cost-plus-fixed-and-award-fee supplemental contract and 11-mo contract extension to Rockwell International Corp. Space Div. for development of the space shuttle orbiter and integration of all elements of the shuttle system. The supplement brought the total value of the contract, which would continue through 30 June 1975, to $943 248 000. Rockwell was to subcontract more than 50% of the dollar value of the contract. (NASA Release 74–195)

*12 July:* Ninety-three research teams in the U.S. and foreign countries had been selected for Earth Resources Technology Satellite follow-on investigations, NASA announced. The studies, selected from 669 proposals, included work in agriculture, forestry, environment, interpretive techniques, land use, marine resources, meteorology, mineral resources, oil, civil works, hazards, and water resources. The investigators would use data from *Erts 1*, launched 23 July 1972, and *Erts–B*, scheduled for early 1975 launch. NASA would fund the domestic studies at a cost of nearly $5.9 million. (NASA Release 74–193)

- President Nixon signed into Public Law 93–344 the Congressional Budget and Impoundment Control Act of 1974, H.R. 7130, which strengthened congressional control over the budgetary process. Major provisions of the legislation would establish House and Senate Budget Committees and a Congressional Budget Office, establish a tight timetable for congressional action on authorization and appropriations bills, change the fiscal year from 1 July to 1 Oct. beginning in FY 1976 (with a transitional three-month period from 1 July 1976 through 30 Sept. 1976), provide for a congressional process to bring appropriation bills into conformance with overall budget totals, increase the flow of budgetary information from the Executive Branch to the Congress, and require congressional agreement on deferred use of funds and reductions in funding levels. (*CR*, 16 July, S12585; *PD*, 15 July 74, 799–801)

*13 July:* President Nixon signed Proclamation 4303 declaring the week of 16–24 July United States Space Week. In signing, President Nixon said: "Just five years ago, two Americans thrilled the world when they took man's first steps on the moon. . . . In that single journey five years ago, man's knowledge of our universe, diligently gathered over the centuries, resulted in a spectacular leap from the earth. . . . In time man may take for granted such wonders as we cannot imagine. . . . But we know that a beginning has been made that will affect the course of human life forever." (*PD*, 22 July 74, 805)

*14 July:* The Air Force launched *Nts 1* Navigation Technology Satellite on an Atlas F booster from Vandenberg Air Force Base into orbit with a

13 762-km apogee, 13 440-km perigee, 7-hr 48-min period, and 125.1° inclination. The spacecraft, designed and built by the Naval Research Laboratory and originally called Timation 3A, would test techniques under consideration for the Dept. of Defense's NAVSTAR Global Positioning System (GPS). *Nts 1* carried receivers and transmitters to broadcast its exact position continuously and an atomic clock to transmit ultraprecise time signals. NAVSTAR, when completed in the mid-1980s, would be a system of 24 satellites providing worldwide, accurate, and instantaneous three-dimensional positions of air, sea, and surface vehicles equipped with a GPS receiver. (Pres Rpt 74; *Av Wk,* 22 July 74, 14; AFSC Release OIP 274.74)

- Gen. Carl A. Spaatz, first Air Force Chief of Staff, died in Washington, D.C., of congestive heart failure at the age of 83. Gen. Spaatz had been one of the first 25 Americans to earn "wings" as the Army's First Aero Squadron in 1916. In 1929 Gen. Spaatz, with Gen. Ira S. Eaker, set a refueling endurance flight record by staying aloft 151 hrs over Los Angeles. For this feat he received the Distinguished Flying Cross. During World War II Gen. Spaatz commanded the largest armada of aircraft and airmen ever assembled under the control of a single commander. President Truman appointed him Chief of Staff when the Air Force became a separate branch of the military service in 1947. After his retirement in 1948 he served as chairman of the Civil Air Patrol. (Johnston, *NYT,* 15 July 74, 1)

*15 July:* Evidence of a possible second black hole, from observations by x-ray astronomy satellite *Uhuru* (*Explorer 42,* launched 12 Dec. 1970) between January 1971 and February 1973, was reported by Center for Astrophysics astronomers in the *Astrophysical Journal.* Black holes were believed to be superdense collapsed stars with gravity so great that even light could not escape. *Uhuru* had detected radiation from the x-ray source Circinus X–1 with characteristics similar to those detected in 1972 by *Uhuru* and in 1973 by *Oao 3* for the suspected black hole Cygnus X–1. Varying periods of x-ray intensity had indicated that Circinus X–1 was possibly binary; the x-ray energy spectrum had showed changes from 2 to 20 kev. No optical counterpart for the x-ray source had been found.

Black holes might be an answer to two basic problems in cosmology, Walter S. Sullivan had written in the *New York Times Magazine* 14 July. All visible material accounted for only 2% of the matter necessary to prevent the universe from flying apart in a continuing expansion; some scientists thought black holes might well make up the deficit. Also, black holes might provide the "glue," the necessary gravity, to keep clusters of galaxies from flying apart. Not all scientists were convinced of the existence of black holes, but the holes' strange effects—such as the slowing of time, the tight curvature of space, and the influence of gravity on light—had been anticipated by Einstein's theory of relativity.

One major dilemma was what happened within a black hole. Some scientists had proposed that a star drawn into a black hole by its gravity might emerge in some other time and place as a quasar and that quasars—exceptionally bright objects of relatively small volume and high density—might be "white holes" into which energy was pouring from somewhere else, perhaps even another universe. (Jones *et al., Astro Journ,* 15 July 74, L71–4; *NYT Magazine,* 14 July 74, 11–35; *A&A 1971, 1972, 1973*)

- The Air Force was developing a new series of global meteorological satellites for the Defense Meteorological Satellite Program (DMSP), *Aviation Week & Space Technology* reported. The new Block 5D spacecraft would be twice the weight of its predecessors, have a payload capacity about four times greater, and carry a broader mix of meteorological sensors to provide uniformly high-resolution photos. Block 5D satellites would employ "selective redundancy"—redundancy in certain critical, rather than all, spacecraft systems. This redundancy would increase probable mission lifetime to two years, more than twice the 9- or 10-mo lifetime of most earlier USAF weather satellites. Rather than choosing a single contractor, the USAF had awarded a series of fixed-price contracts: to RCA Corp. for the spacecraft, Westinghouse Electric Corp. for the primary sensor, Barnes Engineering Co. for the supplementary sensors, and Radiation, Inc., for ground terminals. (*Av Wk*, 15 July 74, 41–47)
- McDonnell Douglas Corp. had been awarded a $13-million contract by Alyeska Pipeline Service Co. to produce 100 000 heat pipes for the trans-Alaska pipeline, the *Seattle Journal of Commerce and Northwest Construction Record* reported. The heat pipes, called cryo-anchor soil stabilizers, were developed by McDonnell Douglas for thermal control of Skylab Orbital Workshop (launched 15 May 1973) and unmanned satellites. They would prevent thawing of permafrost around the pipe supports of elevated portions of the 1285-km pipeline by removing heat from the ground and transmitting it to an above-ground radiator, where it would be dispersed into the atmosphere. (Seattle *Journ Commerce & NW Constr Record*, 15 July 74)
- Federal support to universities and colleges had declined 7%, to $3823 million, during FY 1973 in the first decline since 1970, the National Science Foundation reported. The drop had been largely the result of a $267-million decrease in funds from the Dept. of Health, Education, and Welfare but other significant decreases were posted by NSF, down $40 million; Dept. of Defense, down $11 million; and NASA, down $8 million. Federal funds for academic science activities totaled $2463 million, 5% below 1972. Two science fields had shown an increase in Federal R&D support between 1972 and 1973. Life sciences' $934 million was a 4% increase; engineering's $204 million was up 6%. (NSF *Highlights*, 15 July 74)

*15–16 July:* Western Union Telegraph Co.'s *Westar 1*—first U.S. domestic communications satellite, launched by NASA 13 April—began operations during inaugural ceremonies in New York and Los Angeles. Russell W. McFall, Chairman of the Board and President of Western Union Corp. transmitted the first message, from New York to Los Angeles, using the original Morse key to tap "What hath God wrought!" It was the same message sent by inventor Samuel F. B. Morse 24 May 1844 over the first experimental telegraph, a 65-km line between the Capitol in Washington, D.C., and Baltimore, Md.

Regular commercial operations began 16 July, providing services for *Westar 1*'s first two commercial customers, American Can Co. and Bank of America. (Westar Inaugural Transmission press kit, 15 July 74)

*16–29 July:* NASA launched *Aeros 2* (Aeros-B) aeronomy satellite for West Germany from Western Test Range at 4:51 am PDT on a four-stage Scout launch vehicle into polar orbit with an 879.0-km apogee, 221.3-km perigee, 95.65-min period, and 97.44° inclination. The NASA mission objective—to place the *Aeros 2* satellite in an earth orbit permitting successful achievement of the scientific objectives—was met and the mission was adjudged successful 26 July. The scientific objective was to measure the main aeronomic parameters of the upper atmosphere and the solar ultraviolet radiation in the wavelength band of the main absorption. The long-range objective of the satellite, which carried one Goddard Space Flight Center and four West German experiments, was to provide a model of the ionosphere from which upper atmospheric conditions could be predicted. By 29 July the vehicle and experiments were performing satisfactorily and the West Germans were initiating spacecraft operations.

A July 1965 agreement between the U.S. and West Germany had provided for a series of cooperative satellite projects. The Aeros program, the second in the series, was authorized by a 10 June 1969 agreement and *Aeros 1* was launched 16 Dec. 1972. The 1974 contract between NASA and the West Germany Federal Ministry for Research and Technology (BMFT) for the *Aeros 2* launch provided for reimbursement of NASA for launch vehicle and services. NASA would also provide one experiment, tracking, and backup for data acquisition. BMFT was responsible for development of spacecraft and instrumentation operation and control of the satellite after launch, and data reduction and analysis except for the GSFC experiment. The spacecraft reentered 25 Sept. 1975. (NASA MORs, 30 Nov 72, 8 & 29 July 74; NASA Releases 69-91, 74-183; GSFC *SSR*, 31 Oct 75)

*16–24 July:* U.S. Space Week marked the fifth anniversary of the first manned landing on the moon.

The 1969 launch of *Apollo 11* was commemorated 16 July by ceremonies at Kennedy Space Center Launch Complex 39, Pad A. At 9:32 am EDT, the moment of launch, *Apollo 11* Astronauts Neil A. Armstrong, Michael Collins, and Edwin E. Aldrin, Jr., unveiled a plaque designating the site a National Historical Landmark. Dr. George M. Low, NASA Deputy Administrator, told the thousands of guests: "No matter our achievements, . . . we cannot relax our pursuit of knowledge, for we must face the coming crises of our time. Every effort must be made to see that the opportunity which our technology presents is not lost. We may be sure that the opportunity will not occur soon again, if ever. To fail now would mean a ravaged environment and depleted resources—a plunge into a new and terrible Dark Age from which there would be no return." Unless "we continuously restore the fund of basic knowledge, we shall ultimately exhaust it. . . .

"Some day we will establish scientific outposts on the Moon, and even tap its mineral resources. And in the not-too-distant future astronauts may tread the surface of Mars and the glaciers of the moons of Jupiter in search of extraterrestrial life. . . . The future began here at 9:32 a.m. on July 16, 1969. At that moment on July 16 in the year 2069, launch windows to Mars, Neptune, and Pluto will be open. Will we fly through those windows? I think we will."

In 19 July ceremonies at the San Clemente, Calif., Western White House, Armstrong presented President Nixon a plaque bearing the names of each astronaut who had represented "this country in a flight above the surface of the Earth." President Nixon told Armstrong the greatest contribution to the space program was not the exploration or technology but the "spirit you and your colleagues . . . have had the opportunity to demonstrate."

The Smithsonian Institution hosted 20 July activities marking the day of the landing. Speaking on the Mall in Washington, D.C., Smithsonian Under Secretary Robert A. Brooks said *Apollo 11* had been a "dramatic demonstration . . . that progress in human affairs does not spring full blown from the efforts of one man or group of men. What we saw was the culmination of the efforts of hundreds of scientists and engineers from Leonardo da Vinci . . . to the Apollo 11 astronauts. Each built on what his predecessors had accomplished until the accumulation of knowledge made possible the kind of quantum leap forward represented by the Apollo Program."

Dr. James C. Fletcher, NASA Administrator, called the day of the landing "the day man ended an era of one-planet civilization and began a new age as traveler and colonizer of the universe." Despite difficulties, men would "make more of these steps into the vast but exciting reaches of the solar system."

A 21 July service at the Washington Cathedral in Washington, D.C., dedicated the Space Window donated by Dr. Thomas O. Paine, NASA Administrator at the time of the landing. The window, designed by Rodney Winfield, depicted whirling orange, red, and white stars and orbiting planets on a deep blue and green field. A thin white trajectory, representing a manned spacecraft, emphasized man's minuteness in God's universe. During the ceremony, Armstrong presented a 7.18-g, 3.6-billion-yr-old lunar rock, brought from the moon by *Apollo 11*, to the Cathedral. The rock would be inserted into a hydrogen-filled plexiglass container embedded as the centerpiece of the window. (*PD*, 22 July 74, 815; *NASA Activities*, 15 Aug 74, 5–14; transcript; program)

*17 July:* Dr. Raymond L. Bisplinghoff, Deputy Director of the National Science Foundation since 1970, announced his resignation effective 30 Sept. Dr. Bisplinghoff had served as Director of NASA's Office of Advanced Research and Technology 1962–1963 and as NASA Associate Administrator for Advanced Research and Technology 1963–1965 and held NASA's Distinguished Service Medal. He would become Chancellor of the Univ. of Missouri at Rolla 1 Oct. (NSF Release 74–188)

*19 July:* Astronaut Alan B. Shepard, Jr., the first U.S. man in space and the fifth man to walk on the moon, announced his retirement from NASA and the Navy effective 1 Aug. During his 5 May 1961, 14.8-min Mercury-Redstone 2 mission, Shepard flew his *Freedom 7* spacecraft to a 186.2-km altitude and initiated the U.S. manned space program. He commanded the *Apollo 14* lunar landing mission 31 Jan.–9 Feb. 1971. Shepard would join the Marathon Construction Co. of Houston, Tex., as a partner and Chairman. (NASA Release 74–203; *A&A 1961*)

- NASA announced the appointment of William C. Schneider as NASA Deputy Associate Administrator for Manned Space Flight, effective 21 July. Schneider had served as both Acting Associate and Acting Deputy Associate Administrator for Manned Space Flight from 18 March to 17 May and had been Skylab Program Director from December 1968 until the end of the Skylab program. (NASA Ann, 19 July 74; NASA Release 74–207)

- British Prime Minister Harold S. Wilson and French President Valery Giscard d'Estaing decided at a Paris summit meeting to limit construction of the Concorde supersonic jet airliner to 16 aircraft. The decision, which would end the joint project by 1977, was a compromise between the British Labor Party's desire to cut off state funds entirely and the

French demand that the program be expanded to 19 aircraft. Since many parts had already been manufactured, the decision meant that many of the 30 000 employees of manufacturers Aérospatiale France and British Aircraft Corp. Ltd. might lose their jobs. There were firm orders for only nine of the $40-million aircraft, with options on several others. U.S. airlines had refrained from ordering any of the aircraft because of doubts about its operating economy. (*WSJ*, 22 July 74, 6; UK Embassy PIO, Wash, D.C.)

- Dr. Floyd L. Thompson, former Langley Research Center Director, had been selected to receive the Society of Automotive Engineers' 1974 Daniel Guggenheim Medal, the *Langley Researcher* reported. He would receive the award for his "farsighted development of men and facilities, and for decisive leadership of research that provided technological foundations for manned flight beyond the speed of sound, safe reentry of spacecraft, and successful exploration of space." (*Langley Researcher*, 19 July 74, 1; *A&A*, Dec 74, 72)

- President Nixon announced his intention to nominate James E. Dow to be Deputy Administrator of the Federal Aviation Administration, succeeding Kenneth M. Smith, who had resigned. Dow had been Associate Administrator of FAA since August 1972. The nomination was submitted to the Senate 22 July and confirmed 8 Aug. (*PD*, 22 July 74, 817; 29 July 74, 995; *FonF*, 7 Sept 74)

- Communications Satellite Corp. reported a net income of $21 013 000, or $2.10 a share, for the first six months of 1974—up from $15 223 000, or $1.52 a share, for the same period in 1973. Operating revenues for the period were $63 322 000, up from $55 902 000 in 1973. The increased revenues had resulted from a gain in the number of half-circuits leased by ComSatCorp to its carrier customers for overseas satellite communications. (ComSatCorp Release 74-41)

- Columbia Broadcasting System newsman Walter Cronkite recalled the two most exalting moments in his years of TV space coverage, during an interview with the *Christian Science Monitor*. The first was "Alan Shepard's little 15-minute plop-down into the Atlantic. Rocketry itself was so uncertain . . . I just didn't know whether or not I was going to be able to watch." The second was the first lunar landing. "That was a highly emotional moment for me. The actual touchdown, not so much Neil Armstrong's walk." Cronkite, preparing for the TV documentary "Space: A Report to the Stockholders," said he still believed the major achievement of the moon landing was that "it proves that we truly have the capability at this stage of man's civilization to do any darn thing we want—the most outlandish or the most practical things."

  Of course there were important spinoff values. "Probably one of the most important industrial spinoffs . . . is the command that U.S. industry had gotten in the computer business. . . . I attribute that almost solely to the space program." Cronkite said that recent cutbacks in the manned program were proper. "It is just about where it ought to be—most of the concentration now is in unmanned satellites" and a deliberate, but not crash, program toward the reusable space shuttle. If newsmen were allowed to travel on the space shuttle, he'd go "on any flight they offered me." When asked if he wouldn't be missed by American TV viewers at blastoff, he said, "What do you mean. . . ? I hope to be broadcasting all the time from inside!" (*CSM*, 19 July 74)

**20 July:** A *New York Times* editorial commented on the fifth anniversary of the *Apollo 11* lunar landing: "On 20 July 1969, millions of people all over the globe watched . . . as [*Apollo 11* Astronaut] Neil Armstrong's great left boot groped for the lunar surface and he made his 'one small step for a man, one giant leap for mankind.' " Today, just five years later, many Americans would be hard put to name all three participants of the original mission. "The attention of the world . . . has turned to problems closer to home. Yet when those future centuries look back on this one, the first time that men landed on another celestial body will loom large and they will be remembered long after those in today's headlines have passed beyond recall." (*NYT*, 20 July 74, 30)

**21 July:** The leadership of "NASA's other hero," former Administrator James E. Webb, was recalled by Julian Scheer, former NASA Assistant Administrator for Public Affairs, in the *Washington Post*: The fifth anniversary of the 20 July 1969 *Apollo 11* lunar landing would bring toasts to former astronauts Neil A. Armstrong, Edwin E. Aldrin, Jr., and Michael Collins and others. But it had been Webb's "incredible leadership" which had made the event possible. Webb had "fought the Bureau of the Budget for every nickel" and usually got his way. In Congress, few believed he was not a scientist or engineer. He dazzled them with his knowledge of the technical as well as his management of the vast NASA program structure. "But it was in the internal workings of NASA where his strength lay." He had an uncanny knack for "standing back and looking at the big picture." He never took on a problem without coming up with as many as four options and, for Webb, every option was an opportunity. Abruptly in 1968, Webb resigned. Most of the major Apollo decisions had been made. "Almost shockingly, it was less than a year away from the first manned lunar landing. Had he . . . waited just a while longer, he could have basked in the moment of tremendous glory." (*W Post*, 21 July 74, C6)

- A multibillion dollar competition between the U.S. and France was being waged over which country would produce the new generation of fighter aircraft for the Western European allies, the *New York Times* reported. To the victor would go one of the largest aircraft export sales in history, the possibility of producing 2000 or more aircraft worth between $10 billion and $15 billion. To replace their 10-yr-old, U.S.-built F-104 Starfighter, the Europeans would like a low-cost, high-performance fighter that could meet military requirements through the 1990s. The U.S. was offering versions of the lightweight fighter—the General Dynamics Corp. YF-16 and the Northrop Corp. YF-17, now in competition for an Air Force production contract [see 9 June]—while the French were offering an improved version of the F-1 Mirage. Pentagon officials claimed both the YF-16 and YF-17 were superior to the Mirage in acceleration and maneuverability despite the more powerful engine in the French aircraft.

    The competition would probably be decided as much on economic and political grounds as by military considerations. The French argued that the North Atlantic Treaty Organization allies should "buy European" and maintain European solidarity. The U.S. emphasized standardization of weapons within NATO and the future willingness of the U.S. to maintain forces in Western Europe. Although the immediate potential customers—Belgium, The Netherlands, Norway, and Denmark—would need about 300 aircraft, American officials were envisaging even larger markets in other West European countries and the world market. (Finney, *NYT*, 21 July 74, 3)

**23 July:** The U.S.S.R. launched *Molniya II–10* communications satellite from Plesetsk into orbit with a 40 724-km apogee, 603-km perigee, 12-hr 18-min period, and 62.9° inclination. The comsat would help provide long-distance telephone and telegraph radio communications and transmit U.S.S.R. TV programs to stations of the Orbita network. (GSFC *Wkly SSR*, 18–24 July 74; Tass, FBIS–Sov, 23 July 74, U1; *SF*, 1 Jan 75, 35)

- The Senate Committee on Aeronautical and Space Sciences held hearings on S. 3542, a bill introduced 29 May to authorize $41 700 000 for NASA to launch a seventh Applications Technology Satellite. Sen. Frank E. Moss (D–Utah) said experimenters using *Ats 6,* launched 30 May, had urged another launch because more than the planned one year of operation over the U.S. was needed to complete experiments. Two satellites in orbit at the same time would "greatly improve the experimental information necessary for further decisions on the use of advanced communications satellites." Sen. Barry Goldwater (R–Ariz.) said that ATS satellites held "the promise of improved health care at reduced costs" for people in remote regions of the U.S.

    Dr. James C. Fletcher, NASA Administrator, testified that *Ats 6* was in excellent condition and had a life expectancy of three additional years over the U.S. following its return from the planned year of operation over India. In addition, use of the U.S.–Canada Communications Technology Satellite scheduled for a mid-December launch could provide experimenters additional experiment time and continuity. Dr. Fletcher expressed NASA opposition to the bill, saying it would cost $4 million more than the bill provided, to cover the cost of the Titan IIIC launch vehicle. To avoid further costs, a decision to launch another spacecraft would have to be made by September, when the NASA and industry ATS team were to be disbanded. A seventh ATS satellite could be considered a potential operational vehicle and the user community might consider bearing the costs. (Transcript)

- Marshall Space Flight Center announced the award of two contracts for studies of Skylab results. American Science and Engineering, Inc., received $2 044 860 and the Smithsonian Institution's Astrophysical Observatory received $264 684 to study jointly results from the Apollo Telescope Mount x-ray spectrographic telescope experiment. The telescope had been used during the three Skylab missions (25 May–22 June 1973, 28 July–25 Sept. 1973, and 16 Nov. 1973–8 Feb. 1974) to photograph x-ray producing solar events. (MSFC 74–132)

**25 July:** The 3–19 July *Soyuz 14* mission had prompted concern among U.S. specialists that the U.S.S.R. might be developing a manned spacecraft for military reconnaissance, the *Washington Post* reported. *Soyuz 14* cosmonauts, who had linked up with the *Salyut 3* orbiting space laboratory, had used a special coded channel of communications with Soviet ground stations, surprising U.S. stations that usually monitored Soviet flights. In addition, the *Soyuz 14* flight carried the first all-military crew since the U.S.S.R. began flying multiple-man crews in 1969. All previous flights had at least one civilian aboard. Also, sources reported that at the time of the flight a series of optical sighting targets was laid out on the ground not far from the launch site. The targets were believed to be much larger than the targets usually laid out to check spaceborne optics, prompting speculation that the U.S.S.R. might by trying to learn what man, rather than cameras, could see from space. (Getler, *W Post*, 25 July 74, A1)

- The Navy Sea Systems Div. awarded a contract with a target value of $285 400 000 to General Dynamics Corp. Electric Boat Div. for construction of the lead submarine in the planned 10-submarine Trident (undersea long-range missile system) program. The submarine, which would carry 16 Trident I missiles with multiple independently targeted vehicle (MIRV) warheads, would be faster, quieter, and better equipped than the Poseidon-class submarine. And the Trident I would have far greater range than the Poseidon missile. (DOD Releases 335–74, 336–74; AP, *W Post*, A4)
- Preliminary studies of the use of laser beams to monitor water temperature, turbidity, salinity, and other factors affecting quality were under way at the University of Miami under Kennedy Space Center contract. No method had been known for remote sensing of vertical temperature profiles in bodies of water, but now high-energy, coherent laser beams could penetrate deep into the water. Measuring the return radiation was expected to yield data on a number of factors, offering a potential ecology tool. (KSC Release 121–74)
- John W. Crowley, Jr., NASA Director of Aeronautical and Space Research from the agency's establishment in 1958 until his retirement in 1959, died in Denver, Colo., at the age of 75. Crowley had joined NASA's predecessor, the National Advisory Committee for Aeronautics, at Langley Aeronautical Laboratory in 1921, becoming Chief of the Aeronautics Div. in 1940 and Head of the Research Dept. in 1943. He came to NACA Hq. in 1945 as Acting Director of Aeronautical Research and was appointed Associate Director for Research in 1947. Crowley received NASA's first Distinguished Service Medal in 1959. (*W Post*, 1 Aug 74, C7)

*26 July:* Marshall Space Flight Center announced the award of six contracts, totaling $1.8 million, for development studies of the interim upper stage (IUS) and space tug. The IUS—a predecessor to the space tug, to be used with the space shuttle by both the Dept. of Defense and NASA until the more sophisticated and capable tug was ready for use in 1983—was being developed as the orbit-to-orbit stage (OOS) by the Air Force.

General Dynamics Corp. Convair Div. was awarded a $290 000 contract to study space shuttle and tug compatibility and develop detailed interface requirements. A second contract to General Dynamics, totaling $268 000, was awarded for a definition study of space tug avionics systems with emphasis on areas where transition from the IUS to the tug would require significant hardware or operational changes. United Aircraft Corp. Pratt & Whitney Div. received a $375 371 contract to evaluate and verify the RL-10 high-area-ratio nozzle engine performance for space tug application. Under a $246 000 contract, Martin Marietta Corp. would develop concepts, techniques, and tools for tug fleet and ground operations, mission assignments, payload and orbiter integration, tug scheduling, and physical status. MSFC awarded McDonnell Douglas Corp. a $293 277 contract to study IUS-tug payload requirements compatibility and International Business Machines Corp. a $337 250 contract to study IUS-tug orbital operations. (MSFC Release 74–135)

- NASA announced the selection of four U.S. scientists—Dr. David J. Anderson, Kresge Hearing Research Institute at the Univ. of Michigan; Dr. Joseph Bragin, California State Univ.; Dr. John Oro, Univ. of Houston; and Dr. Alfred R. Potvin, Univ. of Texas—for its 1974 life sciences research program. Each scientist would spend a year at a NASA Center studying life sciences related to space exploration. The program had been recommended by the National Academy of Sciences to increase participation in space- and aeronautics-related research by the scientific community. (NASA Release 74-197)

*28 July:* The economy of Florida's Brevard County was recovering from the impact of Kennedy Space Center layoffs that had followed the end of NASA's Apollo and Skylab programs, the *Washington Post* reported. Employment at the launch facilities had formed the basis for Brevard County's economy and when manpower at KSC was cut in half "the bottom dropped out." Construction came to a near halt, small businesses closed, and personal bankruptcies soared. By March 1971 the Federal Housing Administration held title to 986 repossessed Brevard County homes and the total tripled in the next three years. However, an advertising campaign, begun in 1971 to attract retired persons to the area and its vacant bargain real estate, started an upswing in the economy. Also contributing to the upswing were a pledge that NASA would not let its KSC manpower fall below the current 9500-employee (Civil Service and contractor) level, assurances that the Air Force would maintain its 13 300-man work force at Eastern Test Range for the next five years, construction of a new regional shopping center, and the advent of Disney World 95 km away. (Myers, *W Post*, 28 July 74, E10)

*29 July:* The U.S.S.R. launched *Molniya I-S*—the first Molniya communications satellite in synchronous orbit—from Baykonur Cosmodrome near Tyuratam into transfer orbit and then into synchronous orbit with a 35 790-km apogee, 35 787-km perigee, 23-hr 56-min period, and 0.0° inclination. *Molniya I-S* was the second Soviet satellite in synchronous orbit. The first, *Cosmos 637*, was launched 29 March. Tass announced the purpose of the satellite as "further perfection of communication systems with the use of artificial earth satellites." (GSFC *SSR*, 31 Aug 74; Tass, FBIS-Sov, 31 July 74, U1; *SF*, 1 Jan 75, 35; *SBD*, 5 Aug 74, 100)

*30 July:* Ames Research Center awarded a $400 000 contract to Hughes Aircraft Co. for a design study of an infrared telescope to be carried on the space shuttle above the impeding effects of the atmosphere. The telescope, to operate in orbit attached to the shuttle orbiter, was to have a one-arc-sec pointing accuracy, a 1- to 1.5-m aperture, and cooled optics and sensors. The study would include the telescope facility concept, interfaces with the shuttle orbiter and Spacelab, operating requirements for shuttle and telescope, and ground station requirements including data transfer from the telescope to the ground and control of the telescope from the ground if necessary. (ARC proj off, interview, 29 July 75; *Av Wk*, 19 Aug 74, 18)

- NASA announced the award of a $1.7-million contract to John A. Blume & Associates to study design modifications for the 40- by 80-ft wind tunnel at Ames Research Center. The modifications, which would permit full-scale testing for future aircraft, included repowering the facility from 26.9 mw to 100.7 mw. The new test section under consideration would measure 24.4 by 36.6 m (80 by 120 ft) and increase the speed capability from the present 370 to 555 km per hr. (NASA Release 74-212)

*31 July:* Deputy Secretary of Defense William P. Clements, Jr., approved Air Force plans for initial production of the A-10 close air support aircraft. The Dept. of Defense would release $39 million in FY 1975 funds for long-lead-time items. Approval was granted to buy 52 aircraft from the manufacturer with the provision that contract options to procure a smaller quantity be kept open. Selection of the A–10 for full development followed a competitive prototype flyoff in 1972. Since that time two A–10 prototypes had flown almost 500 hrs in test flights. (DOD Release 346-74)

*During July:* Dramatic changes in a star, altering its chemistry and foretelling its death, had been observed for the first time, Dr. Robert F. Kraft, California Lick Observatory astronomer, reported in *Sky and Telescope.* FG Sagittae, a superstar 8000 light years away in the constellation Sagitta, was undergoing a sudden cooling, abrupt changes in color and brightness, and increases in the number and amounts of heavy elements in its atmosphere. The changes, which occurred in every star that was exhausting its nuclear fuel, might also take place in our own sun, possibly in 10 billion yrs. (*Sky and Telescope*, July 74, 18–22; O'Toole, *W Post*, 10 July 74, A3)

- Industrial research and development spending had totaled $19.4 billion in 1972, an increase of 6% over 1971, National Science Foundation report *Research and Development in Industry 1972* (NSF 74–312) revealed. Company R&D funds had increased 7% and Federal R&D support registered a 6% gain after three years of decreases. The full-time-equivalent number of R&D scientists in industry had increased to nearly 360 000 between January 1971 and January 1972, the first increase since January 1969, when the total was 386 000. Dept. of Defense and NASA had provided 87% of the Federal R&D funds to industry in 1972 and supported 88% of the industrial R&D scientists and engineers working on Federal programs. (Text)

## August 1974

*1 August:* President Nixon transmitted *World Weather Program, Plan for Fiscal Year 1975* to Congress. The plan outlined 1975 U.S. participation in the program seeking to give man "the understanding, tools and techniques necessary to cope with his atmosphere." Goals were to extend the time, range, and scope of weather predictions; assess the impact of atmospheric pollution on environment; study the feasibility and consequences of weather modification; and encourage international cooperation in meeting meteorological needs. He said the U.S. would soon begin continuous viewing of storms over much of the earth's surface with two geostationary satellites and that five more such satellites would be made operational in cooperation with other nations. (*PD*, 5 Aug 74, 1001)

- A 6.4-cm underwater borehole TV camera to provide geologists views of the earth's interior layers had been developed from Skylab technology, Marshall Space Flight Center announced. The camera, designed for MSFC by Sperry Support Services of Alabama, would permit geologists to study geological formations from the interior of a 7.6-cm-dia survey borehole drilled to a depth of 490 m. Images received by the remotely controlled camera would be transmitted along an armored cable to a standard vidicon receiving tube on the surface. (MSFC Release 74–138)

- A digital computing cardiotachometer developed for monitoring beat-to-beat heart rates of astronauts training underwater was being used in two Birmingham, Ala., hospitals, Marshall Space Flight Center announced. The device, designed to operate with a standard electrocardiographic unit, employed an electronic digital system to use the time between two consecutive heart beats to calculate the patient's pulse rate. A numerical display showed the subject's pulse rate 0.3 sec after detecting his second heart beat. (MSFC Release 74–140)

- "Magnetic domain bubbles" being tested as computer memory devices could do mass memory jobs faster, cheaper, and more reliably than conventional devices and could store information for many years for long space journeys, the Air Force Systems Command reported. Dr. Millar Mier of AFSC believed the minute bubbles—densely packed in very thin, magnetic garnet material—would replace tape recorders and computer disc and drum memories in aircraft and spacecraft within a few years. The solid-state bubble bits of information were resistant to radiation and would remain stable during vehicle reentry into the atmosphere or operation in the Van Allen belts. A 10 000-bit test module delivered to the Air Force Avionics Laboratory in May was undergoing tests. (AFSC Release OIP 138.74)

*2 August:* Dr. Dudley G. McConnell was appointed Assistant Associate Administrator for Applications by Dr. James C. Fletcher, NASA Administrator. Dr. Fletcher also announced the promotion of Dr. Harriett G. Jenkins to succeed Dr. McConnell as Assistant Administrator for Equal

Opportunity. Dr. Jenkins had been Dr. McConnell's deputy since February. The appointments would be effective 12 Aug.

Dr. McConnell had spent 12 yrs in laboratory research at Lewis Research Center before coming to NASA Hq. in 1969. In his new position, he would be responsible for developing a coordinated planning capability and a system analysis function for the agency's space applications program.

Before joining NASA in Feb., Dr. Jenkins had been educational consultant for the Response to Educational Needs Project of the Anacostia District of Columbia schools. (NASA Release 74-217)

- The Aeronautical Satellite (Aerosat) Memorandum of Understanding was signed in Paris by Acting Director General Roy Gibson of the European Space Research Organization and Canadian Ambassador Leo Cadieux. U.S. Federal Aviation Administrator Alexander P. Butterfield had signed the Memorandum 9 May. The joint Aerosat program would test, evaluate, and demonstrate the system in preparation for a worldwide operational system of air traffic control in the mid-1980s. The first of two geostationary satellites over the Atlantic Ocean was to be launched in 1977. (ESRO Release, 1 Aug 74; FAA Satellite Div, interview, 21 July 75)

- The last of 130 new solid-state instrument landing systems were being delivered to Air Force and Federal Aviation Administration facilities worldwide. The Air Force Systems Command said the new ILS, "one of the most significant improvements to aircraft landing equipment in 20 years," would become the primary landing approach system at Air Force bases in two years, permitting instrument landings under weather conditions of 30-m vertical and 365-m horizontal visibility. (AFSC Release OIP 136.74)

*4 August:* The *Explorer 51* Atmosphere Explorer, launched by NASA 15 Dec. 1973, had achieved its prelaunch objective of investigating the photochemical processes accompanying the absorption of solar radiation in the earth's atmosphere and was adjudged successful. All 14 onboard scientific instruments were acquiring worthwhile data and, except for minor problems, all spacecraft systems were performing normally. The onboard propulsion system had been used on numerous occasions for perigee altitude adjustment and apogee altitude restoration. The lowest perigee excursion into the lower thermosphere to date had been to an altitude of 130 km. (NASA MOR, 6 Aug 74)

*5 August:* NASA and the Office of Naval Research launched the world's largest unmanned balloon from Fort Churchill, Canada, carrying a 363-kg payload to 47 000-m altitude. The experiment, designed by Goddard Space Flight Center scientist Dr. Robert C. Hartman, measured position and electron spectra between 20 and 800 mev at the 1.2-millibar level. The helium-filled balloon, constructed of half-mil polyethylene, weighed 1360 kg and had a volume of 1.4 cu m and a 0.25-km dia. Launched and tracked with facilities of the Navy's Skyhook program, the balloon rose over Hudson Bay and moved westward 805 km. After 18 hrs, it descended to 30 km, where the payload was separated by radio command and parachuted to the ground. (*Naval Research Reviews*, Nov 74, 13)

*6 August:* An automated blood pressure monitoring system developed for the NASA integrated medical and behavioral laboratory measurement system (IMBLMS) program was being evaluated for use in studies on control of high blood pressure, Johnson Space Center announced. Using Skylab technology, the system combined a blood pressure cuff with a pressure ramp programmer for continuous monitoring and a numerical display. Dr. Ted Andrechuk of Texas Tech Univ. was evaluating the system, which was providing hypertensive subjects with biosensory data for a biofeedback conditioning program. The conditioning would allow the patient to lower his blood pressure permanently. (JSC Release 74-184)

**6, 8, 9 August:** The Senate Aeronautical and Space Sciences Committee conducted hearings on S. 2350 and S. 3484, bills to establish an Office of Earth Resources Survey Systems within NASA and an Earth Resources Observation Administration within the Dept. of the Interior. Both proposals were designed to move the Earth Resources Technology Satellite system from an experimental to an operational phase. Dr. James C. Fletcher, NASA Administrator, opposed the organizational changes proposed by the bills. He argued that S. 2350 would erect barriers in NASA between the ERTS system and related programs in NASA's Office of Applications; keeping these programs together would facilitate hardware development and data reduction and handling. S. 3484 would limit NASA's research and development in earth resources to activities supported and funded by the Dept. of the Interior.

Citing the need for more spectral channels, more rapid handling of data, and more frequent repetitive coverage with ERTS-class satellites, Dr. Fletcher stated that there was no clear-cut distinction between experimental and operational phases in the earth resources program. What was needed was "a continuous evolutionary approach, in which operationally useful data from space are provided to users for research, for experimental demonstrations, and for routine operations when they are ready, while at the same time and with the same satellites necessary improvements to the system can be developed and used." Proponents of the bills argued that ERTS data would be exploited fully only when users could be guaranteed the continuously available data offered by an operational system. Hearings would be continued 18 Sept. (Transcript)

**7 August:** The Air Force Systems Command announced that it would launch an S3-1 satellite late in 1974 as part of a three-launch series to measure atmospheric effects on satellites. The first satellite would be placed in polar orbit with a 4000-km apogee and a 140-km perigee. It would be equipped with ionization gauges and accelerometers to measure air density and to provide data on space vehicle trajectory, slowdown, and decay. Other devices would measure atmospheric heating sources, solar radiation, and atmospheric parameters. With a mass spectrometer in addition, to gather neutral and ion composition data, the experiments would help researchers understand physical and chemical processes in interactions of the sun's radiation with the upper atmosphere, as well as neutral atmospheric effects on the ionosphere. The low-altitude measurements would be used to predict the occurrence and duration of radio blackouts and of atmospheric effects on orbits. (AFSC Release OIP 137.74)

**8 August:** Lewis Research Center awarded an $800 000 cost-sharing contract, of which the Government would supply $500 000, to Pratt & Whitney Aircraft Div. of United Aircraft Corp. in NASA's experimental clean combustor program. Pratt & Whitney was to maintain the already low smoke emissions of its JT8D (Class T4) aircraft engine, while reducing other pollutants, and was to demonstrate the combustor in an engine in 1976.

LeRC was also negotiating a contract with Garrett Corp. AiResearch Aviation Co. for refining Class T1 engines, gas turbine engines with a thrust of less than 35 000 newtons (8000 lbs). A third contract was being negotiated with Detroit Diesel Allison Div. of General Motors Corp. to reduce pollutants from Class P2 turboprop engines. Both companies would be awarded cost-sharing contracts for about $500 000. (LeRC proj off, interview, 30 July 75; Yaffee, *Av Wk*, 26 Aug 74, 56–59)

- The Air Force announced the Aeronautical Systems Div. was awarding a $4 000 000 firm fixed-price contract to Northrop Corp. Aircraft Div. for the YF–17 lightweight fighter aircraft transition program. ASD was also awarding $4 000 000 to General Dynamics Corp. Convair Aerospace Div. for the YF–16 transition program. (DOD Release 368–74)

*9 August:* President Nixon resigned as 37th President of the United States. Vice President Gerald R. Ford was sworn in as the Nation's 38th President. President Nixon had been facing possible impeachment after two years of bitter public debate over his conduct following the 17 June 1972 break-in of the Democratic National Committee headquarters at the Watergate compound in Washington, D.C., by agents of the Committee for the Reelection of the President. (Kilpatrick, *W Post*, 9 Aug 74, 1; Montgomery, *NYT*, 9 Aug 74, 13; *PD*, 12 Aug 74, 1014–1017)

- NASA announced it had awarded RCA Astro Electronics Div. a fixed-price $15 282 143 NASA contract, with an additional incentive amount of $692 638 for performance in orbit, to provide two operational meteorological spacecraft—Improved Tiros Operational Satellite H and I—and parts for ITOS–J. The work, to be performed in Hightstown, N.J., would extend through 15 Feb. 1977. The satellites would support the National Oceanic and Atmospheric Administration's National Operational Meteorological Satellite System through 1978. The noncompetitive procurement was a follow-on to an existing contract. (NASA Release 74–218)

- The *Federal Plan for Meteorological Services and Supporting Research— Fiscal Year 1975*, published for presentation to Congress, called for programs totaling $576.7 million, an increase of $5.7 million over 1974. Published by the National Oceanic and Atmospheric Administration and including proposed activities of all Federal agencies with meteorological programs, the 1975 plan proposed an improved NOAA severe weather warning system; use of new technology for more efficient observation, recording, analysis, and warning in the Dept. of Defense and other agencies; and expanded Federal Aviation Administration weather services to aviation. NASA proposed increases for research with the Nimbus satellite series. (NOAA Release 74–128)

- Thiokol Corp. had been awarded an $8 900 000 contract for strap-on motors for the Thor-Delta launch vehicle, NASA announced. Delivery would run from May 1975 through May 1976. The contract called for 124, TX–354–5 (Castro II) rocket motors, 41 upper-burn nozzle closure plugs, and documentation for the Thor-Delta, which was managed by Goddard Space Flight Center. The Castro II strap-ons had a thrust of about 270 000 newtons (60 000 lbs). (NASA Release 74–219)

- Communications Satellite Corp. submitted its 11th annual report to the President and Congress. FY 1974 had included the launch of *Intelsat-IV F–7* (23 Aug. 1973) and the COMSAT General Corp. award of a $65 900 000 contract to Hughes Aircraft Co. for four high-capacity satellites to be used by American Telephone & Telegraph Co. for domestic communications early in 1976. By 30 June, the number of communications satellite pathways had increased by about 60 to a total of 325, with earth stations in 55 countries. With the addition of 10 more users, the number of countries, territories, and possessions leasing satellite services had exceeded 100. Half circuits leased full time in the global satellite system had increased from 8434 in 1973 to 9627 by 30 June 1974, a gain of about 14%. This voice and record traffic had accounted for the majority of all intercontinental long-distance communications. Net income had risen to $36.3 million, a 45% increase over FY 1973. (ComSatCorp, *1974 Report to the President and the Congress*)

*11 August:* "No other person who has flown in space has captured the experience so vividly," Henry S. F. Cooper said in a review of *Carrying the Fire: An Astronaut's Journeys* by *Apollo 11* Astronaut Michael Collins. Although the book was "long and rambling," touching a great many bases in Collins' career from Air Force test pilot through *Gemini 10* and Apollo, "it is something new under the sun to find an astronaut who isn't afraid to express his feelings." Collins was best at single impressions, "little snapshots," still sharp and clear five years after the lunar landing. Two days out on the mission, " 'The moon I have known all my life, that two-dimensional, small yellow disk in the sky, has gone somewhere to be replaced by the most awesome sphere I have ever seen. . . . This cool, magnificent sphere hangs there ominously, a formidable presence without sound or motion, issuing us no invitation to invade its domain.' "

Collins had cast off the "diffidence astronauts normally wear like spacesuits to protect themselves from prying earthlings. . . . 'I have been places and done things you simply would not believe. I have dangled from a cord a hundred miles up; I have seen the earth eclipsed by the moon, and enjoyed it. I have seen the sun's true light, unfiltered by any planet's atmosphere. I have seen the ultimate black of infinity in a stillness undisturbed by any living thing. . . . I do have this secret, this precious thing that I will always carry with me.' " (*NYT Book Review*, 11 Aug 74, 1)

*12 August:* The U.S.S.R. launched *Cosmos 672*—an unmanned Soyuz spacecraft testing docking systems for the July 1975 Apollo-Soyuz Test Project mission—from Baykonur Cosmodrome near Tyuratam. The satellite entered an orbit with 277-km apogee, 142-km perigee, 88.7-min period, and 51.8° inclination. Maj. Gen. Vladimir A. Shatalov, Soviet chief of cosmonaut training, identified the mission and that of *Cosmos 638* (launched 3 April) at a 9 Sept. Johnson Space Center ASTP press briefing, saying that both had successfully tested the systems. *Cosmos 672* reentered 18 Aug. (GSFC *SSR*, 31 Aug 74; JSC briefing transcript, 9 Sept 74; *SF*, Feb 75, 76)

• Operational utility of satellite communications for maritime service had been demonstrated with *Ats 1* and *3* (Applications Technology Satellites launched 6 Dec. 1966 and 5 Nov. 1967), *Aviation Week & Space Technology* reported. Tests had been made by Exxon Corp. in cooperation with NASA from mid-1973 to early 1974 from an Exxon tanker, using a small General Electric Co. ship terminal and GE earth station. Exxon had reported at a Minneapolis communications conference that the satellites had transmitted voice, teletypewriter, facsimile, and slow-scan TV rapidly, without the delay often caused on high-frequency radio by spectrum congestion. (Klass, *Av Wk*, 12 Aug 74, 56–8)

- Raymond A. Kline was appointed Assistant Associate Administrator for Center Operations. Formerly Assistant Associate Administrator for Organization and Management, Kline would continue to perform management studies and would also oversee the activities of the Automatic Data Processing Management Office. (NASA Ann, 12 Aug 74)

*12–13 August:* Dr. Richard T. Whitcomb, Chief of the Transonic Aerodynamics Branch of the High Speed Aircraft Division at Langley Research Center, received the Aircraft Design Award of the American Institute of Aeronautics and Astronautics 12 Aug. at its 6th Aircraft Design, Flight Test, and Operations Meeting, held in Los Angeles. Dr. Whitcomb received the award for "innovative experimental research resulting in the development of the supercritical wing which permits significant increases in speed and range as well as improved maneuverability of subsonic aircraft without increases in power or fuel consumption."

The Air Transport Assn.'s Octave Chanute Award went to Charles A. Sewell, Grumman Aerospace Corp. chief test pilot, for promoting and flight-testing the concept of spin prevention in a tactical aircraft. A. M. O. Smith, Chief of Aerodynamics Engineering Research at Douglas Aircraft Co., was selected to present the 37th annual Wright Brothers Lecture, on high-lift aerodynamics. (AIAA *News,* 17 July 74; AIAA PIO, interview, 16 July 75)

*13 August:* NASA began "Outlook for Aeronautics," a comprehensive 10-mo planning study to assess aeronautical challenges and opportunities for 1980–2000. Under the leadership of Dr. Leonard Roberts, Director of Aeronautics and Flight Systems at Ames Research Center, representatives of NASA, Federal Aviation Administration, and Dept. of Defense would survey NASA, industry, and Government to determine probable directions of civil and military aviation, technological advances needed, and the role of NASA.

In a 4 Oct. letter to NASA employees Dr. James C. Fletcher, NASA Administrator, said, "We want to produce a study that will assist NASA in presenting the best possible aeronautics program relevant to national needs, for the remainder of this century, to the Administration, the Congress, and the people." In the aeronautical world market, the U.S. had sold more than $235 billion worth of products—not including space products—over the past 25 yrs, maintaining a favorable balance of trade for a nation that might otherwise have been in the red. In preparing this long-range study, "we have the great responsibility of indicating how America's pre-eminent position in areonautics can be maintained and enhanced. The stakes are too high for any other approach." (*NASA Activities,* 15 Oct 74, 2–3, 11–13; NASA OAST, Civil Aircraft Progs Off, interview, 13 Nov 75)

*14 August:* The NASA Delta Project Review Committee, established 31 May to review Thor-Delta launch vehicle project management and launch vehicle performance abnormalities, had completed its investigation, NASA announced. The committee recommended several design and hardware changes, more stringent quality control, strengthening of project management by additional staffing, and more rigorous management and contractual procedures.

The committee endorsed the ongoing redesign of the 1st-stage liquid-oxygen pressurization system and 2nd-stage fuel and oxidizer shutoff valves, revision of the redundant attitude-control-system manufacturing sequence and inspection processes, revision of installation and inspection procedures for the solid-fueled-rocket strap-ons, and a complete investigation of the start-induced forces on the the 1st-stage engine. The schedule for the next Delta launching would be established upon NASA Hq. acceptance of the plan for corrective actions in response to the committee's recommendations. (NASA Release 74-220)

- NASA objectives for *Explorer 50* (IMP-J, launched 25 Oct. 1973)—to make detailed and nearly continuous studies of the interplanetary environment for orbital periods comparable to several rotations of active solar regions and to study particle and field interactions in the distant magnetotail—had been met and the mission was adjudged successful. All 12 instruments were performing well with the exception of an electronic field antenna. Loss of this antenna reduced scientific data 10% for two investigators. All spacecraft housekeeping systems were functioning properly. (NASA MOR, 15 Aug 74)

*15 August:* Elmer S. "Todd" Groo had been appointed Associate Administrator for Center Operations at NASA Hq., effective 4 Sept., NASA announced. Groo, the first to hold the position created in the reorganization announced 5 March, would be responsible for agency-wide planning and direction of institutional resources and operational activities at NASA field installations. Groo had retired in 1973 from IBM World Trade Corp., where he had served as vice president since 1959. (NASA Release 72-221; MSFC Release 74-146)

- Dr. Robert A. R. Parker, scientist-astronaut and program scientist for the Skylab mission, had been named Chief of the Astronaut Office, Science and Applications Directorate, Johnson Space Center announced. Dr. Parker succeeded Dr. Harrison H. Schmitt, who had been named Assistant Administrator for Energy Programs at NASA Hq. in May. The S&AD would serve as an interface between eventual space shuttle payload users and NASA and participate in the shuttle flight research program. Dr. Parker had been named a NASA scientist-astronaut in 1967 and had been a member of the support crews for the *Apollo 15* and *17* missions. (JSC Release 74-186)

- Air Force Systems Command announced plans for a global Radio Solar Telescope Network (RSTN), a four-site space-surveillance linkup for 24-hr operation. Solar and space environmental data would be gathered by RSTN parabolic antennas monitoring frequencies between 245 and 15 400 mhz. Data would be used to forecast atmospheric responses to solar activity, to compensate for communications blackouts of satellite and warning systems, and to provide real-time solar-alert and analysis data. Operational testing would begin early in 1976, with the network to become fully operational over a three-year period. (AFSC Release OIP 132.74)

*16 August:* The Senate passed, by a unanimous vote of 60, H.R. 15572, the Dept. of Housing and Urban Development-Space-Science-Veterans FY 1975 appropriations bill. The bill was passed as reported out by the Senate Committee on Appropriations 15 Aug. plus an amendment to permit NASA to transfer up to 0.5% of its funds between research and development and research and program management programs. The bill appropriated $3.207 billion to NASA, $36 million below the FY 1975 budget request and $60.2 million below the FY 1975 authorization signed 22 June. An additional $4.435 million was appropriated in a Special Energy Bill.

**16 August**

NASA R&D funds totaled $2.327 billion, construction of facilities $140.2 million, and R&PM $740 million. The Senate approved all budgeted R&D items, including $6.2 million for the Large Space Telescope and $8.0 million for SEASAT, both of which had been cut from the House appropriation 26 June. A reduction of $15 million from R&D funds was not directed to any specific program because the Committee suggested that NASA was better able to apply this reduction with a minimum disruption in its priorities. Agreeing with the House, the Committee had urged NASA to reprogram funds to begin work on ERTS–C and said that further reprogramming of funds for Viking would be considered only with the greatest reluctance.

Funds for construction of facilities were up $4.5 million from the House appropriation and down $4.3 million from the authorization. The $740 million for R&PM equaled the House appropriation and was down $9.6 million from the authorization.

H.R. 15572 had been reported out of the Senate Committee on Appropriations for a first time 1 Aug. with a total NASA appropriation of $3.243 billion. It was returned to the Committee 5 Aug. by the full Senate for additional cuts.

The Senate requested a conference with the House to resolve differences in the passed versions. (Sen Com Rpts 93–1056, 93–1091; *CR*, 5 Aug 74, S14313–40; 16 Aug. 74, S15161–5)

- Static test firings of a 6.4%-scale model of the space shuttle to gather acoustical data for design and development activities began at Marshall Space Flight Center. A series of 24 acoustic test firings would continue through November, with 7 additional tests later to study possible sound-suppression methods.

  The test model's two solid-propellant engines, representing the shuttle vehicle's solid-fueled-rocket boosters, were being tested first, to be followed by full system firings, including the three model liquid-propellant engines representing the main engines, which would use liquid hydrogen and liquid oxygen. On each firing, 60 acoustic measurements were taken by specialized microphones to monitor areas potentially sensitive to high sound-pressure levels: vehicle skin panels, thermal protection devices, electronic guidance components, and ground support and related structures. (MSFC Release 74–154; MSFC PIO, interview, 15 July 75)

- Dr. James C. Fletcher, NASA Administrator, announced that Mrs. Ruth Bates Harris, who had been dismissed as NASA Deputy Assistant Administrator for Equal Opportunity 26 Oct. 1973, had accepted an appointment as Deputy Assistant Administrator for Public Affairs for Community and Human Relations effective 19 Aug. In a memorandum to NASA employees, Dr. Fletcher stated that he was firmly convinced of Mrs. Harris's "pride in NASA's program accomplishments in aeronautics and space, her belief that they are relevant to all Americans, and her enthusiasm in conveying this pride and conviction to others." Mrs. Harris would serve as the point of contact between NASA and community groups and between NASA and state and local governments. (NASA Release 74–223; Fletcher memo, 16 Aug 74)

- Dr. James C. Fletcher, NASA Administrator, presented awards at an Ames Research Center ceremony to 50 members of the *Pioneer 10* team for their contributions to the 3 Dec. 1973 flyby of Jupiter. Charles F. Hall, Manager of the Pioneer Project at ARC, received NASA's Distinguished

Service Medal for his "outstanding leadership and dedicated performance."

At another awards ceremony, Dr. Fletcher presented 28 NASA awards to personnel of Jet Propulsion Laboratory and other organizations for contributions to the *Mariner 10* flyby of Venus (5 Feb. 1974) and Mercury (29 March 1974). Project Manager Walker E. Giberson received the Distinguished Service Medal for "distinguished leadership in the conception, organization, and management" of the mission. (ARC Release 74–35; ARC PIO, interview, 18 July 75; JPL Release 711)

- The Navy awarded $1.3-billion contract to Lockheed Aircraft Corp. for the initial production of the submarine-launched Trident I missile and the development of a maneuverable nuclear warhead. The warhead, called the Mark 500, would not be limited to targets on a missile's orbital trajectory and it would be able to evade enemy defensive missiles. (Reuters, *W Post*, 17 Aug 74, A5; *WSJ*, 19 Aug. 74, 5)

*17 August:* The first test flight of an Air Force remotely piloted vehicle was successfully completed, operated from a ground console at Edwards Air Force Base. The RPV—built by Teledyne, Inc., Ryan Aeronautical Div. and incorporating a 25-m wingspan and a single jet engine mounted on an 11-m fuselage—was in the air 1 hr 50 min and reached an altitude of 7770 m. One of two kinds of high-altitude, long-range reconnaissance aircraft produced under the Compass Cope program, the aircraft could carry a 3200-kg payload at a cruising altitude of 20 000 m for 30 hrs. (Miles, *LA Times*, 22 Aug 74, 31; AP, *W Post*, 23 Aug 74, 11)

*19 August:* The prototype Mini-Sniffer, a small 55-kg radio-controlled aircraft equipped to test pollution in the upper atmosphere, made the first of a series of eight flights from Flight Research Center. The vehicle, powered by a 9-kw gasoline engine, maintained an altitude of 15–30 m and performed well at 55–87 km per hr. The purpose of the flight was to test airworthiness at lower altitudes and speeds. Another series of flights would check out the auto-pilot system at low altitude and then the aircraft would be taken to 6100 m, using radar and a long-range radio-control system. The Mini-Sniffer would eventually be fitted with an 11-kw engine which would carry it up to 21 000 m. (FRC *X–Press*, 2 Aug 74, 30 Aug 74; FRC proj off, interview, 24 July 75)

- Naval Weapons Center scientists had discovered that nitrogen lasers could be used to locate downed flyers, *Aviation Week & Space Technology* reported. The sea dye marker routinely included in Navy life vests and rafts was found to fluoresce when illuminated by a nitrogen laser from an aircraft. (*Av Wk*, 19 Aug 74, 47; Sullivan, *NYT*, 14 Sept 74, 26)

*19–29 August:* The backup command and service module, launch escape system, and Saturn IB 1st stage in the Apollo-Soyuz Test Project were subjected to simulated lightning strikes at Kennedy Space Center to determine effects of lightning during launch preparations. Test results would be used in shuttle planning; the shuttle would not be surrounded by a protective mobile service structure on the launch pad. (KSC PIO, interview, 17 June 75; *Spaceport News*, 6 Sept 74)

*20 August:* President Ford nominated Nelson A. Rockefeller, former Governor of New York, as 41st Vice President of the United States. (*PD*, 26 Aug 74, 1050–1; 1063, 1064)

- The International Telecommunications Satellite Organization marked its 10th anniversary. Membership, originally at 14, had grown to 86 countries, which jointly owned the four Intelsat IV satellites and three spare

Intelsat III satellites in the global communications system. The system provided more than 5000 international telephone circuits plus telegraph, telex, data, and TV services; charges had progressively dropped from the initial $32 000 per year for a half circuit to $9000. The permanent INTELSAT Organization had come into being 12 Feb 1973, replacing the interim Consortium that had operated since 20 Aug. 1964. (INTELSAT Release 74–2)

- Dr. Edward G. Gibson, science pilot on the 84-day *Skylab 4* mission (16 Nov. 1973—8 Feb. 1974), announced he would leave the NASA astronaut corps 30 Nov. 1974. Dr. Gibson, a scientist-astronaut since 1965, would join Aerospace Corp. of Los Angeles as a senior staff scientist, specializing in the interpretation of solar data gathered during the 171 days of manned operation in the Skylab program. He had held primary responsibility for the 338 hrs of Apollo Telescope Mount operation on *Skylab 4*, making extensive observations of solar processes. (JSC Release 74–246)

*20–22 August:* More than 70 papers on the Skylab program were presented at the 20th annual meeting of the American Astronautical Society, held at the University of Southern California. Billed as the definitive report on Skylab, the conference addressed Skylab evolution, accomplishments, and applications to future NASA missions. Eight NASA representatives were presented at the 22 Aug. session of the conference as newly elected Fellows of the AAS. (MSFC PIO, interview, 11 July 75; MSFC Release 74–136; WFC Release 74–8; JSC Releases 74–163, 74–246)

*21 August:* A House-Senate Conference Committee favorably reported a compromise version of H.R. 15572, the FY 1975 Dept. of Housing and Urban Development-Space-Science-Veterans appropriations bill that included a $3.207-billion NASA appropriation. NASA funds were $40.394 million below the budget request and $60.194 million less than the FY 1975 authorization. An additional $4.435 million for energy research and development was appropriated separately. The compromise appropriation was close to the one passed by the Senate 16 Aug. and $3.685 million above the $3.203 billion approved by the House 26 June. (NASA Off Budget Ops, Chron History; CR, 21 Aug 74, H8787–90; H Rpt 93–1310)

- The U.S.S.R. cosmonaut corps was being built up at same time the U.S. astronaut force was dwindling, a *Miami Herald* article said. The *Herald* quoted Astronaut Thomas P. Stafford, who had visited the U.S.S.R. during training for the 1975 Apollo-Soyuz Test Project mission, as saying : "Star City (where most of the cosmonauts live and train) looks like a boom town. There's no doubt the Soviets have a vigorous ongoing manned space effort planned for the next five or six years." He said that the number of buildings in Star City had tripled in the previous two years, with more under construction. The total of 75 to 80 cosmonauts was a sharp increase over the estimated 55 of five years ago. In contrast the U.S. astronaut corps had fallen to 34 from a peak of 60 during the Apollo program, with only three training for a flight. Stafford said he did not believe the U.S.S.R. had any new spacecraft under development but rather planned for the continued use of the Soyuz as a workhorse. (*M Her*, 21 Aug 74)

*22 August:* NASA reactivated the main telemetry system on *Explorer 46* Meteoroid Technology Satellite (launched 13 Aug. 1972), to gain additional data on the impact flux of small meteoroids. The telemetry system had been shut down 22 Aug. 1972 after the spacecraft battery had overheated, and a solar-powered backup had since transmitted only primary experiment data. Data received before shutdown had indicated that sensors flown on previous missions had overestimated the micrometeoroid population, but variation in the impact flux necessitated additional data. Data transmission would continue through February 1975. (NASA Release 74–321; NASA OAST PIO, interview, 3 Sept 75)

- The House and Senate, in separate floor actions, agreed to the 21 Aug. conference report on H.R. 15572, the Dept. of Housing and Urban Development-Space-Science-Veterans Appropriations Bill for FY 1975, clearing the measure for the White House. The bill included a NASA appropriation of $3.207 billion. (CR, 22 Aug 74, H8892–9, S15776–8)

*23 August:* Ames Research Center controllers of *Pioneer 11*—launched 5 April 1973 and now 85 000 000 km from Jupiter—increased the spacecraft's spin rate to improve measurements of Jupiter's radiation belts during the flyby, scheduled for December. (ARC Release 74–41)

- Photographs and visual observations made by astronauts aboard the Skylab Orbital Workshop had revealed eddies embedded in warm-water currents flowing poleward from equatorial oceans, Robert E. Stevenson of the Office of Naval Research reported in a *Nature* article. Observations from *Skylab 4* (16 Nov. 1973–8 Feb. 1974) over four ocean areas had confirmed initial evidence from Caribbean photographs taken on *Skylab 2* and *3* (25 May–22 June, 28 July–25 Sept. 1973) that eddies with vortices of 5500- to 37 000-m dia were present wherever other conditions were suitable, regardless of water temperature. Findings were also confirmed by the Navy's Weather Reconnaissance Squadron Four with bathythermographs.

  Formerly, turbulence had been difficult to document because of limited oceanographic techniques, insufficient information about such features, and the lack of photos from orbital platforms which enhanced the ocean and atmosphere manifestations of the eddies. (*Nature*, 23 Aug 74, 638–40)

- A *Science* editorial commented on the Nixon Administration and science: During his terms in office President Nixon, who had resigned 9 Aug., had been preoccupied by foreign policy and Watergate reaction. Science was "at most an afterthought." Because he thought that science should advance the cause of national security and prestige, "grand technology seemed to have had a special fascination for Nixon," most evident in his enthusiasm for the Apollo program and the supersonic transport. He had made a dramatic declaration of war on cancer and set the Nation toward Project Independence, under which the U.S. was to "achieve energy self-sufficiency by 1980." There was "little evidence that Nixon was hostile to science but rather that, except for the technological extravaganzas, he was simply not interested." He lacked rapport with scientists, allying himself instead with proponents and practitioners of the quasi-science of modern management.

  The Presidential tapes (tape recordings made secretly of conversations in the Oval Office) showed a President "badly informed, even indifferent," helping to explain why the White House science advisory machinery was deemed expendable. "But in the present situation it takes no special wisdom to see how R&D decisions will affect how the government deals with serious energy and food problems and how important these actions will be in future economic developments. The Nixon tapes illustrate why good presidential advisers and good mechanisms to transmit their advice are necessary in every sphere of policy; this should not be overlooked among the lessons of Watergate for the new Administration." (*Science,* 23 Aug 74, 675–8)

*24 August:* A probe into the next 50 yrs was the subject of a series of articles in *Saturday Review/World.* Soviet atomic scientist Andrei D. Sakharov envisioned flying cities—artificial earth satellites with important industrial functions. The satellites would hold nuclear and thermonuclear installations with radiant cooling of heat exchangers, to avoid heating the earth, and would also serve as cosmic research laboratories and way stations for long-distance flights. Sakharov also foresaw greater attempts to establish communication with other civilizations by searching for and transmitting interplanetary signals on all known wavelengths and searching outer space for space stations of extraterrestrial civilizations. Powerful telescopes set up on the moon would permit scientists to see planets orbiting the nearest stars.

*Apollo 11* Astronaut Neil A. Armstrong said that in 50 yrs every interesting crater and cranny of the moon would have been visited and revisited and lunar exploration would be replaced by lunar exploitation. Underground lunar colonies, protected from extreme temperatures, would consist of pressurized compartments connected by tunnels. One day man might want to build an atmosphere on the moon to avoid the inconveniences of living in a near vacuum, but in the nearer future the use of the moon's vacuum for activities such as materials processing was a major reason for being there. Craters on the dark side of the moon, always facing away from the earth, could hold radio and x-ray telescopes for transmissions, unadulterated by earth noise, to and from outer space. Deep craters, which received no sunlight and had temperatures nearing absolute zero, would be ideal for low-temperature laboratories where scientists could investigate the point where molecular motion nearly stops. Lunar agriculture would be underground under lights in modified lunar soil fed with a nutrient solution. Although by 2024 the only extraterrestrial colonies would be on the moon, men probably would have visited Mars and a few selected celestial bodies. Men exploring space might well use a miniaturized heart-lung machine, circumventing the need for breathing and spacesuits.

Dr. Wernher von Braun, former Marshall Space Flight Center Director, predicted a worldwide satellite search would be started for additional deposits of metal ore, coal, oil, shale, and natural gas. Satellite data would also provide key data for a badly needed worldwide food supply management system. Von Braun predicted nuclear lasers in geosynchronous orbit would handle millions of TV channels simultaneously and billions of telephone conversations and would lead to worldwide video-telephone service. Advanced meteorological satellites would provide accurate two-week weather forecasts and disaster warnings, saving millions of dollars in property damage. (*Saturday Review/World,* 24 Aug 74)

- Maj. Alexander P. de Seversky (U.S. Army Air Corps), aviation pioneer, died of a respiratory ailment in New York at the age of 80. Maj. de Seversky had begun his career as a pilot in the Czarist air force in Russia during World War I. After coming to the U.S. in 1918 he contributed to military and commercial aviation, helping to develop the first fully automatic synchronous bombsight, gyroscopically stabilized flight instruments, inflight fueling techniques, and a variety of landing gear for seaplanes and aircraft. He also led the effort to design, build, and test an all-metal fighter aircraft, the P-35, the first aircraft to fly faster than 480 km per hr. In 1931 he founded the Seversky Aircraft Corp., the forerunner of Republic Aircraft Corp.

During World War II de Seversky became one of the country's leading advocates of strategic air power, saying that the U.S. could not win a "limited war fought with traditional forces." He argued that "the whole military philosophy must be changed or we shall forever trail the rest of the world not only in conventional aircraft, but in intercontinental ballistic missiles and even in man-made earth satellite development." (Fraser, *NYT*, 26 Aug 74, 32; AP, Orlando, Florida, *Sentinal Star*, 26 Aug 74, B4)

*26–29 August:* The U.S.S.R. launched *Soyuz 15*, carrying crew members Lt. Col. Gennady V. Sarafanov and Col. Lev S. Demin, from Baykonur Cosmodrome near Tyuratam at 12:58 pm local time (3:58 pm EDT) 26 Aug. in an attempt to dock with the orbiting *Salyut 3* research station (launched 25 June). *Soyuz 15* entered orbit with a 239-km apogee, 175-km perigee, 88.5-min period, and 51.7° inclination. Tass reported the purpose of the mission was "to continue scientific research and experiments in space started July 3 . . . during the flight of the transport ship 'Soyuz 14' and the station 'Salyut 3.'"

Tass reported 28 Aug. that on the second working day the cosmonauts felt well and had made "experiments to perfect the technique of piloting the ship in different flight regimes." During maneuvering *Soyuz 15* approached the *Salyut 3* space station many times, the cosmonauts controlling spaceship functions, observing the approach, and inspecting the station. Tass then reported the crew was "concluding the flight and . . . preparing the spacecraft for the return to the earth."

After a 48-hr 12-min mission, *Soyuz 15* softlanded safely at 1:10 am 29 Aug. Baykonur time (4:10 pm EDT 28 Aug.)—the first Soviet night landing—48 km southwest of Tselinograd in Kazakhstan despite a "difficult meteorological situation." On-the-spot medical examinations showed the cosmonauts to be in "good health."

Early press reports quoted Western experts as saying the flight probably ended early because of a failure in the docking mechanism of the spacecraft, damage to or failure of the spacecraft during the attempted linkup, or a serious deterioration of the Salyut orbital station. In response to questions about the impact of the *Soyuz 15* mission on the planned July 1975 U.S.–U.S.S.R. Apollo-Soyuz mission, Dr. James C. Fletcher, NASA Administrator, said in a 29 Aug. statement, "NASA personnel engaged in the Apollo-Soyuz Test Project . . . were informed in advance by Soviet personnel of the 'Soyuz 15' flight and its approximate timing." Dr. Fletcher said that neither *Soyuz 14* (launched 3 July) nor *Soyuz 15* "carried the ASTP docking mechanism. We have no reason to doubt this information."

At a 9 Sept. press briefing at Johnson Space Center, Maj. Gen. Vladimir A. Shatalov, Soviet chief of cosmonaut training, confirmed through his translator that a hard dock had been one objective of the *Soyuz 15* mission and that there were problems with the docking system. "In our previous flights . . . we had worked out the automatic rendezvous and docking, but it worked only up to 150 meters. . . . From that point, it was purely manual." At an 11 Sept. JSC press briefing, Shatalov said that during the *Soyux 15* flight the automatic system "worked well" up to 30 to 50 m. "But when going beyond 50–40, the range rate exceeded the necessary or the intended parameters." He also said the approaches had been repeated "a number of times and each time the automatic system worked up to 30–40 meters. . . . At no time in these approaches was it intended to use manual docking." He emphasized that they were satisfied with the automatic system to the point it had worked. He also said the second major objective of the mission was "to perfect this night landing."

*Soyuz 15* was the second spacecraft to visit *Salyut 3*, the Soviet's third orbiting station. A two-man crew, launched 3 July aboard *Soyuz 14*, had spent nearly 16 days aboard the station performing scientific experiments. (Tass, FBIS–Sov, 27–29 Aug 74, U1; NASA Release 74–242; Wren, *NYT*, 29 Aug 74, 1; *W Post*, 30 Aug 74, A32; *SF*, Dec 74, 470; Transcripts, 9, 11 Sept 74)

*26 August:* Charles A. Lindbergh—the first man to fly solo nonstop from New York to Paris, in 1927—died of cancer on Maui, Hawaii, at the age of 72. He was buried three hours later in a simple funeral near the Kipahulu church on Maui. In Washington, President Ford said: "Nearly half a century has passed since his courageous solo flight . . . but the courage and daring of his feat will never be forgotten. For years to come, we will also remember the selfless, sincere man himself . . . , one of America's alltime heroes and a great pioneer of the air age that changed the world."

For the transatlantic flight, airmail pilot and barnstormer Lindbergh had helped design a Ryan monoplane, calculating every ounce. When the *Spirit of St. Louis* landed at Le Bourget airfield, Paris, at 10:22 pm 21 May 1927 after a 33-hr 29-min flight, Lindbergh was greeted by 25 000 wildly cheering persons. British, French, and Belgian governments awarded the flyer their highest decorations for valor. He received the first U.S. Distinguished Flying Cross and the Medal of Honor and was promoted from lieutenant to colonel in the Army Air Corps Reserve.

Following goodwill flights throughout the U.S. and Latin America, sponsored by the Guggenheim Fund for the Promotion of Aeronautics, Lindbergh served as adviser to Pan American World Airways and to the predecessor of Trans World Airlines, helping to lay out transatlantic, transcontinental, and Caribbean air routes for commercial aviation. He served as a member of the National Advisory Committee for Aeronautics 1931–1939, heading the special committee that recommended to Congress expansion of U.S. aircraft research facilities and establishment of what became the Ames and Lewis Research Centers.

After the kidnaping and murder of his first son and the subsequent publicity, Lindbergh moved to Europe in 1935, where he studied the military forces of the European countries. His conclusion that the German Luftwaffe would be superior to the combined forces of the Allies, his acceptance of the Service Cross of the German Eagle from the German government, and his opposition to U.S. participation in World War II provoked the U.S. public to question his patriotism and he resigned his commission in the Army Air Corps. During the war he was a consultant to bomber aircraft manufacturers Ford Motor Co. and United Aircraft Corp. He traveled to the Pacific war area in 1944 to study the Navy's Corsair F–4U under flying conditions and there, as a civilian, flew 50 missions against the Japanese.

Following the war, his commission restored and promoted to the rank of brigadier general, Lindbergh was a technical consultant for Government and private industry in missile and space flight programs. As a consultant for Pan Am he worked on design specifications for the Boeing 747. Lindbergh devoted the later years of his life to writing and speaking on world conservation policies. (*NYT*, 27 Aug 74, 1; 18; Crawford, *W Post*, 27 Aug 74, C3; *PD*, 2 Sept 74, 1068–9; James, *NYT*, 22 May 27, 1; Affidavit Exec Apptmt, 6 Nov 31; letter, NACA Chm Bush to Lindbergh, 7 Feb 40)

- Dr. Alan M. Lovelace, Acting Deputy Assistant Secretary of the Air Force for Research and Development, had been appointed NASA Associate Administrator for Aeronautics and Space Technology effective 1 Sept., NASA announced. Dr. Lovelace would replace Gen. Bruce K. Holloway (USAF, Ret.), who would retire from NASA 30 Aug. after serving as Acting Associate Administrator for Aeronautics and Space Technology and Assistant Administrator for DOD and Interagency Affairs. Dr. Lovelace had been Director of the Air Force Materials Laboratory in 1967 and had become Director of Science and Technology, Air Force Systems Command, in 1972. He had received numerous awards, including the AFML Charles J. Cleary and Arthur S. Flemming awards for his work in fluorine and polymer chemistry, the National Civil Service League Career Service Award, and the Air Force Decoration for Exceptional Service. (NASA Release 74-236)

- John M. Thole had been appointed Deputy Associate Administrator for Space Science effective 1 Sept., NASA announced. Thole would succeed Vincent L. Johnson, who had retired. Former project manager for the Applications Technology Satellite at Goddard Space Flight Center, Thole had joined GSFC in 1960 and had held a variety of project management positions since that time. (NASA Ann, 26 Aug 74)

*26 August–20 September:* Thirty NASA engineers and technicians met with U.S.S.R. Academy of Sciences counterparts in Moscow as joint preparations for the Apollo-Soyuz Test Project's July 1975 flight continued.

On 8 Sept. another group of 47 specialists and engineers headed by U.S. ASTP Technical Director Glynn S. Lunney arrived in Moscow for 2 weeks of meetings with the Soviet group headed by U.S.S.R. Technical Director Konstantin D. Bushuyev. The project directors announced plans for joint prelaunch tests at the U.S. launch site 1–8 Feb. 1975 and at the U.S.S.R. launch area 5–13 May 1975. Flight crews and technical specialists from both countries would participate. Soviet specialists also announced plans to fly a manned Soyuz mission to test ASTP-related systems. NASA would be informed of further details before the flight to permit tracking.

Thirteen U.S. flight controllers were also spending two weeks in Moscow beginning 15 Sept., meeting with Soviet controllers for familiarization and training, including briefings on Soviet Mission Control Center operations, Soyuz spacecraft systems, and the mission flight plan. (JSC ASTP Press Communiqué, 20 Sept 74; JSC *Roundup*, 13 Sept 74; JSC Release 74-250)

*27–29 August:* A Skylab Life Sciences Symposium at Johnson Space Center reported medical results of the three Skylab missions. Conference Chairman Richard S. Johnston, Director of Life Sciences at JSC, said that nothing in the medical results would preclude man from spending up to a year in space. Dr. G. D. Whedon of the National Institutes of Health stated that weightlessness could threaten capable musculo-skeletal functions during space flights lasting up to three years unless measures could be developed to prevent significant losses of calcium, phosphorus, and nitrogen. Continued loss of calcium could cause astronauts to break their backs or legs and could promote formation of kidney stones.

Dr. William E. Thornton, life scientist in the JSC Astronaut Office, reported that by Mission Day 3 all Skylab crewmen had lost more than 2000 cc (2 qts) of extravascular fluid from calf and thigh; at recovery, a sharp reversal was noted. The Thornton team suggested that the reason for the shift was the intrinsic and unopposed lower limb elasticity, which forced venous blood and other fluids toward the head. Other investigators reported that such cardiovascular changes as the heart's decreased ventricular size and the suppression of red blood cell production was the body's accommodation to weightlessness. These physiological changes had been the expected response to a lack of exertion; they reversed after the astronauts' return to earth. (JSC *Roundup*, 30 Aug 74, 1–2; JSC proj off, interview, 12 Aug 75; Altman, *NYT*, 1 Sept 74, 14)

*28 August:* Marshall Space Flight Center had negotiated a restructured contract with TRW Inc. for the first of three High Energy Astronomy Observatory spacecraft to be launched between 1977 and 1979, NASA announced. Modifications had increased the original contract value by $9 451 139, to a new total of $86 397 579. The contract covered design, development, manufacture, test, checkout, and delivery of the spacecraft. Period of performance was from 1 July 1974 through 15 April 1980.

The original HEAO program, approved in 1971 and suspended in 1973 because of budget cuts, had called for two large unmanned observatories. The new contract called for a series of smaller scale missions using three satellites in circular orbits at an altitude of 416 km to obtain data on high-energy radiation from space—celestial x-rays, gamma-rays, and cosmic rays.

Contracts totaling $23 332 700 had also been signed for five experiments to be carried by the second spacecraft, HEAO-B. The awards included $1 350 000 to American Science and Engineering, Inc., to design and build the x-ray telescope. (NASA Release 74–240; MSFC Release 74–179)

- NASA adjudged the mission of *Skynet IIA* communications satellite, launched 18 Jan. for the United Kingdom, unsuccessful. NASA's primary objective of launching the comsat into orbit for transfer into a synchronous equatorial orbit permitting the spacecraft to accomplish its operational mission had not been achieved. *Skynet IIA* had been placed in an extremely eccentric orbit when the Thor-Delta launch vehicle's 2nd-stage attitude control system failed; preliminary analysis had isolated the problem to a failure in the regulator board of the DC-to-DC converter in the 2nd-stage electronics package. (NASA MOR, 4 Sept 74)

- Martin Marietta Corp. had been awarded a $26 453 600 NASA contract for support of the external tank effort of the space shuttle program, NASA announced. The consolidated facilities contract, which would run through 31 Aug. 1978, provided for acquisition of plant equipment at Michoud Assembly Facility, rehabilitation of existing facilities, and construction, modification, maintenance, and repair of facilities. (NASA Release 74–239)

*29 August:* The NASA Aircraft Office was established under the Associate Administrator for Center Operations to provide a focal point for aircraft operations and management activities. The Flight Activities Office, Flight Research Center, was abolished, and the Administrative Standardization Office at Langley Research Center would report directly to the NASA Aircraft Office. Robert W. Sommer, former Director of the Flight Activities Office, was appointed Acting Director of the new office. (NASA Ann, 29 Aug 74)

- Communications Satellite Corp. announced it had signed a contract to provide technical and operational management for the global commercial satellite system owned by the International Telecommunications Satellite Organization. Under the contract, to run through 11 Feb. 1979, ComSatCorp would recommend and implement research and development programs; advise on procuring space segment facilities; negotiate, place, and administer all contracts for space segments; arrange for satellite launch services and support; prepare and coordinate system operations plans; and prepare and coordinate frequency plans for earth stations. (ComSatCorp Release 74-44)

*30 August–14 September:* NASA launched *Ans 1* Netherlands Astronomical Satellite from Western Test Range on a four-stage Scout launch vehicle at 7:08 am PDT (10:08 am EDT). The satellite entered an elliptical polar orbit with a 1167-km apogee, 254-km perigee, 99.0-min period, and 98.1° inclination—significantly different from the planned circular orbit—because of a malfunction in the 1st stage of the Scout. Changes in the spacecraft program and scientific instrument operations program were made to accommodate to the actual orbit. Although the orbit hindered the observation program, the primary objective—to increase the scientific knowledge of stellar ultraviolet and x-ray sources—could be achieved.

The spacecraft was pointed at the sun, the solar panels were deployed, and fine pointing in a three-axis stabilized condition was established. All three experiments—a hard x-ray experiment provided by the U.S. and a UV telescope and soft x-ray experiment provided by The Netherlands—were turned on by 8 Sept. In-orbit checkout of experiments and spacecraft was completed 13 Sept. and the operational phase of the mission began 14 Sept.

Principal effect of the orbit on the scientific program was the loss of observing time. The higher apogee placed the spacecraft in the Van Allen radiation belts longer than planned, creating a more intense radiation background than anticipated. The UV telescope could operate effectively for 70% of each orbit and observe 10th-magnitude stars only when the spacecraft was below 500 km. Only 5th-magnitude or brighter stars were observable most of the time.

The U.S. x-ray experiment could make effective observations about 25% of the total possible observing time; the experiment could operate below 700 km but not over the poles or the South Atlantic Anomaly. The soft x-ray experiment consisted of a large-area detector, which was operating similarly to the U.S. experiment, and a grazing-incidence telescope, which was seriously affected by the radiation background. Although much of the acquired data was not usable, most future data might be recovered through improved analysis techniques.

Despite the problems, the orbit would provide nine months of full sunlight, rather than the six months expected, permitting reobservations of objects observed during the first three months. In addition, aerodynamic drag would lower the apogee over a period of time, decreasing the passage time through the Van Allen belts.

*Ans 1* was launched under a Memorandum of Understanding signed 5 June 1970 by NASA and the Astronomical Netherlands Satellite Program Authority (ANSPA). NASA provided the x-ray experiment, spacecraft technical support, the Scout launch vehicle and services, and limited tracking and data acquisition. The Netherlands was responsible for the spacecraft design, fabrication, and testing; two experiments; and primary spacecraft command, control, and tracking and data acquisition. *Ans 1* was the first satellite to be controlled completely from the European Space Research Organization's operations center at Darmstadt, Germany. The ANS program was managed for NASA by Goddard Space Flight Center under the direction of the Office of Space Science. Langley Research Center had responsibility for the Scout launch vehicle. (NASA MORs, 21 Aug, 6 Nov 74; GSFC *SSR*, 31 Aug 74; GSFC proj mgr, interview, 15 July 75; ESRO Release, 23 Aug 74)

*31 August:* A galaxy six times larger than any previously known in the universe had been discovered by an international team of astronomers using The Netherlands' Westerbork radio telescope, the *Washington Post* reported. Named 3C 268, the galaxy was 18 million light years in length and 1.8 billion light years from the earth. It was composed of an optical galaxy—a group of stars emitting visible light—at the center, with enormous gaseous radio components spreading like giant propeller blades on either side. (AP, *W Post*, 31 Aug 74, A13)

*During August:* Flight Research Center began wake vortex studies with a Boeing 747 to test different configurations and mechanical devices for suppressing the formation or reducing the strength of vortices. Wind-tunnel and water-channel tests had indicated that engine-induced turbulence and span-load tailoring by flap modification would permit reduction of standard separation distances between aircraft. Removing the outboard flap on a model of a Boeing 747 and adding a trailing spline reduced the wing-tip vortex so that the separation between a Boeing 747 and a following Learjet or DC-9 could be reduced from 8 to less than 3 km. During inflight tests, lowering the inboard flap 30° dissipated the normal wing-tip vortex as a second vortex flowing in a counter direction formed. Flight testing would continue into September. (FRC *X-Press*, 13, 27 Sept 74; NASA Release 74-230)

- The Air Force Space and Missile Systems Organization (SAMSO) awarded a $42 847 797 contract to Rockwell International Corp. Space Div. to design, develop, fabricate, launch, and operate three prototype satellites in the Navigation System Using Time and Ranging (NAVSTAR) Global Positioning System (GPS). With a Naval Research Laboratory NAVSTAR technology satellite, the prototypes would provide positioning signals to support development and test of user equipment for all military services. Plans called for refurbished Atlas F space boosters to launch the spacecraft into 20 200-km orbits in 1977. (ASFC *Newsreview*, Aug 74, 16)
- Federal research and development support of space programs had declined steadily in proportion to civil projects from 1969 to 1975, the National Science Foundation reported in its *Analysis of Federal R&D Funding*

*by Function, Fiscal Years 1969–1975.* Energy and environment programs had outpaced other major civil programs, while space programs had been reduced more than any other R&D function in the period. In FY 1975, the space budget was 13% of the planned $19.6-billion Federal R&D funding, a decline from 23.9% in 1969. Manned space flight, including the space shuttle, would account for more than half the 1975 total. (NSF Report 74-313)

- The combined calculated weight at launch of the U.S.S.R.'s *Soyuz 4* (launched 14 Jan. 1969) and *Soyuz 5* (launched 15 Jan. 1969) had indicated that the spacecraft had carried propulsion systems far in excess of requirements for an earth orbital mission, David Woods reported in *Spaceflight*. The 1.97-km-per-sec velocity requirement of such active docking vehicles as *Soyuz 4* had matched the total velocity requirement of 1.94 km per sec of *Apollo 11* (launched to the moon 16 July 1969), indicating that these early Soyuz spacecraft had been capable of manned lunar missions, Woods deduced. (*SF*, Aug 74, 300–302)

## September 1974

*1 September:* In an effort toward accurate earthquake prediction, antennas at Jet Propulsion Laboratory and Goldstone Tracking Station—200 km apart and on either side of the San Andreas Fault—were receiving radio signals from quasars outside the galaxy. As part of JPL's Astronomical Radio Interferometric Earth Surveying (ARIES) project, techniques developed for spacecraft navigation could measure the precise distance between the antennas by monitoring the arrival times of identical radio signals, accurate to one tenth of a billionth of a second. Changes in the times could indicate three-dimensional shifts in the earth's crust up to an initial accuracy of 10 cm, with an eventual accuracy goal of 2 cm. Scientists hoped that ARIES techniques would help prove the theory that earthquakes were preceded by a swelling of the earth's surface up to one meter over hundreds of square kilometers and provide a major new tool in earthquake research and prediction. (JPL Release 710; JPL proj off, interview, 3 Dec 75)

*1–13 September:* The Air Force SR–71 Blackbird aircraft sped from New York to London in a record 1 hr 56 min on 1 Sept., piloted by Capt. Harold B. Adams. Flown to England for display at the Farnborough International 74 air show through 8 Sept., the strategic reconnaissance aircraft built by Lockheed Aircraft Corp. could survey more than 260 000 sq km of the earth's surface an hour from above 24 400 m. On 13 Sept., the SR–71 set another record, flying from London to Los Angeles in 3 hrs 47 min. (DOD Release 410–74; AP, *W Post*, 14 Sept 74, A11)

*3 September:* First wind-tunnel tests of a third-generation X–24 lifting body had been completed by Arnold Engineering Development Center, the Air Force Systems Command announced. A model of the proposed X–24C equipped with phase-change paint was exposed to a mach 6 airflow in a wind tunnel. By photographing the progression of the melt line formed by the paint, engineers determined where heat built up and how quickly, around a new canopy designed for flights up to six times the speed of sound.

The X–24A lifting body had flown test flights from 1969 to 1971 and the X–24B was continuing flights from Edwards Air Force Base in a NASA–USAF research program into concepts for future hypersonic aircraft and returning spacecraft. The X–24C was similar to the B model with the exception of the new canopy and omission of one vertical stabilizer. (AFSC Release OIP 158.74)

- President Ford signed H.R. 11864, the Solar Heating and Cooling Demonstration Act of 1974, into P.L. 93–409 and signed H.R. 14920, the Geothermal Energy Research, Development, and Demonstration Act of 1974, into P.L. 93–410.

P.L. 93–409 provided for demonstration within three years of the practical uses of solar heating technology and for development and demonstration within five years of the practical use of combined heating and cooling technology. It amended the National Aeronautics and Space Act of 1958 to direct NASA to carry out solar heating and cooling research but provided that, within 60 days of a law creating the Energy Research and Development Administration or other energy R&D agency, functions given to NASA and the National Science Foundation could be transferred to the new agency.

P.L. 93–410 authorized guaranteed loans to finance commercial ventures in geothermal energy and provided for coordination of Federal activities in the development of geothermal energy. (*PD*, 9 Sept 74, 1099; texts)

*4–5 September:* Plans for the Laser Geodynamic Satellite (LAGEOS) were completed at a two-day Preliminary Design Review by officials of NASA and the Smithsonian Institution, at Marshall Space Flight Center. LAGEOS, to be launched in 1976, would provide information on earthquake-producing motions and strain in the surface of the earth. (MSFC Release 74–165; NASA Release 73–261; MSFC PIO, interview, 6 June 75)

*5 September:* The European Space Research Organization announced selection of Communications Satellite Corp. as its U.S. partner in the Aeronautical Satellite (Aerosat) program [see 9 May]. Aerosat was to deploy two satellites over the Atlantic Ocean for communications between aircraft in transatlantic flight and the ground. (ESRO Release, 5 Sept 74)

• Maj. Gen. Vladimir A. Shatalov, U.S.S.R. chief of cosmonaut training, stated that experiments conducted during the *Salyut 3* (launched 25 June), *Soyuz 14* (3 July), and *Soyuz 15* (26 Aug.) flights had no bearing on the Apollo-Soyuz Test Project's scheduled July 1975 flight. In a Tass interview before leaving Moscow for Johnson Space Center for joint training of ASTP crews, Shatalov announced that the U.S.S.R. would launch a manned spacecraft a few months before the Apollo-Soyuz launch, to test hardware in flight. (*Pravda*, FBIS–Sov, 11 Sept 74, U1–2)

*6 September:* Western Union Telegraph Co.'s *Westar 1* (launched 13 April) made America's first mail delivery via satellite, sending a "mailgram," combination telegram and letter, from New York to Los Angeles. A 60-word mailgram to the West Coast cost $1.35, rather than the $6.00 for a telegram. Regular, one-business-day transmission of mailgrams by satellite from New York to Los Angeles would begin in late September, with mailgram traffic from the East Coast to Texas and parts of the West Coast beginning late in the year. (NASA Hq *WB*, 7 Oct 74; AP, B *Sun*, 6 Sept 74, A3)

• President Ford signed H.R. 15572 into Public Law 93–414, making appropriations for Dept. of Housing and Urban Development, space, science, and veterans activities for FY 1975. The bill included a $3.206-billion NASA appropriation and $666.3 million for the National Science Foundation. [See 21 and 22 Aug.] (P.L. 93–414)

*7–27 September:* Apollo-Soyuz Test Project cosmonauts arrived in the U.S. 7 Sept. for training with American crews at Johnson Space Center 8–27 Sept. President Ford received the U.S. and U.S.S.R. crews at the White House before taking them to visit an Alexandria, Va., picnic. At the picnic the President said the planned July 1975 joint mission was not only a "technological achievement" but was "far broader in its implications and ramifications as far as the world is concerned."

The joint session at JSC included flight simulations, training in communications and docking procedures, and study of spacecraft systems.

(PD, 16 Sept 74, 1101–02, 1146; NASA OMSF PAO, interview, 21 Aug 75; JSC *Roundup*, 13 Sept 74, 1)

*8 September:* The Apollo command and service module for the Apollo-Soyuz Test Project, which had been delivered to Cape Canaveral Air Force Station by a C–5A transport, arrived at Kennedy Space Center. The spacecraft completed the ASTP flight hardware with exception of the docking module, which would arrive at KSC in late October after thermal and vacuum tests at Johnson Space Center. (KSC Release 132–74)

- President Ford as a Congressman had been a "staunch defender of space spending" whenever critics suggested cuts in NASA's budget, the *San Diego Union* reported. As minority leader, Ford had declared that the space program could be justified by its spinoffs. In April 1972, Ford had defended a $200-million space shuttle appropriation because he had believed that its loss would end the space program. (Macomber, *SD Union*, 8 Sept 74, 19)

*8–12 September:* Dr. James C. Fletcher, NASA Administrator, visited the U.S.S.R. at the invitation of the Soviet Academy of Sciences. Interviewed by a Tass correspondent, Dr. Fletcher compared implementing the Apollo-Soyuz Test Project with scaling a high mountain from which new vistas of Soviet-American scientific and technical cooperation would open.

With President of the Soviet Academy of Sciences Mstislav V. Keldysh, Dr. Fletcher visited the space flight control center near Moscow, from which the ASTP mission would be controlled in July 1975. Dr. Fletcher also visited the cosmonaut training center and the Apollo-Soyuz laboratory center with the head of the Intercosmos Council, Academician Boris N. Petrov, and discussed Soviet-American cooperation in exploration and peaceful uses of outer space with Deputy Chairman of the U.S.S.R. Council of Ministers Vladimir A. Kirillin. (NASA Off of Admin, Daily Appointments Calendar; Tass, FBIS–Sov, 12 Sept 74, U1; 13 Sept 74, U1)

*9 September:* NASA and the Maritime Administration demonstrated that satellites could be used for high-quality ship-to-shore communications in L-band frequencies. Test transmissions from the National Maritime Research Center in New York were relayed by NASA's *Ats 6* Applications Technology Satellite (launched 30 May) to a commercial container ship en route from Virginia to France. *Ats 6* could provide two-way communications between ground stations and mobile units operating in a 1080-km-wide area and would permit transmittal of cargo manifests and documentation before ship arrival in port, transmittal of payroll data for shore processing, and remote monitoring of ship machinery performance. (NASA Release 74–255; GSFC PIO, interview, 24 June 75)

- Johnson Space Center announced award of a 12-mo, $300 000 contract to Martin Marietta Corp. Denver Div. to study technical and operational concepts for manned and automated assembly of space systems from the space shuttle orbiter. Martin was asked to consider use of manipulative devices, an astronaut maneuvering unit, and simple docking systems. (JSC Release 74–251)

- NHK Technical Research Laboratory of Japan had developed a TV receiving system that could bring satellite-to-home transmission within five years, *Broadcasting* journal reported. With General Electric Co., NHK was constructing an experimental 12-ghz broadcast satellite designed to transmit signals within the FM band. An AM–FM converter attached to a standard receiver would transform the image into a standard TV signal.

The experimental satellite was scheduled for launch in early 1977. (*Broadcasting*, 9 Sept 74, 68)

*9–13 September:* A Lighter-than-Air Workshop sponsored by NASA, the Dept. of Transportation, the Federal Aviation Administration, and the U.S. Navy attracted 200 persons from nine nations to Monterey, Calif., to explore the technical and economic feasibility of LTA vehicles. The workshop was conducted by the Flight Transportation Laboratory of Massachusetts Institute of Technology. Among the speakers, Ames Research Center representatives argued that LTA vehicles would be economically competitive with aircraft on established freight routes and would be suitable for heavy-load low-level transportation. Nuclear power units, agricultural products, and ore could be transported to and from underdeveloped countries. (MIT, FTL Report R75–2; NASA Release 74–229; ARC PIO, interview, 24 June 75; O'Lone, *Av Wk*, 30 Sept 74, 42–46)

*9–14 September:* A review of "Technology in the Service of Man" at Lewis Research Center featured a ferrofluid separator for recycling scrap metals. Based on the magnetized fluid concept devised for taking showers in space, the recycling process could separate nonmagnetic solids of different densities—like copper, brass, and zinc—by magnetically controlling the apparent density of the ferrofluid. The method could also detect metals in waste materials at incinerators and retrieve titanium used in aircraft construction and weaponry.

The review, attended by 11 000 persons, included presentations on technology for safer aircraft, supersonic and hypersonic aircraft, energy-efficient flight, low-speed aircraft, quieter airport communities, protection of the marine environment, preservation of the earth's atmosphere, exploration of the planetary environment, and energy sources and use. (NASA Tech Brief B73–10463; LeRC PIO, interviews, 24 June, 24 July 75; Biggins, Newport News, Va, *Times-Herald*, 11 Sept 74; *NASA Activities*, Oct 74, 11–12)

*10 September:* *Pioneer 10* data returned from the 3 Dec. 1973 encounter with the planet Jupiter appeared to support the theory that Jupiter was a spinning ball of liquid hydrogen without a detectable solid surface, NASA reported. At most, the planet probably had only a small rocky core deep below the heavily clouded atmosphere. Jupiter's great red spot probably was the vortex of a gigantic storm that had raged along a 40 000-km front for 700 yrs.

Analyses by NASA experimenters and other scientists had shown that the *Pioneer 10* information, the first measurements made from closer than the earth, contradicted some previous theories and verified others. As early *Pioneer 10* reports had indicated [see 25 Jan., 2 and 9 April], Jupiter's turbulent interior was much hotter than expected, its magnetic field much larger than some predictions, and its radiation belts far more intense. The planet was a source of high-energy particle radiation, the only one in the solar system besides the sun. In its atmospheric circulation patterns, cyclones and anticyclones (ascending and descending) were stretched around the planet, rather than being circular as on the earth. This "weather-stretching," probably caused by heat radiation from the planet's interior and its 35 200-km-per-hr rotation speed, accounted for Jupiter's planet-girdling, alternating brown-red and gray-white cloud bands.

Density of the four largest moons decreased from the density of rock for the innermost Io and Europa to the density of a water-ice and rock mixture for the outer Ganymede and Callisto. Io had a tenuous atmosphere. Ganymede was known to have an atmosphere and the other two large moons probably had also.

Meanwhile *Pioneer 11*, the sister spacecraft launched 5 April 1973, was drawing toward Jupiter for its December flyby, before going on to Saturn. (NASA Release 74-238)

- Safety was of prime importance to NASA management and NASA-Soviet joint working groups planning the Apollo-Soyuz Test Project mission, Dr. James C. Fletcher, NASA Administrator, said in a letter to Sen. William Proxmire (D-Wis.). The response to a Sen. Proxmire letter expressing concern for ASTP crew safety in light of past failures of the Soyuz and Salyut spacecraft explained that NASA had requested and received from the Soviets a detailed explanation of the *Soyuz 11* (6-30 June 1971) failure, in which three cosmonauts were killed, and of the test program to correct the problem. NASA officials had also asked for information on the 26-29 Aug. *Soyuz 15* mission, which ASTP technical personnel were expected to discuss during joint meetings 9-23 Sept. (Letter, Proxmire to Fletcher, 3 Sept 74; letter, Fletcher to Proxmire, 10 Sept 74)
- The Air Force Systems Command's Arnold Engineering Development Center (AEDC) announced that it had refined computerized wind-tunnel techniques for predicting aircraft loss-of-control characteristics. AEDC correlated wind-tunnel data from a model F-15 aircraft mounted on a computer-controlled support with NASA drop-test data of a remotely controlled model. The computer predicted model response, such as stalls, to forces acting on it; activated the high-pitch rig to place the model in the appropriate position; and checked the prediction. This use of the computerized mounting system was the first without a human in the control cycle. (ASFC Release OIP 174.74)

*10-12 September:* California Institute of Technology scientist Charles T. Kowal discovered a possible 13th moon of Jupiter. In photos taken with the Hale Observatories' 122-cm Schmidt telescope at Mount Palomar, Calif., Kowal detected a 10- to 16-km moon as a pinpoint of reflected light 11 million km from the planet.

Photos taken 22 Sept. from the Univ. of Arizona's Steward Observatory and again 15 and 16 Oct. from Mount Palomar supported Kowal's findings. Dr. Kaare Aksnes of the Smithsonian Astrophysical Observatory determined the object belonged to the intermediate group of moons orbiting Jupiter at a distance of about 11 million km. (Alexander, *LA Times*, 20 Sept 74; Sullivan, *NYT*, 23 Oct 74, 19)

*11 September:* A declassified feasibility study by Goodyear Aerospace Corp. had indicated that reflector satellites 200-900 m in diameter could illuminate ground areas continuously at night, the *Huntsville Times* reported. Pointing out commercial uses, an official from Marshall Space Flight Center, which had managed the Project Able study, told the Times that potential applications included reflecting microwaves from space for conversion to electrical power, beaming rays on Florida's orange groves should frost threaten the fruit crop, and lighting Alaska during its extended darkness. Launched on a Saturn V booster or on the space shuttle, a 640-m-dia, 20 000-kg reflector satellite could provide twice the light of the full moon and raise the temperature of the area covered by several degrees. (Casebolt, *Huntsville Times*, 11 Sept 74)

- A new artificial limb had been developed by Kennedy Space Center and Rancho Los Amigos Hospital in California, KSC announced. The strapless device was attached directly to the amputee's bone through a carbon collar that remained at skin level. Originally developed for rocket motors, the high-purity carbon material would not cause infection in patients. Previously, the body's rejection of foreign materials and the failure of the skin to form a hygienic seal had prevented direct attachment to the bone. A quick-release ball connector for the limb was based on a device that held rocket umbilical attachments in place until liftoff. (KSC Release 134–74)

- Pages of the *Wall Street Journal* were being printed via satellite in a joint test program, Dow Jones & Co., Inc., and Communications Satellite Corp. announced. A Dow Jones composition plant in Massachusetts transmitted high-resolution facsimile pages to *Intelsat-IV F–7* communications satellite in synchronous orbit 35 900 km above the Atlantic Ocean. The satellite relayed the data to the Dow Jones production facility in New Jersey for production of press plates. The test program was the first time the entire process from composition to printing had been conducted via satellite transmission. (Dow Jones-ComSatCorp Joint Release 74–45; ComSatCorp PIO, interview, 1 Aug 75)
- Trace materials in Cleveland's airborne particulate matter were not present in dangerous levels, but lead might approach these levels, scientists from Lewis Research Center reported at the Earth Environment and Resources Conference in Philadelphia. With the City of Cleveland's Air Pollution Control Division, LeRC had conducted a two-year study using neutron activation and gas chromatography to determine levels of 60 trace materials. Processed by computer, levels of trace materials were related to wind direction in the form of maps. Sources of pollution could then be established by knowing which industry produced which elements and compounds and in what ratios. (LeRC Release 74–54)
- Lockheed-California Co. had received a $25-million letter contract from Rockwell International Corp. for structural testing of the space shuttle orbiter, the *Huntsville Times* reported. (*Huntsville Times*, 12 Sept 74)

*13 September:* Albert P. Loening, developer and manufacturer of the first successful amphibian plane—the Loening amphibian—in 1923, died at the age of 88. He had been a founding stockholder and director of Grumman Aircraft Engineering Corp., now Grumman Corp. (NYT, 15 Sept 74, 59)

*14 September:* A Federal grand jury had charged Charles R. Riser of Maryland with mail fraud in setting up Andromeda Space Covers, a mail-order firm selling fake space exploration items, the Baltimore *Sun* reported. Fifteen persons had paid for golf balls driven across the moon by Astronaut Alan B. Shepard, Jr., during the January 1971 *Apollo 14* mission, but he had used only two balls and these were still on the moon. The assistant U.S. attorney on the case said that astronauts might be called to state whether signatures on other souvenirs were authentic. (Erlandson, B *Sun*, 14 Sept 74, B20)

*15 September:* Five embryonic stars in the nebula of Orion had been discovered by astronomers using infrared detectors attached to the Hale Observatories' 254-cm telescope on Mount Wilson, the *Chicago Sun-Times* announced. California Institute of Technology and Hale scientists had reported that the stars, probably less than 10 000 yrs old, were about 1600 light years from earth. (UPI, *Chicago Sun-Times*, 15 Sept 74)

- Space engineers in India were testing subsystems for their first satellite launch vehicle, in preparation for eventual launch of a 40-kg satellite into near-earth orbit. All instruments and payload had been developed at Vikram Sarabhai space center. (Delhi Overseas Service, FBIS—India, 17 Sept 74, U4)

*15–22 September:* Arab League telecommunications representatives voted to establish a satellite communications system, at a meeting in Beirut, Lebanon. Three proposals were under consideration: Hawker Siddeley Group Ltd. had proposed a spacecraft similar to the Orbiting Test Satellite (OTS) being developed by the European Space Research Organization. Messerschmitt-Boelkow-Blohm had recommended a three-axis stabilized spacecraft. A consortium of Hughes Aircraft Co., Nippon Electric Co. of Japan, and Thomson-CSF of France had proposed a spacecraft similar to Telesat Canada's spin-stabilized 12-transponder Anik satellites.

Meanwhile a number of nations also were moving toward domestic satellite systems. Algeria had begun installation of the first of 14 earth terminals for its domestic satellite system following award of a $9.6-million contract to GTE International, a subsidiary of General Telephone & Electronics Corp. The system would begin operations soon after 1 Jan. 1975, when Algeria would lease one transponder from the International Telecommunications Satellite Organization. Brazil would begin leasing an Intelsat transponder for its domestic service 1 June 1975. Indonesia had begun development of two satellites of its own; initial terminals in a 25-station system would be easily convertible from operation with an Intelsat satellite to later operation with an Indonesian comsat. Malaysia was purchasing three ground terminals to use Intelsat comsats. The Phillippine government was considering a similar plan and, with Malaysia, possibility of switching from an Intelsat to Indonesia's planned satellite.

Saudi Arabia had installed two terminals with 11-m-dia antennas for domestic service, which, with the completion of two larger stations in mid-1975, would be moved to remote locations for use with an Intelsat or the planned Arab League satellite. (*Av Wk*, 23 Sept 74, 23; Johnsen, *Av Wk*, 30 Sept 74, 22; Hughes contract officer, interview, 27 June 75)

*16 September:* Two scientists had predicted that in 1982 a rare alignment of all nine planets on the same side of the sun would produce major earthquakes, *Newsweek* reported. In their book *The Jupiter Effect,* John Gribbin of the British magazine *Nature* and Stephen Plagemann of Goddard Space Flight Center warned that increased solar activity caused by the alignment, which occurred every 179 yrs, would alter wind directions in the upper atmosphere. The frictional effect exerted on the solid earth by atmospheric circulation would change with the winds. The consequent braking of the earth's rotation rate would trigger earthquakes in areas, like California, under geological stress. (*Newsweek*, 16 Sept 74, 57)

- Dr. John E. Naugle, NASA Deputy Associate Administrator, would receive the National Civil Service League's Career Service Award for Sustained Excellence, NASA announced. Dr. Naugle had joined NASA in 1959 as head of the Nuclear Emulsion Section at Goddard Space Flight Center and, in 1967, had become Associate Administrator for the Office of Space Science at Headquarters. Under his leadership, scientists had mapped the entire surface of Mars, collected the first detailed information about Jupiter and its moons, and advanced the use of unmanned spacecraft as astronomical observatories. Dr. Naugle would receive the award, given to Federal officials with at least 10 yrs outstanding public service, on 9 Oct.

Dr. Carolyn Huntoon, Johnson Space Center, would receive the League's Special Achievement Award for her design of a Skylab experiment that had measured biochemical reactions to weightlessness. (NASA Release 74-252)

- L/G Duward L. Crow (USAF, Ret.) became NASA Assistant Administrator for DOD and Interagency Affairs and Special Assistant to Dr. James C. Fletcher, NASA Administrator. Succeding Gen. Bruce K. Hollc way (USAF, Ret.), who had retired from the agency, Gen. Crow had been Assistant Vice Chief of Staff for the Air Force before coming to NASA. (NASA Release 74-258)

*17 September:* The channeled scabland of eastern Washington had been formed 20 000 yrs ago by the greatest flood known to have occurred, the *New York Times* quoted U.S. Geological Survey as reporting. From ground scars and photographs from *Erts 1* Earth Resources Technology Satellite (launched 23 July 1972), scientists had traced the flood for 885 km. An ice dam blocking drainage of western Montana into Idaho had collapsed, pouring 2000 cubic km of water down the valley of the Spokane River. Scars visible in satellite photographs were formed by erosion of surface soil and underlying basalt. As water drained, gravel ridges 6 to 9 m high and 3 km long had been formed. The largest channel was the Grand Coulee, 80 km long and 270 m deep. (Sullivan, *NYT*, 17 Sept. 74; USGS PIO, interview, 12 June 75)

- NASA authorized two flight kits for Apollo-Soyuz astronauts: an official flight kit, containing selective items for memorabilia, and an astronaut preference kit, containing 20 personal items selected individually. Special flight items having particular significance for the mission would be suggested by the Director, Johnson Space Center, and approved by the Administrator. (NASA NMI 8020.19B)

- The congressional Joint Committee on Atomic Energy issued the panel report *Transportation of Radioactive Material by Passenger Aircraft.* The panel recommended that the transport index (the level of radiation in millirem per hour one meter from the surface of packages of radioactive material) be reduced from 10 to 1 and that the maximum permissible radiation at the package surface be lowered by one fourth to one half. Total maximum permissible quantities of curies of radionuclides per package should be at least 10 times lower. Improved packaging, installation of high-level radiation monitors in baggage compartments, and Federal monitoring of packages in transit from supplier to loading were also recommended. (Joint Com Print)

*17–18 September:* The second in a series of balloon flights carrying development instruments for NASA's planned High Energy Astronomy Observatory satellites reached an altitude of 38 000 m, descending 56 hrs after launch from Sioux City, Iowa. The balloon carried a French-Danish experiment to study isotopic composition of primary cosmic rays. The first flight, carrying the same instrument from Sioux City 27 Aug., had reached 31 000 m and descended 160 km west of the launch site after 12 hrs aloft. A third would be launched from Palestine, Tex., 17 Dec., carrying a Univ. of San Diego x-ray experiment; a fourth was scheduled for spring 1975. The flights were being coordinated by Marshall Space Flight Center.

Three HEAO satellites were scheduled for launch into earth orbit 1977–1979 to study celestial x-rays, gamma rays, and cosmic rays. (NASA Release 74-271; MSFC proj off, interview, 12 Nov 75)

*17–20 September:* Telephone communications via satellite to remote areas of Mexico were demonstrated during the International Congress of Electrical and Electronic Communications. *Westar 1* (launched 13 April) and a portable antenna-earth station developed by Hughes Aircraft Co. relayed signals from Isla de Cedros through *Westar 1* to Western Union Telegraph Co.'s earth station in Texas; signals were then transmitted by terrestrial facilities to Mexico City. The experiment showed satellites could span terrain where conventional communications were difficult to construct and uneconomical to operate. (Western Union Release, 17 Sept 74; Western Union PIO, interview, 6 June 75)

*18 September:* Dr. Kurt H. Debus, Director of Kennedy Space Center since its establishment in 1962, announced his retirement. Born in Germany, Dr. Debus had begun his career in U.S. military and NASA space programs in 1945, when with Dr. Wernher von Braun he had joined the Army's ballistic missile program. He had supervised the launch of the first Redstone ballistic missile in 1953 and the flight-testing of Jupiter and Pershing missiles and Jupiter C and Juno launch vehicles until the Army team was transferred to NASA in 1960. He had directed the launch of the first U.S. satellite, *Explorer 1*, on 31 Jan. 1958, and NASA launch programs from the flight of Alan B. Shepard, Jr., in 1961 to preparations for space shuttle operations. Dr. Debus had received the Louis W. Hill Space Transportation Award from the American Institute of Aeronautics and Astronautics as well as the Army's highest civilian decoration, the Exceptional Civilian Service Medal, and NASA's Distinguished Service Medal. (KSC Release 135–74; Debus letter to KSC employees)

- The Senate Aeronautical and Space Sciences Committee continued hearings on S. 2350 and S. 3484, bills to develop an operational Earth Resources Technology Satellite (ERTS) system and to establish an Office of Earth Resources Survey Systems within NASA and an Earth Resources Observation Administration within the Dept. of the Interior [see 6, 8, 9 Aug.]. Frank G. Zarb, Office of Management and Budget Associate Director for Material Resources, Energy and Science, warned against immediate development of an operational ERTS system, citing budgeting restraints and marginal value of current ERTS data. Since an advanced satellite offering higher spatial resolution would increase the usefulness of ERTS data, Zarb contended that the third satellite, ERTS–C, should be experimental. His view was supported by S. Benedict Levin, Executive Vice President of Earth Satellite Corp., and by John V. Granger, Acting Director of the Bureau of International Scientific and Technological Affairs for the State Dept. (Transcript)

- Soviet astronomers had detected a black hole several light years away in the center of the globular cluster Omega in the constellation Centaurus, Tass reported. Using the interaction effect between gravitation fields of globular clusters and stars, Soviet astronomer Kiriol Ogorodnikov and his assistant Yelena Naumova of the Univ. of Leningrad had determined the value of the mass of the cluster Omega Centauri to be one trillion times that of the sun. Since this value far exceeded the admissible density of all luminaries in the cluster, the scientists concluded that a black hole existed within the cluster. (Tass, FBIS–Sov, 19 Sept. 74, U1)

- West German scientist Gisbert Winnewisser announced discovery of the organic molecule vinyl cyanide in outer space, a possible link to the chemistry of living organisms. The discovery had been made with Australian Frank Gardner, using the Australian Commonwealth Science and

Industrial Research Organization (CSIRO) telescope. A member of the Max Planck Institute of Radioastronomy, Winnewisser announced the find at a meeting of astronomers in Wurzburg, West Germany. (Agence France Presse, *W Post*, 20 Sept. 74; Turner, Nat'l Radio Astron Obs, W Va, interview, 6 Aug 75)

- Grigory Khozin of the Soviet Academy of Sciences' Institute of U.S. Studies expressed resentment of criticism by some U.S. press members and others of Soviet motives and technical ability in the cooperative Apollo-Soyuz Test Project. Writing in the weekly *Literaturnaya Gazeta*, Khozin particularly censured suggestions of early Soviet reluctance to cooperate in space and of Soviet spacecraft docking difficulties, made in the recent book *U.S.-Soviet Cooperation in Space* by former U.S. Ambassador to the U.S.S.R., Foy D. Kohler. He accused "certain U.S. circles" of being "glad to cast a shadow" on ASTP and to "smear the forthcoming joint experiment precisely because it is a joint one [and] promotes the establishment of normal relations between countries." Khozin declared that the U.S.S.R., as well as U.S. space program leaders and "ordinary Americans," favored further close cooperation in space. (FBIS–Sov, Sept 74, B2–3; Smith, *NYT*, 19 Sept 74)

*18 September–21 October, 24 November–14 December:* Two sets of critical surface-simulation tests were successfully conducted on the Viking lander proof-flight vehicle in a Martin Marietta Corp. thermal vacuum chamber under Martian temperature and pressure conditions. The tests verified the ability of Viking systems to withstand Martian temperatures and verified lander operations and ground support systems from the acquisition of a simulated Martian soil sample to analysis, interpretation, and reporting of data by Viking science teams. The spacecraft performed tasks—using the surface sampler, imaging system, meteorological sensors, x-ray spectrometer, seismometer, and thermal control systems—in temperatures ranging from 178 to 308 K ($-135°$ to 95°F). (NASA prog off, interview, 21 Oct 75; *Av Wk*, 4 Nov 74, 44–45)

*19 September:* The U.S.S.R. launched eight Cosmos satellites on a single booster from Plesetsk, in its second eight-satellite launch of the year [see 23 April]. Orbital parameters were:

*Cosmos 677*—1466-km apogee, 1398-km perigee, 114.5-min period, and 74.0° inclination.
*Cosmos 678*—1533-km apogee, 1466-km perigee, 116.0-min period, 74.0° inclination.
*Cosmos 679*—1511-km apogee, 1466-km perigee, 115.8-min period, 74.0° inclination.
*Cosmos 680*—1492-km apogee, 1466-km perigee, 115.5-min period, 74.0° inclination.
*Cosmos 681*—1472-km apogee, 1466-km perigee, 115.3-min period, 74.0° inclination.
*Cosmos 682*—1466-km apogee, 1453-km perigee, 115.1-min period, 74.0° inclination.
*Cosmos 683*—1467-km apogee, 1434-km perigee, 114.9-min period, 74.0° inclination.
*Cosmos 684*—1467-km apogee, 1416-km perigee, 114.7-min period, 74.0° inclination.

The U.S. press speculated that the satellites were part of an operational system for worldwide naval communications. (GSFC *Wkly SSR*,

19–25 Sept 74; Tass, FBIS–Sov, 20 Sept 74, U1; *SBD*, 23 Sept 74, 101; 21 Oct 74, 52)
- Dr. Richard T. Whitcomb, Head of Langley Research Center's Transonic Aerodynamics Branch, had been selected to receive the 1974 Wright Brothers Memorial Trophy for development of the area-rule concept of aircraft design and invention of the supercritical wing, the National Aeronautic Assn. announced. Dr. Whitcomb would receive the award 13 Dec. (NAA Release, 19 Sept 74)
- A high-temperature metal alloy and a self-lubricating bearing material, both developed at Lewis Research Center, had been chosen by *Industrial Research* magazine as 2 of the 100 most significant products of the year, NASA announced. Composed principally of tungsten, nickel, and aluminum, the alloy was three times stronger at 1480 K (2200°F) than any existing commercial cast-nickel-base alloy. The bearing material—a product of glass, metal, and calcium fluoride—functioned at temperatures up to 1170 K (1650°F). Both products would be used for more efficient and cleaner turbine engines, electric-power generating systems, and other high-temperature devices. (LERC Release 74–51; *Industrial Research*, Oct 74)

*20–23 September:* *Mariner 10* (launched 3 Nov. 1973) encountered the planet Mercury a second time, with its closest approach on 21 Sept. at 48 069 km. The objective of the second flyby was to extend photographic coverage of the sun side of Mercury. The spacecraft obtained some 500 TV science photos of Mercury, with 350 additional TV pictures to support an optical navigation experiment. Encounter TV pictures were transmitted in real time at the rate of 117.6 kilobits per second. New photo mosaics would correlate incoming and outgoing portions of Mercury photographed during the first encounter (25 March–5 April) and provide additional views of the south polar area. In the two encounters, *Mariner 10* had photographed 45% of Mercury's surface. Ultraviolet airglow scans were also made to obtain more information about Mercury's atmosphere.

The spacecraft did not pass through either sun or earth occultation, and communications with the Deep Space Network stations were uninterrupted. The second flyby was adjudged successful 3 Oct. A third flyby was planned for March 1975.

Preliminary results, presented at a 23 Sept. press briefing at Jet Propulsion Laboratory, indicated that scarps and evidence of volcanism observed on the first flyby were pervasive and not localized. The scabby hills detected on the first encounter in an area opposite the Caloris Basin did not reappear, suggesting that such terrain was peculiar to that site. (NASA MOR, 3 Oct 74; NASA Release 74–248; NASA proj off, interview, 30 June 75; Sullivan, *NYT*, 24 Sept 74)

*21 September:* The prototype West German, British, and Italian multirole combat aircraft (MRCA) made its first public flight, lifting off from Manching, West Germany. The aircraft—product of three nations' combining technological and avionic resources to develop one model to fulfill six combat roles, from low-level attack to high-altitude air superiority—demonstrated its variable-geometry wing in the short flight. Fully extended, the wing made short takeoff and landing possible. Moved back at a 25° angle to the fuselage, it provided top speed of more than mach 2. Final decision on whether to produce the MRCA would follow 12 mos of testing. (Morris, *LA Times*, 22 Sept 74, 1A)

*22 September:* The North American Air Defense Command had predicted more than 10 000 pieces of space clutter would be circling the earth by the year 2000, the *Los Angeles Times* reported. More than 3200 man-made objects were already orbiting the earth, with one object falling back to earth daily. More than 77% of the total had resulted from explosions of satellites already in orbit, NORAD said. Space exploration by countries other than the U.S. contributed to increasing space clutter. (UPI, *LA Times*, 22 Sept 74)

*23 September:* A four-engine DC-8 transport using a NASA-developed two-segment approach had reduced the area impacted by noise above desirable tolerances by 53% in a normal landing approach, NASA announced. Being evaluated in service by United Air Lines, the aircraft was the second in regular airline service to use a steeper glide path, inclined 5.5° to the ground instead of 3°. When the aircraft intersected the normal flight path, the glide path angle was decreased, with the landing made as usual. Both higher altitudes of the glide path and reduced engine power setting contributed to noise reduction. (NASA Release 74-226)

- Japan Rocket Development Assn., an industry group, had proposed a three-phase, $9-billion Japanese program to launch 65 satellites between 1978 and 1987, *Aviation Week & Space Technology* reported. Nineteen satellites, each weighing under 300 kg, would be launched in the $3.3-billion Phase 1 for engineering, meteorological, geographical survey, communications, navigation, and ionospheric research missions. Of 25 spacecraft in the $2.8-billion Phase 2, some would have earth resources survey missions. Weighing up to 600 kg, they would be launched by N-2 rockets incorporating McDonnell Douglas Delta 1st stages and two 4500-kg, Japanese-built upper stages. The first prototype N-2 was scheduled for summer 1975. The 21 Phase 3 satellites, some weighing 1000 kg, would be orbited at a cost of $2.6 billion, and many would be launched by a Japanese O-model launch vehicle still being designed. The program also called for launch trials for Japanese M-X boosters. (*Av Wk*, 23 Sept 74, 21)

- Personnel from the National Oceanic and Atmospheric Administration, operator of the solar observatory at Johnson Space Center since 1970, would move to Marshall Space Flight Center, JSC announced. NOAA would continue at MSFC as co-investigator in an experiment conducted with the Apollo Telescope Mount on the Skylab Workshop during its 1973-1974 mission. The experiment had gathered solar radiation data in the x-ray region of the solar spectrum. The JSC-NOAA solar observatory would reopen for space shuttle operations. (JSC Release 74-253)

- A mainshaft seal for turbine engines was being developed at Lewis Research Center to reduce wear, maintenance, and fuel consumption. By creating a thin gas film, the self-acting lift-pad seal eliminated contact between the seal bore and the rotating shaft shoulder except at startup and shutdown. The seal functioned at higher pressures, speeds, and temperatures than conventional contact seals and, with its low wear and leak rate, would be used in the NASA space shuttle liquid-oxygen pump. (LeRC Release 74-55)

- Westinghouse Corp. subsidiary TCom Inc. (for Tethered Communications) was testing the aerostat, an inexpensive alternative to earth-orbiting satellites, *Time* reported. Developed from signal-relay balloons used for reaching isolated U.S. outposts in Vietnam, the aerostat was a helium-filled balloon anchored at altitudes of 3000–5000 m. In the Bahamas, an aerostat at a 3500-m altitude picked up signals from Miami 180 km away, trebling the range of Florida stations. With its 1800 kg of electronic equipment, the aerostat could receive and broadcast up to 4 TV channels, 2 command radio stations, and data from 5000 to 10 000 microwave circuits. (*Time*, 23 Sept 74)

- President Ford outlined "Project Interdependence," a global approach to the energy problem, in his address to the ninth annual World Energy Conference in Detroit. The President proposed that all nations increase production, reduce consumption, and eliminate waste of energy; that a cooperative spirit exist; that poor nations receive special attention; and that fuel prices reach a level providing strong incentives to producers without disrupting consumer economics.

  Describing U.S. mobilization to achieve long-term goals, the President said the energy program would require "a commitment in excess of the successful one made by John F. Kennedy to put a man on the Moon in the last decade." (*PD*, 30 Sept 74, 1181–6)

- The Global Atmospheric Research Program Atlantic Tropical Experiment (GATE) ended its operations off the coast of Senegal [see 15 June–23 Sept.]. (NOAA Release 74–154)

*24 September:* Flight Research Center's highly instrumented F–111 aircraft, to be used to test the integrated propulsion control system (IPCS), made its first baseline flight to gather data on the engine's inlets and exhaust controls. Data would be used by NASA in support of the program and by Boeing Co., which would build the system. (FRC *X–Press*, 27 Sept 74, 2; FRC PIO, interview, 17 June 75)

- NASA had designated Marshall Space Flight Center as lead Center for NASA activities under the Solar Heating and Cooling Demonstration Act signed into law 3 Sept., MSFC announced. The Center would work under the direction of the NASA Hq. Office of Energy Programs. (MSFC Release 74–175)

*25 September:* A helical rotary-screw expander that could be used to tap geothermal energy of hot aquifers under deserts for low-cost energy was described at a National Science Foundation conference at California Institute of Technology. The invention of Pasadena engineer Roger S. Sprankle was the first electrical power-generating system to use the entire amount of energy available in geothermally heated water and might reduce production costs 50%. Tested by Jet Propulsion Laboratory and Hydrothermal Power Co., the Sprankle system's rotary screws filtered impurities to make maximum use of the steady flow of pumped water. Once the geothermally heated water flowed through the screw expanders, the liquid dropped in pressure and temperature as water vaporized. The increasing mass flow of vapor turned rotors with an output shaft linked to an electric generator. (Caltech-JPL Release, 25 Sept 74)

*26 September:* The U.S.S.R. announced the mission of the *Salyut 3* space station (launched 25 June) had been "completely fulfilled." The "main program of work" had ended 23 Sept. and the recoverable module containing research materials and experiments had separated from *Salyut 3*. Engines were discarded before entry into the atmosphere, and the parachute system was activated at an altitude of 8.4 km to land the module in a preset area of the U.S.S.R. *Salyut 3* plans had been for a

90-day mission but the station would continue in orbit for more experimentation.

Equipment tested during the 90-day flight included power supply systems with revolving panels of solar batteries, an electromechanical stabilizing system, a high-precision control system, an autonomous navigation system, thermoregulating life-support and radio communication systems, and engine systems.

A two-man crew aboard *Soyuz 14* (launched 3 July) had successfully docked with the orbiting station 5 July for 15 days of scientific experimentation. However, the press reported Western observers doubted that *Salyut 3* had accomplished its entire mission, since the *Soyuz 15* cosmonauts (launched 26 Aug.) had failed to dock with the station. (Tass FBIS–Sov, 27 Sept 74, U1–2; *W Post*, 27 Sept 74, A22; *SBD*, 27 Sept 74, 132)

- The House Committee on Science and Astronautics' Subcommittee on Aeronautics and Space Technology, reporting on almost a year's investigation of NASA's tracking and data acquisition program, said the program was effective and efficient and that planned consolidation was progressing well. The consolidation, to be completed by 1976 with a reduction of stations from 25 to 15, would combine the Manned Space Flight Network and the Space Tracking and Data Acquisition Network into one Spaceflight Tracking and Data Network (STDN). The Subcommittee recommended that the FY 1976 NASA Authorization request include a detailed analysis of the need for an additional subnet of three 64-m antennas for the Deep Space Network and report on the status of relations and agreements with the U.S.S.R. The Subcommittee warned that NASA's T&DA capabilities were not designed for the workload that future operational systems would require; solutions should be studied by the President's Science Adviser. NASA was asked to furnish an analysis of lease versus purchase of satellites for the Tracking and Data Relay Satellite System (TDRSS) in its FY 1975 authorization request. (Com Print)

*27 September:* *Ats 6* Applications Technology Satellite (launched 30 May) and *Sms 1* Synchronous Meteorological Satellite (launched 17 May) had caused substantial interference for radio telescopes in the U.S., Canada, and Great Britain, the *Washington Post* reported. Stars in the southwest sky with radio frequencies near those of the satellites could not be studied. Dr. Frank Kerr, spokesman for the National Academy of Sciences Committee on Radio Frequencies, described failed communications between NASA and radio astronomers. The astronomers had learned that *Ats 6* would interfere with study of the sun, pulsars, and quasars only when it was too late to change the satellite's transmitter and that *Sms 1* would overlap radio telescope signals only after the satellite had begun sending back weather pictures 17 May. (O'Toole, *W Post*, 27 Sept 74)

- DeElroy E. Beeler, Flight Research Center Director for Center Development, left NASA to join private industry after 28 yrs of service. Beeler had joined the National Advisory Committee for Aeronautics in 1941, specializing in high-speed aeronautics. He was appointed project engineer for the X-1, the world's first aircraft to reach supersonic speeds, and transferred in 1946 to what was to become FRC. There he was instrumental in research planning, design, and flight investigations of advanced research aircraft projects through X-15 and XB-70. In 1954

Beeler was named FRC's first Director of Research and in 1958 he became Deputy Director, a position he held until 1973. (FRC *X-Press,* 27 Sept 74, 2; NASA biog, 25 July 63)

- Facilities at Marshall Space Flight Center were being modified to consolidate Concept Verification Test (CTV) simulation in a central location. The CTV project would determine by testing in simulators whether concepts generated during definition studies for space activities, especially Spacelab experiments, were valid, before commitments to proceed with final design and hardware fabrication. (MSFC Release 74–178)

*29 September:* A robot with an artificial intelligence for making independent decisions in environments hostile to man was being developed at Jet Propulsion Laboratory and California Institute of Technology to prove the feasibility of using robots for planetary exploration. The planned rover would be equipped with metal arms and hands, a visual system of two TV cameras and a laser, wheels for legs, and thousands of instructions programmed into its computer brain. It would be able to analyze a scene, extract information from it, make some choices, and move around obstacles. (NASA Release 74–260)

*30 September:* NASA's *Explorer 45* (SSS-A Small Scientific Satellite launched 15 Nov. 1971) was turned off. The spacecraft had been investigating the causes of worldwide magnetic disturbances associated with large solar flares. Major accomplishments of *Explorer 45* included the first measurement by satellite of man-made stimulated electromagnetic emissions in the magnetosphere; first measurements of alpha particles in the equatorial region of the magnetosphere, where they were 100 times more numerous than expected; first detailed analysis of the interaction of electromagnetic waves with charged particles; measurement of the earth's ring current, which produced the magnetic storm main phase; and measurement of the closest approach to the earth of the boundary of the magnetosphere at about 25 000-km altitude. (*Goddard News,* Nov 74)

*30 September–4 October:* The fourth simulated Spacelab mission of the airborne science–shuttle experiment system simulation (ASSESS) program was conducted at Marshall Space Flight Center using an Ames Research Center Learjet. The Lear 4 mission was the first in which substitute experiment operators were trained as crewmen. Their performance would be compared with that of the principal investigator and his team during previous missions. During 10 nighttime flights into the base of the stratosphere, the crew measured infrared radiation at selected wavelengths from targets believed to be forming new stars. (MSFC Release 74–185)

*30 September–5 October:* A seminar to predict the probable state of the world in the year 2000 was held by the Smithsonian Institution for NASA as part of NASA's "Outlook for Space" study begun during June. Experts from various fields discussed a new age-oriented global culture, the probability of widespread famine, the prolonged disruption of technological economies by high oil prices, and the spread of nuclear power beyond control of the great powers. Seminar participants concluded that the intensification of nationalism would continue, particularly in response to economic stresses. Speakers from the high-technology sector, however, stressed that the U.S. would not have to suffer from scarcity of resources, although resource extraction would be more expensive and would require new technology.

Satellite reconnaissance to find mineral resources and to monitor threats to the global environment would still be necessary. The group also proposed surveillance to track all asteroids crossing the earth's orbit so that the impact of an asteroid might not be mistaken for a nuclear attack. Communications satellites were considered both useful for uniting countries and dangerous in their potential to increase threats of demagoguery. Many participants agreed that the time for uncontrolled free enterprise was passing and that we would require independent agencies for both planning and technology assessment, exempt from the special interests of particular government departments. Specialists also proposed that NASA's planning and analytical abilities be applied to national concerns in their full complexity, taking into account world economic, political, and social problems. (Sullivan, *NYT*, 8 Oct 74, 16; GSFC proj off, interview, 13 Aug 75)

- The International Aeronautical Federation held its 25th Congress in Amsterdam with the theme "Space Stations, Present and Future." Among speakers, Astronaut Thomas P. Stafford and Cosmonaut Aleksey A. Leonov, U.S. and Soviet commanders in the Apollo-Soyuz Test Project, stressed how well joint work and understanding had advanced. A paper by Daniel J. Shramo of Lewis Research Center, with B. R. Foushee and P. J. O'Leary of General Dynamics Corp. Convair Div., described the Centaur primary upper stage for solar system exploration and its potential adaptation as an upper stage for the space shuttle.

  A Ukranian Academy of Sciences research team reported on experiments in the formation of melts in weightlessness with radiant heating and on prospects of using solar energy for space engineering. Comparative studies of metal solidification had suggested that a more uniform distribution of short-grained structures under weightlessness might become the basis of space technology, with the added attraction of using the sun as an energy source. The team had built a prototype solar plant that could be mounted on a spacecraft or used in a vacuum chamber on the earth.

  Soviet medical investigations during the flights of *Soyuz 12* (launched 27 Sept. 1973), *Soyuz 13* (18 Dec. 1973), and *Soyuz 14* (3 July) were reported by Dr. Oleg G. Gazenko and others. Findings indicated that prolonged exercises, full and partial spacesuits, and gravity-simulating fittings had reduced effects of weightlessness during the space flight and after the crew's return. Although reports of initial adaptation symptoms were similar to American findings, the Soviet paper also noted that cosmonauts had reduced vestibular symptoms by deliberately slowing down movements, especially head movements.

  Prof. Hilding A. Bjurstedt, Head of the Dept. of Aviation Medicine at the Karolinska Institute in Stockholm, received the 1974 Daniel and Florence Guggenheim International Astronautics Award of the International Academy of Astronautics for studies of physiological responses to environmental stress.

  Leonard Jaffe, NASA Deputy Associate Administrator for Applications, was elected IAF President. (IAA Ann Rpt, 28 Aug 74; Parks, *SF*, Feb 75, 42–50; Program, 25th IAF Congress; *NASA Activities*, 15 Dec 74, 21)

*During September:* Large-scale verification tests of the "rectenna" receiving system for microwave energy transmission were run at Jet Propulsion Laboratory for NASA. A microwave energy transmission system was being developed to determine the possibility of transmitting energy through space. The tests established a new record for microwave power transmission, delivering one kilowatt over a 1.6-km range. The 10-yr program had increased DC-to-DC efficiency of transmissions from 17% to 50% and efficiency of the rectenna in converting radio frequency to DC from 25% to 78%. (NASA Gen Mgt Rev Rpt, 15 Oct 74, 57–60)

- Interstellar space flight appeared possible, British Interplanetary Society engineers and scientists concluded after the first 15 mos of a study to establish the feasibility of flight outside the solar system. Costs would be high, but so far none of the engineering difficulties seemed insoluble. The chief limitation that might delay such a project was the low natural abundance of helium 3 for the nuclear microbombs proposed for propulsion.

The Project Daedalus study had taken the star Barnard, six light years away, as a theoretical target. The journey was to be limited to 30–40 yrs, a working lifetime. To achieve the necessary 51 000-km-per-sec velocity, BIS scientists proposed a starship driven by a series of controlled nuclear explosions—each equivalent to 90 tons of TNT—occurring at the rate of 250 per sec. Microbombs, small pellets of nuclear fuel, would be compressed and heated to detonation point by electron beams. The rocket-powered boost phase would last 5 yrs, followed by a 35-yr coast. During flyby of Barnard, 10–20 probes would be released to search for inner planets and moons, sending back information to the mother ship for transmission to the earth.

NASA had studied the nuclear-pulse propulsion concept for interplanetary exploration through a General Dynamics Corp. contract in 1963, but had suspended the study because of costs and the Nuclear Test Ban Treaty. (Gatland, *SF*, Sept 74, 356–358; *A&A 1963, 1964, 1965*)

## October 1974

*1 October:* U.S.S.R. Supreme Soviet Presidium Chairman Nikolay V. Podgorny presented the highest Soviet award, the Order of Lenin, to Cosmonauts Pavel R. Popovich, Yury P. Artyukhin, Gennady V. Sarafanov, and Lev S. Demin. Popovich and Artyukhin were honored for their 15-day mission in the *Salyut 3* orbital scientific station and *Soyuz 14* transport craft (launched 3 July to join the *Salyut 3*, launched 25 June). Sarafanov and Demin were recognized for their *Soyuz 15* flight (launched 26 Aug.), during which they experimented in maneuvering and docking with *Salyut 3* before landing on land at night. Popovich, already a Hero of the Soviet Union, was also awarded the Second Gold Star Medal. Artyukhin, Sarafanov, and Demin were awarded the title Hero of the Soviet Union, Gold Star Medals, and Badges of Pilot-Cosmonaut of the U.S.S.R. (FBIS–Sov, 2 Oct 74, U1)

*1–7 October:* Astronaut-physician Dr. F. Story Musgrave and Dr. Dennis R. Morrison of Johnson Space Center's Bioscience Payloads Office spent one week in a mockup Spacelab at JSC in a shakedown test of operational procedures and experiments. Spacelab, a scientific space laboratory, was being developed and built by the European Space Research Organization for orbital missions aboard NASA's space shuttle. A rehearsal for future tests and a means of perfecting operational procedures, this first test also evaluated proposed space shuttle ground support and flight crew operational procedures, data-handling techniques, man-machine integration, and biomedical demonstrations representative of Spacelab medical experiments. (JSC Release 74–255; JSC *Roundup*, 11 Oct 74, 1)

*2 October:* Roy E. Anderson, General Electric Co. Research and Development Center engineer, demonstrated the potential of geostationary satellites for search-and-rescue missions. Standing on the NASA Hq. steps in Washington, D.C., and using an ordinary walkie-talkie aided by an antenna built on the frame of a golfer's umbrella, Anderson beamed a message to GE's Radio-Optical Observatory in Schenectady, N.Y., via NASA's *Ats 3* Applications Technology Satellite. The inexpensive, 1.4-kg antenna increased the normal eight-kilometer range of the walkie-talkie more than 10 000 times. Anderson suggested that a global search-and-rescue system, covering all but the polar regions, would require only six geostationary satellites. The necessary transponder could be attached to any communications satellite and would draw only a small portion of the satellite's power. (GE Release, 2 Oct 74; AP, B *Sun*, 3 Oct 74, A3)

- Rockwell International Corp. Space Division, prime contractor for the space shuttle orbiter, signed a $16.5-million supplemental agreement with NASA. This 27th supplemental agreement since the contract was awarded 9 Aug. 1972 brought the estimated contract value to $995 million. It incorporated authorization for sneak circuit analysis, vertical flight-test support, midfuselage mockups, integration of all shuttle elements, and other changes. (JSC Release 74–258)

**2 October**
- Skynet IIB, United Kingdom military communications satellite, arrived at Kennedy Space Center to be prepared for November launch by NASA on a Thor-Delta launch vehicle. NASA had launched *Skynet A* and *B* in 1970. *Skynet IIA*, launched 18 Jan. 1974 into too low an orbit, had reentered the atmosphere. (*Spaceport News*, 18 Oct 74)

*3 October:* A spherical object from *Cosmos 654* satellite was reported recovered in Ohio. *Cosmos 654*, launched 17 May 1974, had been reported by the press as a Soviet ocean surveillance satellite. The debris was believed to have been a portion of the staging system used to move the craft from its initial low, near-circular orbit into a higher, longer life orbit with a 1022-km apogee and 915-km perigee. There the nuclear systems required for ocean surveillance radar equipment and sensors could decay naturally before the orbit decayed years in the future. (*SBD*, 3 Oct 74, 164–5; GSFC *SSR*, 31 Oct 74)

*3, 4, 9 October:* The House Committee on Science and Astronautics' Subcommittee on Space Science and Applications held hearings on H.R. 14978 and H.R. 15781, bills to establish offices in the Dept. of the Interior and NASA to operate an Earth Resources Survey System. Hugh T. Loweth, Deputy Associate Director for Energy and Science of the Office of Management and Budget, testified that more experimentation was needed before the Earth Resources Technology Satellite program entered the operational phase, but witnesses from industry and other agencies of Government recommended accelerating the program immediately. Dr. Rocco A. Petrone, NASA Associate Administrator, and Charles A. Mathews, Associate Administrator for Applications, opposed adoption of the bills, stressing NASA's desire for "an evolutionary approach" that minimized an artifical distinction between experimental and operational satellites and concentrated instead on immediate dissemination of data and simultaneous improvement of the system [see 6, 8, 9 Aug. and 18 Sept.]. (Testimony)

*4 October:* Dr. Lowell J. Paige was named Acting Deputy Director of the National Science Foundation by NSF Director H. Guyford Stever. He would succeed Dr. Raymond L. Bisplinghoff, who had announced his resignation 17 July. (NSF Release 74–206)

*5 October:* Scientists using the 305-m dish antenna at the Arecibo Observatory in Puerto Rico had discovered a pulsar—a pulsating radio source—orbiting a companion celestial object in the constellation Aquila, the *New York Times* reported. Both pulsar and the object weighed as much as the sun but were only a few kilometers in diameter. It was the first pulsar discovered circling another object, a phenomenon which made it possible for scientists to estimate its mass. Its orbit had been calculated from the slight variations in its pulse rate. While it was possible that the companion was a black hole, scientists thought it more likely that both were neutron stars and a search was continuing to find out if the companion was a very weak pulsar. (Sullivan, *NYT*, 5 Oct 74, 50)

*5–13 October:* Dr. James C. Fletcher, NASA Administrator, visited Japan to discuss U.S. and Japanese space programs and evaluate opportunities for cooperation. He met with representatives of government agencies, the scientific community, and Japan's Federation of Economic Organizations and visited Tanegashima Space Center and the Uchinoura Space Center at Kagoshima. (NASA Release 74–276; *NASA Activities*, 15 Nov 74, 23–4)

*7 October:* NASA adjudged the mission of *Noaa 3* (ITOS–F, launched 6 Nov. 1973) successful. NASA mission objectives—to launch the spacecraft into sun-synchronous orbit of sufficient accuracy for the operational mission, to evaluate and check out the spacecraft in orbit, and to turn operational control over to the National Oceanic and Atmospheric Administration—had been accomplished. (NASA MOR, 8 Oct 74)

- Robert Hotz, in an *Aviation Week* editorial, praised the "stabilizing and peace-keeping functions" of the large U.S. military space program. Scoring the "official government hypocrisy about our non-military space program," Hotz argued the Eisenhower Administration's attempt to separate civilian and military roles in space had "persisted to this day and badly impaired and obscured the truly peaceful role played by military space vehicles." Air Force boosters and military-trained astronauts and managers had played key roles in NASA programs. Military reconnaissance satellites and early warning satellites added up to "one of the biggest forces in deterring major aggression and keeping as much peace as there is left in the world." In contrast, claimed Hotz, the Soviet Union "never made any pretense of separating its military and scientific space activities" and had gone beyond reconnaissance satellites into "offensive space weapons." (*Av Wk*, 7 Oct 74, 11)

*8 October:* Astronaut Henry W. Hartsfield was presented the Gen. Thomas D. White Space Trophy for 1973 by Secretary of the Air Force John L. McLucas in a ceremony at National Geographic Society Headquarters. Hartsfield received the trophy—presented annually to the Air Force employee who made the most significant contribution to U.S. progress in space—for his part in saving the $2.5-billion Skylab program. Hartsfield, serving as capsule communicator during the *Skylab 2* mission (25 May–22 June 1973), played a key role in preparing and evaluating crew procedures and equipment for transporting, installing, and deploying the parasol sunshade to protect the Skylab Orbital Workshop after it had been damaged during its 14 May 1973 launch. Hartsfield also had been instrumental in developing procedures for installation by the *Skylab 3* crew (launched 28 July 1973) of a compact, lightweight replacement package for the failing gyroscopes in the Skylab attitude control system. (NGS *Release*, 8 Oct 74)

- The House agreed to H. Res. 988, the Committee Reform Amendments of 1974, by a 359-to-7 vote, reorganizing House committee jurisdictions and procedures. The name of the Committee on Science and Astronautics was changed to Committee on Science and Technology effective 3 Jan. 1975, and the Committee's jurisdiction was broadened to include all research and development in non-nuclear energy, environment, weather, and aviation, with special oversight over all nonmilitary Federal R&D. Legislative responsibility for the National Oceanic and Atmospheric Administration would be shared with the Committee on Merchant Marine and Fisheries. Responsibility for Federal Aviation Administration civil aviation R&D programs would be added to that for NASA's aeronautical R&D. Major emphasis on the civilian national space program would continue. (*CR*, 8 Oct 74, D1210–11, H10146–69; 17 Feb 75, 534–5; House Com Staff interview, 24 March 76)

- President Ford signed an $82.1-billion FY 1975 Dept. of Defense appropriation bill, almost $5 billion below the Administration request. The conference committee bill had been approved by the House 23 Sept. and the Senate 24 Sept. Research, development, test, and evaluation funds included $445 million for the B-1 advanced strategic bomber, clearing the way for construction of a fourth developmental prototype during FY 1975. The Navy VFAX fighter program was allocated $20 million for adaptation of the air combat fighter to be selected by the Air Force—either the General Dynamics Corp. YF-16 or the Northrop Corp. YF-17. Other funds included $81.4 million for six A-10 close-air-support R&D aircraft (in addition to $138 million for purchase of 25 production aircraft); $175 million for F-15 air superiority fighter R&D, and $8.6 million for Navy V/STOL development. (PD, 14 Oct 74, 1250; CR, 23 Sept 74, D1125; 24 Sept 74, D1130; Av Wk, 23 Sept 74, 17-8)

*8–9 October:* Marshall Space Flight Center hosted its Second Annual Research and Technology Review. Dr. William J. Patterson of MSFC's Materials and Processes Laboratory reported the development of polyimide laminates to replace epoxy printed wiring boards for high-reliability electronic circuits. The polyimides could overcome cracking that had caused Saturn and Skylab electronic problems. A contract has been negotiated to demonstrate the practicability of commercial production. Lewis L. Lacey and Iva C. Yates, Jr., of MSFC's Space Sciences Laboratory reported Skylab experiments had demonstrated that dispersed immiscible liquid emulsions were much more stable in the low-gravity environment of space than on the earth and that processing immiscible metals in low $g$ could lead to new alloys with unique electrical properties. Apollo-Soyuz Test Project experiments were planned to develop techniques for future alloy-processing experiments on Spacelab.

Donald Stone of MSFC's Electronics and Control Laboratory reported a scanning laser radar for automatic rendezvous and docking had been tested for possible use on the space shuttle, space tug, and solar electric propulsion stage. Plans called for delivery of the radar to a simulator of the space tug flight system for evaluation under conditions similar to actual use. (Proceedings; MSFC Release 74-182)

*8–10 October:* The Spacelab Operations Working Group held its first meeting, at Marshall Space Flight Center. Representatives of four NASA Centers, NASA Hq., and the European Space Research Organization began planning technical documentation, requirements review, operations, and safety for the European pressurized laboratory that would be carried on the space shuttle. (NASA, *Spacelab Newsletter* 74-7)

*9 October:* The House Committee on Science and Astronautics' Subcommittee on Aeronautics and Space Technology reported on prospects for *Advanced Nuclear Research* in the wake of the February 1972 termination of the NERVA nuclear rocket program and the FY 1974 reductions in nuclear power and propulsion research brought by NASA budget cuts. Concerned lest the $1.5 billion invested in these programs since NASA took charge of them in 1958 be entirely lost, the Subcommittee had held hearings 7 Feb. 1974 to determine how to maintain a research capability and to ensure that results of the previous R&D would be used toward solving critical energy problems. The Subcommittee was "disappointed" that the AEC–NASA Space Nuclear Systems Office had been abolished.

Witnesses from industry, NASA, and AEC described work being done in thermionic technology, plasma-core reactor research, high-temperature gas-cooled reactors, and medium-power reactor systems. The Subcommittee concluded that, though funding in these fields was modest, "the potential benefits [were] enormous in scope and magnitude." While the traditional R&D areas were becoming less well defined for the various agencies, the Subcommittee found the greater portion of the funds for the programs should go to AEC rather than to NASA.

The Subcommittee recommended: (1) NASA should continue, and expand as its budget permitted, its advanced nuclear research in potential energy applications. (2) NASA should continue complementary arrangements with other Federal agencies, especially AEC. (3) The Joint Committee on Atomic Energy should evaluate the Subcommittee findings. (4) NASA and AEC should jointly investigate and report how nuclear power and propulsion R&D had been used and how data was stored. And (5) NASA and AEC should expand industrial participation. (Com Print)

- Marshall Space Flight Center held a final review of the Zero-Gravity Atmospheric Cloud Physics Project Phase A Definition Study. The study established the feasibility of atmospheric cloud physics research on Spacelab and recommended preparation of a Phase B study. (MSFC Orbital & Space Environmental Br, interview, 6 Aug 75; MSFC Release 74-181)

*9–10 October:* The fourth flight of the balloon-borne ultraviolet stellar spectrometer (BUSS) provided the best test results yet of the candidate telescope experiment for the space shuttle. The National Scientific Balloon Facility of the National Center for Atmospheric Research (NCAR) at Palestine, Tex., launched the 91-m tall, 440 000-cu-m, helium-filled balloon, which carried a telescope, Ebert-Fastie spectrometer, detector, star-pointing-and-control system, and telemetry and command system to 40 000-m altitude. During five hours at that altitude the instruments observed seven stars in UV wavelengths. The stars included one giant star and three super giants. Dr. Yoji Kondo of Johnson Space Center's Astrophysics Section, principal investigator for the experiment, said the flight was free of environment and hardware problems and its success marked a new stage in balloon-borne astronomical observation after past high failure rates. Data would be combined with other observations for better information on the structure, atmosphere, and evolution of a variety of stars. The mission also evaluated state-of-the-art detector and tracking performance, flexibility of attaching instrumentation to a general-purpose telescope platform, and developing real-time, man-in-the-loop data acquisition, evaluation, and ground operations. (Kondo interview, 1 Oct 75; JSC Release 74-254; Bulban, *Av Wk*, 2 Dec 74, 43-5)

*9-25 October:* Johnson Space Center Flight Operations Aircraft Div. completed the first mission of Project Airstream, gathering high-altitude gaseous and particulate samples from the upper atmosphere in the Western Hemisphere with the NASA high-altitude WB-57F aircraft. Sampling missions, flown for the Atomic Energy Commission and the Dept. of Transportation, allowed AEC to determine the global distribution of atmospheric nuclear weapon test debris, both gaseous and particulate. Nine NASA crewmen alternated in the 14 flights of the first mission, logging 51 500 km in all. (JSC Release 74-269)

*10–30 October:* NASA launched *Westar 2* (Westar-B), the second U.S. commercial domestic communications satellite, for Western Union Telegraph Co. from Eastern Test Range at 6:53 pm EDT on a Thor-Delta launch vehicle. The satellite was placed in a close-to-planned transfer orbit with a 36 107-km apogee, 231-km perigee, and 24.76° inclination. The apogee kick motor, fired 13 Oct., circularized the orbit with 35 766-km apogee, 35 244-km perigee, 23-hr 42-min period, and 0.5° inclination. By 25 Nov. the comsat was operating normally at its station at about 90° W longitude above the equator.

The launch had been twice delayed, once for investigation of the entire Delta program and once for a last minute check of a connector pin. When launched, the vehicle performed as planned with two minor exceptions (both due to human error): the 2nd-stage cold-gas redundant altitude-control system switched from primary to backup nozzles because the guidance system had been programmed with too low a moment threshold and the umbilical end fitting was damaged because a tether had been improperly installed. Neither problem compromised the NASA mission objective: to place the satellite in an orbit of sufficient accuracy for transfer into a stationary synchronous orbit while retaining sufficient stationkeeping propulsion to meet the mission lifetime requirements. The objective was met, and the mission was adjudged successful 30 Oct.

*Westar 2* joined *Westar 1*, launched 13 April 1974 and later placed in geosynchronous orbit at 99° W longitude. The Westar Satellite Communications System—consisting of *Westar 1* and *2*, a third satellite for later launch, and five ground stations—had the potential for general-purpose telecommunications throughout the continental U.S., Alaska, Hawaii, and Puerto Rico. Western Union planned to make the system available to all users, largely through leasing the voice, TV, and data transmitting services. NASA would be reimbursed for the launch. (NASA MORs, 1, 25 Nov 74; NASA Releases 74-147, 74-265; Westar Release, 12 Sept 74; *SBD*, 11 Oct 74, 214)

*10 October:* An electric rocket engine that short-circuited aboard *Sert II* Space Electric Rocket Test satellite in 1970 had been restarted in space and successfully operated for six weeks, scientists at Lewis Research Center reported. Launched 3 Feb. 1970 to demonstrate that electric propulsion could be used for future space missions, *Sert II* was to have operated an ion engine for six months in space. Engine thruster 2 shut down after less than three months of operation; thruster 1 performed for five months. Both failures were attributed to a small chip apparently eroded from a molybdenum grid and lodged between two grids at the back of the engine. William R. Kerslake, *Sert II* experiment manager, believed the chip was jarred loose in summer 1974 when the spacecraft was spun up by its cold-gas thruster system to obtain a better sun angle for solar arrays. Since then, the thruster had been operated successfully on 17 occasions for short periods, obtaining up to 80% of its maximum thrust. Electrical Propulsion Branch Chief Robert C. Finke said operation after $4\frac{1}{2}$ years proved "the long term storability of this thruster design" and indicated that the system could be "confidently incorporated into future missions requiring several years of thruster operation." (LeRC Release 74-60)

• NASA had deleted a backup Viking lander spacecraft and would substitute the proof test orbiter for one of two flight orbiters to help offset significant cost increases and meet target dates for the two 1975 missions to explore the planet Mars, NASA reported to Speaker of the House Carl Albert. Technical problems in the guidance-and-control computer, biology instrument, and lander proof-test capsule qualification had delayed completion of the landers, made the schedule "extremely tight," and increased costs an estimated $48.8 million over the $89 million in the FY 1975 budget. NASA notified the House that it planned to use for Viking $48.8 million available from the Skylab program after its successful completion. (Text in *Viking Proj Hearings*, House Subcom on Space Science, 21–2 Nov 74)

- NASA was negotiating two 11-mo study contracts for assessing methods of producing electric power from coal. Funded by the National Science Foundation, Dept. of Interior, and NASA, the program would compare costs and impact on the environment of such energy systems as a potassium Rankine topping cycle, advanced steam plants, open and closed cycle gas turbine systems, supercritical carbon dioxide systems, magnetohydrodynamic systems, and fuel cells. Westinghouse Corp. and General Electric Co. had been selected to make the studies, to provide data for a Lewis Research Center energy systems model for evaluating variables. (NASA Release 74–273)

*11 October:* A modified Piper Seneca aircraft completed the first test flight of a new general-aviation wing, the GAW–1 developed at Langley Research Center. A derivative of the supercritical airfoil, also developed at LaRC, the new wing showed potential for lift increases up to 30%. The lift-to-drag ratio was increased about 50%. In addition to the low-speed airfoil, the new wing design had a 25% reduction in area, tapered planform, full-span Fowler flaps, and spoilers for roll control instead of conventional ailerons. (Mgr, NASA Gen Aviation Tech Off, interview, 25 July 74; NASA Release 74–227)

- A NASA design study for the Dept. of Housing and Urban Development's modular integrated utility system (MIUS) program had revealed the MIUS design could save 33% over conventional practices in providing utility services to a 648-living-unit garden apartment complex in Houston, Johnson Space Center reported. JSC had reported this result and similar savings from other studies at a recent two-day conference on technology transfer in Houston. (JSC *Roundup*, 11 Oct 74)

- President Ford signed into law the Energy Reorganization Act of 1974 (P.L. 93–438), abolishing the Atomic Energy Commission and establishing in its place three new Federal entities: the Energy Research and Development Administration, the Energy Resources Council, and the Nuclear Regulatory Commission. The President also signed Executive Order 11841, activating the Energy Resources Council and appointing the Secretary of the Interior its Chairman. P.L. 93–438 was to become effective not later than 120 days after signature. (*PD*, 14 Oct 74, 1271–3; *CR*, 8 Oct 74, H10134)

*15–30 October:* The United Kingdom's *Ariel 5* (UK–5) satellite was launched on a NASA Scout vehicle at 10:47 am local time (3:47 am EDT) by a team of Italian engineers at the San Marco launch facility off the coast of Kenya. Orbital parameters were 569.5-km apogee, 502.0-km perigee, 95.4-min period, and 2.86° inclination. After completion of checkout, routine experiment and spacecraft operations began 30 Oct.

*Ariel 5*—a 132-kg, solar-powered, 16-sided polygon designed by the U.K.—was the fifth satellite launched in a cooperative program between NASA and the U.K. Science Research Council. The mission of its one-year operating life was to study x-ray sources both within and beyond our galaxy, using five experiments from U.K. and one from Goddard Space Flight Center. Instruments were designed to follow up findings of

*Uhuru (Explorer 42)*, launched 12 Dec. 1970, and *Oao 3*, launched 21 Aug. 1971. *Uhuru* had extended the number of known celestial x-ray sources from about 30 to 200 and found evidence of a black hole companion to Cygnus X-1, later supported by *Oao 3*, sounding rocket, and ground-based observations. Four *Ariel 5* pointing experiments focused on specific x-ray sources, and two scanning experiments swept the entire galaxy with every revolution. The GSFC scanning experiment—first of its kind to fly on an orbiting scientific satellite—consisted of two x-ray pinhole cameras, each with a 90° field of view, providing together a sweep of the entire sky. After evaluation, *Ariel 5* results would be available to the world scientific community.

Though all experiments obtained data during initial tests, two were drawing excessive current. The problem was resolved by outgassing. A slight delay in shutter closing on Experiment A, a star sensor, had little or no effect on the mission.

The U.K.–U.S. cooperative satellite series had begun with *Ariel 1* in 1962 and continued through *Ariel 4*, launched 11 Dec. 1971. *Ariel 5* was launched under a 2 Nov. 1970 SRC–NASA Memorandum of Understanding, with a 27 Dec. 1973 letter modification shifting the launch site from Wallops Flight Center to San Marco to enhance scientific data return. The Memorandum tasked SRC with spacecraft design, fabrication, and testing, five experiments, primary command and control, and data reduction and analysis. NASA was responsible for one experiment, spacecraft technical support, the Scout launch vehicle, and tracking, command, and data acquisition. SRC financed all aspects of the Ariel program. The launch by members of the Univ. of Rome's Aerospace Research Center was the seventh from San Marco, but the first payload for a country other than the U.S. or Italy (NASA MORs, 11 Oct 6, Nov 74; NASA Release 74–274)

*15 October:* Capt. Jaques-Yves Cousteau's research vessel *Calypso* left Galveston, Tex., on a nine-month voyage to carry out oceanographic and weather experiments for NASA in cooperation with Texas A&M Univ. Researchers would measure chemical content of seawater, temperature, depth factors, water colors, and pollutants in different regions for correlation with weather and water appraisals collected by *Ats 3* and *Nimbus 5* satellites (launched 5 Nov. 1967 and 10 Dec. 1972), by Nimbus-F after launch in 1975, and by high-flying U–2 aircraft equipped with a coastal zone color sensor being developed for launch on Nimbus-G in 1978.

One of NASA's Lockheed U–2 earth resources survey aircraft was based at Hickam Air Force Base, Hawaii, during October for stratospheric sampling in a semiglobal Ames Research Center study of the effect of ozone, nitric oxide, and pollutants on the climate. (NASA Release 74–283; ARC Release 74–47)

• The Nobel Prize in physics would be shared by two radio astronomers, the Swedish Academy of Sciences announced in Stockholm, marking the first recognition by the Nobel committee of achievements in astronomy. Sir Martin J. Ryle, Astronomer Royal of Britain and professor at Cambridge Univ. Cavendish Laboratory, was cited for development of a way to use several small radio telescopes to achieve results which would otherwise require an antenna several kilometers in diameter. Dr. Ryles's system accumulated data over a period of days and correlated it through computers to piece together larger pictures. Dr. Antony

Hewish, professor of radioastronomy at the Cavendish Laboratory and leader of the research group that discovered pulsars in 1967, would share the prize. (Rensberger, *NYT*, 16 Oct 74, 1)

*16 October: General Aviation Programs,* forwarded to the House Committee on Science and Astronautics by the Subcommittee on Aeronautics and Space Technology, reported on an oversight review based on hearings 14 and 15 May and a visit to general-aviation manufacturers in Kansas 31 May and 1 June. Noting that general aviation made up about 97% of the civil aircraft fleet, flew 79% of the hours and 71% of the miles, and carried up to one third of all intercity passengers—the Subcommittee recommended: continued close NASA-industry cooperation, NASA assistance to Federal agencies to ensure realistic regulations, individual attention to special general-aviation needs, NASA noise and emission research, concerted industry effort to use new technology, NASA effort to increase public visibility of its program, NASA-industry measures toward better understanding of respective roles in research and development, and an annual General Aviation Manufacturers Assn. evaluation of the NASA program. (Com Print)

- Aerospace industry employment would reach 968 000 by December—the highest level since May 1971—but then drop by 8000 by June 1975, the Aerospace Industries Assn. of America predicted in its semiannual survey. The increase over previous estimates was attributed to an unexpected rise in production and delivery of aircraft, while the missile and space sectors of the industry remained essentially constant. Commercial transport deliveries were expected to slow down in the first half of 1975. (AIA Release 74-16)

*17 October:* Marshall Space Flight Center requested bids on constructing an x-ray telescope test facility for NASA's High Energy Astronomy Observatory (HEAO) program. Bid opening was scheduled for 27 Nov. and contract award by 9 Dec. The facility was to be used for x-ray verification testing and calibration of x-ray mirrors, telescope systems, and instruments. (MSFC Release 74-211)

*18 October:* The first few space shuttle flights would land at Flight Research Center, Dr. James C. Fletcher, NASA Administrator, announced in a talk before the Antelope Valley Board of Trade, Lancaster, Calif. First horizontal test flights would be made from FRC and first orbital test flights would return there because of added safety margins and good weather conditions. FRC would also be a secondary landing site for operational shuttle flights when weather or other conditions made it desirable. (NASA Release 74-279)

- Plans almost to double spending for the European Space Research Organization's Ariane launch vehicle program in 1975 were made at an Ariane Program Board meeting in Paris. Spending authority would exceed $71 million, up from 1974's $37-million budget, and obligational authority was proposed at about $110 million instead of $71 million proposed earlier, in authorizations expected to be approved at the 26 Nov. Board meeting.

The Board—made up of officials from ESRO and 10 European countries financing Ariane development—met to review budget and development following France's confirmation that she would continue to contribute 62.5% of the financing. The development schedule would still call for flight trials of the launcher in 1979 and 1980.

**18 October**

The Board also was informed that the Société Européene de Propulsion (SEP) had sold a production license for the Ariane's Viking rocket motors to the Indian Space Research Organization for ISRO's SLV-3 vehicle, under development for launching future educational broadcast satellites. (*Av Wk*, 28 Oct 74, 19)

- Federal scientists and engineers decreased by three percent in 1973, reducing employment in that category to the lowest level since 1968, the National Science Foundation reported. Nonprofessional scientific and technical personnel declined by six percent, reducing the number below even the 1968 level. The 1973 drop was the second decrease in Federal civilian scientists and engineers since comparable figures were first kept in 1954.

  Within NASA, total scientific and technical personnel decreased by almost five per cent. While the number of scientists increased marginally, the number of engineers decreased by more than five percent and the nonprofessionals decreased by more than seven percent. (NSF *Highlights*, 18 Oct 74)

- The Dept. of Defense announced award of a $29 509 671 cost-plus-award-fee contract to General Dynamics Corp. Electronics Div. to develop, produce, and operate the Phase I control and user system segments of the NAVSTAR Global Positioning System [see During August]. (DOD Release 793-74)

**22 October–12 November:** The U.S.S.R. launched *Cosmos 690* biological satellite from Plesetsk into an orbit with 366-km apogee, 216-km perigee, 90.3-min period, and 68.8° inclination. The satellite carried white laboratory rats, steppe tortoises, insects, and fungi in a test of the combined effects of weightlessness and radiation. A cesium-137 gamma radiation source within the spacecraft provided an average radiation dose of 32 rads per hr on command from the earth. Beginning on the 10th day of the flight, the rats were exposed to doses of gamma-quantums ranging between 200 and 1000 rads over 24 hrs. After return to the earth 12 Nov., the rats were found to be more pathological and less active than after corresponding control experiments. The change in the hemopoietic and lymphoid organs was more sharply pronounced than in the *Cosmos 605* experiments (31 Oct.–22 Nov. 1973). While experiment results were processed, research was continuing on methods of antiradiation protection by deflecting particles with an electric field. Czechoslovak and Romanian specialists cooperated with the U.S.S.R. on the *Cosmos 690* project as part of the Intercosmos program. (GSFC *Wkly SSR*, 17–23, Oct 74; *Pravda*, FBIS–Sov, 3 Jan 75, U1–4)

**22 October:** Dr. George M. Low, NASA Deputy Administrator, signed a NASA and Soviet Academy of Sciences agreement on news coverage of the Apollo-Soyuz Test Project flight, scheduled for July 1975. The agreement brought a protest from Associated Press President and General Manager J. Wes Gallagher because it excluded American newsmen from the Soviet launching. In a 23 Oct. letter to Gallagher, Dr. James C. Fletcher, NASA Administrator, stated that, while the Soviets were adamant on this question, they had otherwise agreed to provide "the most complete, comprehensive release ever to the U.S. news media of real-time information related to a Soviet space mission." The agreement, Dr. Fletcher noted, covered joint U.S.–U.S.S.R. activities while recognizing the right of each nation to establish policies for their respective independent activities. The U.S. would not exclude Soviet newsmen from the U.S. liftoff.

The Public Affairs Agreement provided that, beginning with the Soyuz launch and ending with the Apollo recovery, onboard information from both spacecraft, voice communications between the spacecraft, and ground communications with both spacecraft, as well as ground-based and onboard TV, would be exchanged immediately. The first real-time TV coverage of a Soviet launch would be released to the West. Factual information on joint activities would be released immediately. Lines from the U.S. public affairs center in Moscow would communicate with technical specialists in the Soviet mission control center. For the first time, Soviet commentators would be at the Soviet control center and launch site during mission operations to describe countdown, launch, and mission events as they occurred, for immediate release. (NASA Int Aff, interview, 30 Sept 75; AP, NYT, 12 Oct 74, 2; 24 Oct 74, 9; NASA press conf transcript, 12 Nov 74)

*23 October:* An eight-week series of joint tests on the actual flight docking system to be used in the Apollo-Soyuz Test Project began in Moscow. Specialists from the U.S. and the U.S.S.R. confirmed that the docking units sealed hermetically. (ASTP Press Communiqué, JSC Release, 20 Dec 74; Moscow Domestic News Service, FBIS–Sov, 22 Nov 74, U2; 31 Dec 74, U4)

- The satellite relay system planned for the July 1975 Apollo-Soyuz space mission was tested by Johnson Space Center and Goddard Space Flight Center communications engineers. *Ats 6* Applications Technology Satellite (launched 30 May) and the Apollo S-band, steerable, high-gain antenna were tested together for the transmission of voice, television, and simulated telemettry data in preparation for the first spacecraft-to-satellite-to-ground routing employed in the manned space flight program. Signals were transmitted between JSC and the ground station in Rosman, N.C., via the *Ats 6* in a simulation of mission conditions. (JSC Release 74–264; JSC *Roundup*, 8 Nov 74)

- Detection of a "plasma mantle" by the European Space Research Organization's *Heos 2* Highly Eccentric Orbit Satellite (launched by NASA 31 Jan 1972) was announced by West German space scientists at the Max Planck Institute for Extraterrestrial Physics in Munich. Scientists had become aware of the new belt of electrically charged particles circling the earth while assessing data collected by a specially developed "plasma analysator" in the satellite. The belt of particles emitted from the sun and trapped in the earth's magnetic field was said to be between 10 060 and 20 100 km thick. (B *Sun*, 24 Oct 74, A3; *A & A 1972*)

- "The fact that fewer and fewer young people are entering the field of aeronautics" was endangering the U.S. aeronautical position, Dr. James C. Fletcher, NASA Administrator, said in a speech before the NASA-University Conference on Aeronautics at the Univ. of Kansas. The number of college students in junior classes in aerospace engineering had dropped 75% during 1968–1973, although civil aeronautics export business had increased fivefold in the past 10 yrs. Total civil and military aerospace exports were expected to reach $7 billion in 1974, with a "self-evident" importance to the U.S. balance of payments. U.S.-designed and -produced transport aircraft formed 80% of the non-Communist world's transport fleets.

Because of the increasing system complexity, costs, foreign competition, and complex environmental and economic problems, it was "clear that the support of the research and technology critical to the future health of civil aeronautics will depend on heavier government involvement." NASA was "prepared to support a stronger aeronautics research and technology effort" and had organized a long-range study group to evaluate the "Outlook for Aeronautics" in the 1980–2000 period. A report was due in the summer of 1975.

Among possible technology goals, Dr. Fletcher was convinced that aircraft fuel savings of 50% within the next 20 yrs could be reached, through weight reductions by composite structures and fly-by-wire and active controls and through supercritical aerodynamics, advanced engine techniques, and operational methods. NASA would also give special attention to technology for alternative fuels, such as hydrogen. (Text)

- The first of two C–130 transport aircraft with experimental boron-epoxy composite reinforced wing midsections was delivered to the Air Force to begin three years of tests. The joint NASA–USAF flight evaluation program was to demonstrate the fatigue resistance and weight reductions possible when the metal wing structures were reinforced with a composite of boron filaments embedded in an epoxy resin. Lockheed Aircraft Corp. would make periodic inspections and report to NASA and the Air Force. (NASA Release 74–290)

*24 October:* The U.S.S.R. launched *Molniya I–28* communications satellite from Plesetsk into orbit with 39 707-km apogee, 645-km perigee, 717.7-min period, and 62.8° inclination. Tass reported that the satellite carried equipment to relay TV programs and to establish remote multichannel radio communications. (GSFC *Wkly SSR*, 24–30 Oct 74; *Av Wk*, 4 Nov 74, 13; Tass, FBIS–Sov, 25 Oct 74, U1)

- U.S.S.R. flight controllers for the Apollo-Soyuz Test Project began two weeks training at Johnson Space Center in preparation for the July 1975 mission. Some 20 Soviet controllers attended sessions on Apollo trajectory, spacecraft systems and communications, docking module, Mission Control Center operations, and contingency flight planning. (JSC Release 74–264; NASA Release 74–284; JSC *Roundup*, 8 Nov 74)

- The Air Force successfully test-fired an intercontinental Minuteman I missile dropped from a C–5A transport aircraft flying 6100 m above the Pacific Ocean. The first successful air-drop firing followed drops of two inert missiles in a step toward development of an ICBM with mobility in the air and on land. (Finney, *NYT*, 26 Oct 74; *Aerospace Daily*, 29 Oct 74)

*25 October:* Marshall Space Flight Center announced it had awarded a $169 000 contract for the study of hydrogen as a possible energy source, to the Institute of Gas Technology, Chicago, Ill. The 10-mo study was to evaluate methods of producing large quantities of hydrogen and select a few of the most promising for follow-on studies. It was also to identify industrial and residential uses and estimate costs of developing technology. (MSFC Release 74–207)

*26 October:* President Ford signed the Solar Energy Research, Development, and Demonstration Act of 1974 (P.L. 93–473) establishing the Solar Energy Coordination and Management Project. A NASA Associate Administrator would be one of six members of the project, which was tasked with determining and evaluating the resource base and conducting research, development, and demonstration of solar energies and technologies. NASA would cooperate with other Federal agencies in the project by providing management capability and development of technologies. NASA was further authorized to carry out programs assigned it by the project. (P.L. 93-473)

- The first prototype B-1 intercontinental strategic bomber was unveiled at Palmdale, Calif., by the Air Force and Rockwell International Corp. Secretary of Defense James R. Schlesinger told on-lookers that the new aircraft was needed to complement the U.S. arsenal of land-based and submarine-based missiles. Although smaller and lighter than the B-52 it was meant to replace, the B-1 could fly farther and faster, carry almost twice the payload, and use a shorter runway. Powered by four 133.4-kilonewton-thrust (30 000-lb) turbofan engines and lifted by variable geometry, or "swing" wings, the bomber could fly efficiently at speeds in excess of mach 2. The projected cost of production models—$76.4 million each—made the future of the aircraft controversial and uncertain. (DOD Release 505-74; Getler, *W Post*, 27 Oct 74, A6; Wright, *NYT*, 27 Oct 74, 43)

*26 October-4 November:* An agreement for NASA to submit life sciences experiments for flight aboard the next available Soviet biological satellite was negotiated during the fifth joint U.S.-U.S.S.R. Working Group on Space Biology and Medicine in Tashkent, U.S.S.R. The first U.S. experiments would be passive specimen modules, completely autonomous from spacecraft power, life support, and data-recording systems, and would require no operational commands from the ground during flight. NASA would prepare descriptions of the proposed experiments in December and deliver flight hardware by 15 Aug. 1975. U.S. investigators would also participate in pre- and postflight studies of blood and tissue samples from animals flown aboard the next Soviet biological satellite.

The protocol was confirmed by NASA and the Soviet Academy of Sciences 4 Dec.

Representatives also discussed preparations for the joint July 1975 Apollo-Soyuz Test Project mission; exchanged results from Apollo, Skylab, Cosmos, Salyut, and Soyuz programs; and discussed pre- and postflight examinations, vestibular training, and crew reaction to flight. (NASA Gen Mgt Rev Rpt, 16 Dec 74, 42-3; NASA Release 74-237; NASA Int Aff, interview, 17 Oct 75; Moscow Domestic News Service, FBIS-Sov, 29 Oct 74, B10; 8 Nov 74, U1)

*27 October-2 November:* The 67th Annual General Conference of the Fédération Aéronautique Internationale (FAI) convened in Sidney, Australia. Astronaut Charles Conrad, Jr., was presented the 1973 FAI Gold Space Medal for his role as *Skylab 2* commander, and Astronaut Alan L. Bean was presented the Yuri Gagarin Gold Medal for his role as *Skylab 3* mission commander. Both men also received the De la Vaulx Medal for their successive space duration records. (NAA *News.* Oct 74, Dec 74, 1)

*28 October-9 November: Luna 23*, second Soviet lunar mission of 1974, was launched from Baykonur Cosmodrome near Tyuratam into a parking orbit with 246-km apogee, 183-km perigee, 88.7-min period, and 51.5° inclination, where the "automatic station" was separated and launched on a near-to-planned trajectory to the moon. After a 31 Oct. flight path correction, the spacecraft on 2 Nov. entered a lunar orbit with 104-km apogee, 94-km perigee, 117-min period, and 138° inclination. On 4 and 5 Nov. the orbit was changed to 105-km apogee and 17-km perigee.

The spacecraft landed in the southern part of the moon's Sea of Crisis 6 Nov., but rugged terrain damaged a device designed to take rock samples from a 2.5-m depth. The planned drilling and collection of lunar soil had to be abandoned for a reduced equipment-testing program. Work with the station was discontinued 9 Nov. (GSFC *Wkly SSR*, 24–30 Oct 74; Domestic News Service, FBIS–Sov, 29 Oct 74, U1–2; 10 Nov 74, U1; Tass, FBIS–Sov, 4 Nov 74, U1; *Av Wk*, 4 Nov 74, 13; *SF*, April 75, 152)

*28 October:* The U.S.S.R. launched *Meteor 19* weather satellite from Plesetsk into orbit with 904-km apogee, 842-km perigee, 102.4-min period, and 81.2° inclination. Tass reported that the spacecraft carried meteorological equipment to photograph cloud formations and snow cover and to obtain data on the thermal energy reflected and radiated by the earth and the atmosphere. (GSFC *Wkly SSR*, 24–30 Oct 74; Tass, FBIS–Sov, 29 Oct 74, U1; *Av Wk*, 4 Nov 74, 13)

*29 October:* The Air Force launched three satellites on a single Titan IIID booster from Vandenberg Air Force Base—including *S3–1*, launched to measure air density, obtain upper-atmosphere composition data, and provide information on atmospheric effects on spacecraft orbits [see 7 Aug.]. The first satellite entered orbit with a 279-km apogee, 152-km perigee, 88.8-min period, and 96.7° inclination and reentered 19 March 1975. The second entered orbit with a 542-km apogee, 540-km perigee, 95.3-min period, and 96.1° inclination. The third entered orbit with a 3711-km apogee, 147-km perigee, 125.4-min period, and 97.0° inclination; it reentered 26 May 1975. (Pres Rpt 74; AFSC PAO, interview, 23 March 76; AFSC Release OIP 137.74)

• The Apollo-Soyuz Test Project docking module arrived at Kennedy Space Center aboard a C-141 transport aircraft, from the Space Division of Rockwell International Corp., Downey, Calif. The module virtually completed the array of hardware being processed at KSC for the July 1975 launch. (KSC Release 154–74)

• Intention to nominate Dr. Robert C. Seamans, Jr., as Administrator of the new Energy Research and Development Administration was announced by President Ford as part of a major reshuffling of key energy posts. Dr. Seamans—former NASA Deputy Administrator, Secretary of the Air Force, and President of the National Academy of Engineering—was reported by the press as selected over former NASA Administrator Thomas O. Paine and two other candidates to head the new agency replacing the Atomic Energy Commission. Former astronaut and Atomic Energy Commissioner William A. Anders was to be nominated Chairman of the new Nuclear Regulatory Commission and the former AEC Chairman, Dr. Dixy Lee Ray, to the new position of Assistant Secretary of State for Oceans and International Environmental and Scientific Affairs. Dr. Seaman's nomination was sent to the Senate 2 Dec. and confirmed 12 Dec. Anders' and Dr. Ray's were sent to the Senate 4 Dec.; Anders' was confirmed 12 Dec. and Dr. Ray's 13 Dec. (PD, 4 Nov 74, 1382–3; 9 Dec 74, 1546; *CR*, 4 Dec 74, S20620; 12 Dec 74, S21317; 13 Dec 74, S21497; O'Toole, *W Post*, 16 Oct 74, A10; 30 Oct 74, A1)

• Iran signed an agreement to build a ground station to receive data directly from NASA's experimental earth resources satellites. Under the Memorandum of Understanding signed with NASA in Tehran, the new site would join six others already in the system—three in the

U.S. and one each in Canada, Brazil, and Italy. Iran would bear the full cost of building the facility and would make its data and tapes available to NASA in return for free access to the first two Earth Resources Technology Satellites. Thereafter, NASA had the option to establish a cost-sharing arrangement. (NASA Release 74-289)

- NASA's 16th Annual Awards Ceremony was held in Washington, D.C. Dr. James C. Fletcher, NASA Administrator, presented NASA's highest award, the Distinguished Service Medal, to Robert L. Krieger, Director, Wallops Flight Center; Norman Pozinsky, Hq. Director, Network Development and Engineering; Martin L. Raines, Johnson Space Center Director of Safety, Reliability, and Quality Assurance; Lee R. Sherer, Director, Flight Research Center; and John M. Thole, Deputy Associate Administrator for Space Science.

The NASA Distinguished Public Service Medal, the agency's highest award to non-Government personnel, was presented to Jack M. Campbell, President of the Federation of Rocky Mountain States and former Governor of New Mexico; Harry Dornbrand, Executive Vice President, Fairchild Space and Electronics Co.; and Jesse L. Greenstein, California Institute of Technology astrophysicist.

The Group Achievement Award went to 12 teams at Centers and Headquarters, the Exceptional Service Medal to 51 persons, and the Exceptional Scientific Achievement Medal to 16. (Program; *NASA Activities*, 15 Nov 74, 20)

*30 October:* The first Skylab crew—Charles Conrad, Jr., Dr. Joseph P. Kerwin, and Paul J. Weitz—who flew on *Skylab 2*, 25 May–22 June 1973—was awarded the Haley Astronautics Award at the Conference on Scientific Experiments of Skylab, Huntsville, Ala. The annual award was given by the American Institute of Aeronautics and Astronautics "for outstanding contributions by test personnel who undergo personal risk in the advancement of space flight." The three-day conference, 30 Oct.–1 Nov., was sponsored jointly by the AIAA and the American Geophysical Union. (MSFC Release 74-205; MSFC PAO, interview)

*31 October:* The U.S.S.R. launched *Intercosmos 12* from Plesetsk into an orbit with 697-km apogee, 241-km perigee, 94.0-min period, and 74.0° inclination. Tass reported that the satellite, intended to study the earth's atmosphere and ionosphere and the flow of micrometeorites, carried equipment made by specialists from Bulgaria, Hungary, East Germany, Romania, and Czechoslovakia, as well as the U.S.S.R. The spacecraft reentered 11 July 1975. (LC Science Policy Research Div., interview, 6 Aug 75; GSFC *Wkly SSR*, 31 Oct–6 Nov 74, 10–16 July 75; Tass, FBIS–Sov, 1 Nov 74, U1)

*During October:* The Space Shuttle Operational Management Assessment Team was formed to make "an appraisal of the technical aspects of the program and an assessment of the ability of management to complete the development of the shuttle within the cost and performance targets." Composed primarily of industry representatives and named for its chairman, Willis M. Hawkins of Lockheed Aircraft Corp., the Hawkins Committee was to prepare a formal briefing no later than 1 Nov. Inflation levels of 8–9%, well over the 5% on which FY 1975 funding was based, had increased costs throughout the shuttle program and compromised the ability of NASA and the prime contractors to reach their program milestones. (Team charter; Covault, *Av Wk*, 23 Sept 74, 20–2)

During October

- The European Space Research Organization rejected a plan to install a launch pad for the McDonnell Douglas Thor-Delta launch vehicle at the Kourou Space Center in French Guiana. The ESRO decision, which followed a four-month feasibility study, cited the high cost of installing the pad—set at approximately $50 million—as the reason for the rejection. The press speculated that, although proponents of the proposal contended the Thor-Delta launch capability would complement the European Ariane launcher under development, the plan might have run into opposition from Ariane supporters who feared competition.

  NASA, when queried by ESRO officials, had voiced no objections to the plan even though the equatorial launch site would have increased the vehicle's orbital weight capacity by about 20% over launches from the U.S., making the site an economical alternative for commercial customers. (*Av Wk*, 4 Nov 74, 21; GSFC Delta Proj Off, interview, 5 Dec 75)

- David Baker's annual review of the NASA budget in *Spaceflight* concluded that the FY 1975 NASA budget marked "a significant turning point," because the cuts were not as drastic as they had been in recent years, and, most important, there was a major increase for new programs—from $10.5 million in 1974 to $37.7 million. This led Baker to label "a trifle premature" his prediction of a year earlier that NASA might be relegated indefinitely to the role of serving "as a launch vehicle for satellites and space probes of other agencies, organizations, and nations." (*SF*, Oct 74, 362–5)

## November 1974

*1 November:* Project da Vinci—oft-delayed manned-atmospheric-research balloon flight—was terminated after one third of its projected 36-hr duration, because of threatening weather. The joint project of the Atomic Energy Commission, National Geographic Society, and the Army's Atmospheric Sciences Laboratory was intended primarily to investigate atmospheric behavior at 1200–4300 m, where less scientific research had been done than at very high and very low altitudes. Floating free and staying with a single parcel of air as long as possible, the four crewmen studied atmospheric behavior over mountains, plains, forests, and cities. They had hoped to travel in their 21-m helium balloon from the launch site at Las Cruces, N. Mex., to a landing near Lubbock, Tex. The project safety officer terminated the flight over Wagon Mound, N. Mex., but not before enough data had been gathered to plan future flights. (ERDA PIO, interview, 4 Aug 75; AEC Release T-425; AP, *NYT*, 3 Nov 74, 87)

*2 November–3 January 1975:* NASA's *Pioneer 11* interplanetary probe, launched 5 April 1973, flew past Jupiter three times closer than *Pioneer 10* had a year earlier and sent back the first polar images of the planet. The spacecraft then headed on toward a September 1979 encounter with Saturn.

*Pioneer 10*, launched 2 March 1972, had been the first spacecraft to reach Jupiter—3 Nov. 1973–9 Jan. 1974. The success of that mission permitted NASA scientists to attempt a somewhat riskier approach with the backup *Pioneer 11*, a clockwise trajectory by the south polar region and then straight up through the intense inner radiation belt by the equator and back out over the north pole. Although the spacecraft went within 42 000 km of the planet at 12:22 am EST 3 Dec., its peak speed of 171 000 km per hr and the trajectory's high latitude prevented serious radiation damage.

The Jupiter mission of *Pioneer 11* was to map the magnetic field; measure distributions of high-energy electrons and protons in the radiation belts and look for auroras; find a basis for interpreting radio emissions from Jupiter; detect and measure the bow shock and magnetospheric boundary and their interactions with the solar wind; verify the thermal balance and determine temperature distribution of the outer atmosphere; measure the hydrogen-helium ratio in the atmosphere; measure the structure of the ionosphere and atmosphere; measure the brightness, color, and polarization of Jupiter's reflected light; perform two-color visible-light imaging; and increase the accuracy of orbit predictions and masses of Jupiter and its moons.

The interplanetary mission was to map the interplanetary magnetic field; study the radial gradient of the solar wind and its fluctuations and structure; study the radial and transverse gradients and arrival directions of high-energy charged particles (solar and galactic cosmic rays); investigate relationships between the solar wind, magnetic field, and cosmic rays; search for the boundary and extent of the heliosphere (solar atmosphere); determine the density of neutral hydrogen; and determine the properties of interplanetary dust.

Hurtling toward and then away from Jupiter, *Pioneer 11* sent data back to the earth at 2048 bits per sec throughout most of the month before and after encounter. Picture-taking of Jupiter began 18 Nov., at 14 800 000 km out from the planet. From 25 Nov. through 8 Dec., imaging and polarimetry observations were made 23 hrs a day.

On 26 Nov. *Pioneer 11* crossed the detached bow shock wave (7 772 000 km out) at the juncture of the solar wind and Jupiter's magnetic field and entered the magnetosphere (6 923 000 km out). In five hours the spacecraft passed undamaged through this first field of intense radiation. On 1 Dec. it crossed the orbit of Callisto—outermost of the five inner moons—and began returning its best pictures of Jupiter. On 2 Dec. *Pioneer 11* crossed the orbits of Jupiter's four closest moons and entered the inner radiation belt. During this last leg of the approach to the south polar region, the spacecraft continued the imaging and sensing of the planet and moons and made a closeup, two-hour scan that included the red spot portion of the surface.

Less than 12 hrs before the spacecraft passed behind Jupiter, an intense peak of electron radiation built up a 4000-v electric charge within the spacecraft. The charge created an electric arc that signaled heater lines to turn on—drawing spacecraft power and weakening the signal to earth. In about three hours the problem was identified and override commands from earth switched the heaters back off.

At midnight on 3 Dec. *Pioneer 11* entered Jupiter radio occultation—a blackout that lasted 42 min while the spacecraft swung behind the 142 744-km-dia planet—moving from the south pole straight up over the north pole. Since *Pioneer 11* radio signals required 46 min to cross the 737 million km separating the earth and Jupiter at the time of encounter, the spacecraft had already circumscribed the planet before the earth lost contact with its signal. During occultation the spacecraft passed through the most intense radiation (up to 40 000 times as much in the earth's radiation belts), reached its greatest speed (171 000 km per hr) and came closest to the planet (within 42 000 km of the cloudtops). At 1:24 am EST—62 min after periapsis—the signal reached earth, indicating that the spacecraft had emerged from behind the planet and that all systems had survived the most critical phase of the mission. Temporary failures had lost some data, but no permanent damage was sustained.

Following the Jupiter encounter, Dr. James C. Fletcher, NASA Administrator, renamed the spacecraft *Pioneer-Saturn* as it headed toward the planet Saturn. The exact target point had not been selected but scientists considered an interesting possibility was to target the spacecraft to pass between Saturn and its rings and then close by Titan, one of Saturn's moons and a possible location of life. The five-year, 2.4-billion-km trip to Saturn would carry *Pioneer-Saturn* about 160 000 000 km above the orbit planes of most planets.

*Pioneer 11* was the first spacecraft to use the gravity of an outer planet to throw it to another planet farther out, came the closest to Jupiter, was the fastest spacecraft yet launched, and was the first spacecraft to photograph Jupiter's polar regions. *Pioneer-Saturn* was expected to become the first spacecraft to pass far from the ecliptic plane and the first to reach Saturn.

*Pioneer 11* confirmed, but somewhat modified, the findings of *Pioneer 10*. The model of a relatively flat, disclike magnetosphere was challenged by new evidence suggesting a large, blunt magnetosphere of irregular and fluctuating shape. Temperature measurements of about 125 K at cloudtops supported *Pioneer 10* readings, which were lower than those indicated by earth-based observations. Imaging revealed a decrease in the banding structure of the planet with increased latitude and a more mottled structure across the poles. (NASA Releases 74-292, 74-300; Hall, Opp, *et al.*, *Science*, 2 May 75, 445-77; NASA Pioneer Mgr, interview, 12 Nov 75; NASA press conference transcript, 19 Nov 74; Perlman, *SF Chron*, 3 Dec 74; O'Toole, *W Post*, 3 Dec 74, A1; Sullivan, *NYT*, 4 Dec 74, 27; Mead, *San Jose Mercury*, 5 Dec 74; *NYT*, 9 Dec 74)

*4 November:* President Ford accepted a segment of crystal grown in the orbiting Skylab Workshop during January, from Dr. James C. Fletcher, NASA Administrator, and Massachusetts Institute of Technology Chairman Howard W. Johnson. Johnson said the indium-antimonide crystal, produced in an MIT experiment, was the purest crystal with the longest life yet produced by man. Using it to carry power current, a computer could be the size of a postage stamp. (*NASA Activities*, 15 Nov 74, 17; *PD*, 11 Nov 74, 1428; UPI, *W Post*, 5 Nov 74, 1)

*4-15 November:* The formal phase of the Spacelab Preliminary Requirements Review—first major milestone in the Spacelab program—was completed in Noordwijk, The Netherlands, and Bremen, West Germany. Representatives of NASA and the European Space Research Organization approved 1077 review items, withdrew 56, and marked 100 for more study. Avionics and mechanical interfaces with the space shuttle were selected for particular attention before the next major milestone—the Systems Requirements Review. (*Spacelab Newsletter* 74-8; Spacelab/CVT Program Approval Document, Attach A, 1)

*5 November:* Former Mercury astronaut John H. Glenn, Jr., the first American to orbit the earth, was elected U.S. Senator from Ohio by a wide margin over his Republican opponent, Ralph J. Perk. Glenn had withdrawn from a 1964 senatorial bid after a fall affected his inner ear. In 1970 he lost the Democratic primary to Howard M. Metzenbaum, who in turn lost the general election to Robert Taft, Jr. Metzenbaum, appointed to fill the vacated seat of William B. Saxbe in September 1973, lost to Glenn in the 1974 Democratic primary. (Farrell, *NYT*, 6 Nov 74, 37)

*7 November:* The appointment of Richard G. Smith as Deputy Director of Marshall Space Flight Center, effective 18 Nov., was announced by Center Director William R. Lucas. Smith would also continue as acting director of MSFC's Science and Engineering Directorate. Since 1960, Smith had held positions of increasing responsibility in the Science and Engineering Directorate, including Saturn Program Manager from 1970 to 1973. (MSFC Release 74-218)

- Aircraft designer Clarence L. (Kelly) Johnson announced his retirement as Senior Vice President of Lockheed Aircraft Corp., effective in mid-January 1975. During his 43 yrs with Lockheed, Johnson had contributed to the design of some 40 aircraft—about half of them his original designs—including the P-38 "Lightning" fighter of World War II, the P-80 jet fighter, the Constellation series of airliners, the F-104 mach 2 fighter, the U-2 reconnaissance plane and its successor SR-71, and the C-130 transport. In 1964 President Johnson awarded Johnson the National

Medal of Science, citing his achievements and products "as both incomparable and virtually incredible." (Getler, *W Post*, 8 Nov 74, A12)

*11 November:* A shortage of unsymmetrical dimethylhydrazine (UDMH) rocket fuel was being strictly managed to prevent delays in military and NASA launches. The fuel was produced from raw materials found to have an agent that could cause cancer in humans. Methods had been devised to produce the fuel by removing humans from the step endangering health, but adequate supplies could not be produced before the end of the year. Stopgap measures taken by the Air Force, including strict control and allocation of existing supplies, was preventing the shortage from interfering with any critical launches. (NASA Dir Launch Vehicle & Propulsion Prog, interview, 2 Oct 75; *Av Wk*, 11 Nov 74, 14)

*12 November:* The U.N. Committee on the Peaceful Uses of Outer Space adopted a draft convention that would make compulsory the registration of all spacecraft. The treaty required approval by the General Assembly and signing by member nations before taking effect. (Dept of State, Off Treaty Affairs, interview, 11 Aug 75; *SBD*, 16 Oct 74, 20)

- NASA contract awards to small business during FY 1974 totaled $181 247 000—8.6% of all contracts—NASA announced. This percentage was the highest since 1962. (NASA Release 74–293)

*13 November:* "The only way to face the crisis of our times is through the advancement of science and technology," Dr. George M. Low, NASA Deputy Administrator, said in an address to the Cal Tech Management Club. "The space program is at the forefront of the search for knowledge, . . . of producing knowledge, . . . of developing the technological capability which comes from this knowledge." Yet, "among even the higher echelons of the American business community, lack of understanding rather than comprehension of space ventures is the prevailing frame of mind." Dr. Low called for communication to the American people of how the space program increased understanding of the universe and man's place in it, improved life on earth, and kept the country secure through scientific, technological, and economic strength, as well as international cooperation. (Text)

*15 November:* NASA launched three satellites—*Noaa 4* (ITOS-G) meteorological satellite for the National Oceanic and Atmospheric Administration, *Intasat* ionospheric satellite for Spain, and *Amsat Oscar 7* amateur radio communications satellite—from Western Test Range at 9:11 am PST on one Thor-Delta launch vehicle. The three spacecraft entered almost identical, near-to-planned, polar orbits. It was the first triple launch for a Thor-Delta.

*Noaa 4* entered orbit with a 1458-km apogee, 1444-km perigee, 115-min period, and 101.7° inclination, as the fourth operational Improved Tiros Operational Satellite in the joint NASA and Dept. of Commerce program to provide global cloud-cover observations. Delayed 29 Oct. by faulty electrical connectors on the Thor-Delta and again 13 Nov. when a 2nd-stage hydraulic pump posed a problem, the launch went smoothly on 15 Nov. *Noaa 4* joined *Noaa 2*, launched 15 Oct. 1972, and *Noaa 3*, launched 6 Nov. 1973.

The 345-kg, three-axis-stabilized, earth-oriented satellite moved in a circular, near-polar, sun-synchronous orbit. Day and night coverage of the earth's cloud cover was provided by two very-high-resolution radiometer instruments and by two scanning radiometer sensors. Two vertical-temperature-profile radiometers made continuous radiance measurements for determination of the atmospheric vertical temperature profile over every part of the earth's surface at least twice daily. One secondary sensor, the solar proton monitor, studied solar activity and its effects on the earth's environment. Goddard Space Flight Center was responsible for design, development, launch, and in-orbit checkout of the spacecraft. NOAA's National Environmental Satellite Service (NESS) would take over operation of the satellite after checkout. NASA mission objectives—to launch the spacecraft into sun-synchronous orbit of sufficient accuracy for operational mission requirements and to conduct an in-orbit evaluation and checkout of the spacecraft—were accomplished and the mission was adjudged successful 24 Jan. 1976.

*Intasat*, the first Spanish satellite, separated from the Delta 2nd stage 24 min after launch and entered orbit with a 1468-km apogee and perigee, 115-min period, and 101.7° inclination. The 20-kg, 12-sided, magnetically oriented, spin-stabilized spacecraft would measure ionospheric electron content, irregularities, and scintillations in a joint space research program of NASA and the Instituto Nacional de Técnica Aeroespacial. INTA provided the spacecraft and would acquire experiment data. NASA provided pickaback launch and tracking.

*Amsat Oscar 7*—the Radio Amateur Satellite Corp.'s Orbiting Satellite Carrying Amateur Radio—separated from the Delta 2nd stage 77 min after liftoff and entered orbit with a 1458-km apogee, 1454-km perigee, 115-min period, and 101.7° inclination. The seventh in a series begun with *Oscar 1* in 1961, the solar-powered, octahedral, 29.5-kg *Oscar 7* joined *Oscar 6*, launched 15 Oct. 1972, to serve as a space communications relay station for radio amateurs and as a teaching aid, giving first-hand experience in space science for U.S. schools. *Oscar 7* was built and prepared for launch by the Radio Amateur Satellite Corp. (NASA MORS, 18 Oct 74; 19, 20, 29 Nov 74; 26 Jan 76; NASA Release 74–322; *Goddard News*, Dec 74; *SBD*, 30 Oct 74, 301; 14 Nov 74, 79)

- NASA announced selection of Boeing Aerospace Co., Lockheed Missiles & Space Co., and Martin Marietta Corp. for negotiation of parallel contracts of about $300 000 each for preliminary design and program definition studies for the support systems module (SSM), a major element of the Large Space Telescope (LST). The LST was scheduled for launch into orbit on the space shuttle in the 1980s. The SSM, attached to the structure housing the telescope, would provide electric power, data storage, communications, and automated attitude sensing, pointing, and control. The unit would also contain some of the elements for orbital servicing and retrieval of the telescope by the space shuttle. (NASA Release 74–308)

*16 November:* In an attempt to communicate with another civilization in the universe, a powerful three-minute radio signal was beamed from the Arecibo Observatory in Puerto Rico toward Messier 13, a globular cluster of stars at the edge of the Milky Way, 25 000 light years from the earth. The transmission—devised by the staff of the National Astronomy and Atmosphere Center, which included the Observatory and was operated by Cornell Univ. for the National Science Foundation—was a coded message representing life on the earth. The Observatory's new 450-kw radar transmitter, sponsored by NASA and completed in October, had increased the planetary ranging and surface mapping capabilities a thousandfold. When focused by the Observatory's newly resurfaced 305-m dish the energy was multiplied, sending a beam to Messier 13 brighter than any other source in the Milky Way. (Sullivan,

*NYT*, 20 Nov 74, 77; O'Toole, *W Post*, 17 Nov 74, A1; Pres Rpt 74, 86; NASA Release 74–309; NASA, Chief Plan Astron, interview, 15 Dec 75)

**19 November:** Wind-tunnel tests of 4% scale models of the space shuttle orbiter had begun at Air Force Systems Command's Arnold Engineering Development Center, the Air Force announced. The tests of heat buildup on the forward half of the orbiter during the critical period of reentry and at other points in the flight trajectory were scheduled to continue through the remainder of the year. The 4% model, the fourth size to be tested in the AEDC tunnels, reflected the latest design details on the forward half of the orbiter. (AFSC Release OIP 207.74)

**20 November:** NASA and the Federal Government as a whole were charged with race and sex discrimination in a class action suit filed by the National Association for the Advancement of Colored People (NAACP) Legal Defense Fund in U.S. District Court in Washington, D.C. The suit was initiated by two employees at Johnson Space Center. The NAACP Legal Defense Fund labeled as "lame excuses" NASA's claims that work forces were dropping throughout the agency and that it was difficult to find qualified blacks and women in its specialized areas of work. (AP, *NYT*, 21 Nov 74; Robinson, *W Post*, 21 Nov 74)

**21 November–20 December:** NASA launched *Intelsat-IV F–8* for the Communications Satellite Corp. on behalf of the International Telecommunications Satellite Organization. The 1387-kg satellite was launched from Eastern Test Range at 6:44 pm EST on an Atlas-Centaur vehicle into a transfer orbit with a 35 899.2-km apogee, 536.5-km perigee, 10-hr 39.8-min period, and 26.0° inclination.

NASA's primary objectives were to conduct design, performance, and flight readiness reviews for the Federal Communications Commission, to ensure compatibility of Intelsat spacecraft with NASA launch vehicles and launch environmental conditions, and to launch *Intelsat-IV F–8* into a transfer orbit that would permit the spacecraft apogee motor to inject the spacecraft into a synchronous orbit for communications service.

The apogee kick motor was fired by ComSatCorp at 6:48 pm EST on 23 Nov., placing *Intelsat-IV F–8* in a near-circular orbit with a 35 780-km apogee, 35 630-km perigee, 23-hr 51.9-min period, and 1.8° inclination. The spacecraft then drifted to its station at 174° E longitude over the Pacific Ocean, arriving 12 Dec. NASA adjudged the mission a success 20 Dec.

*Intelsat-IV F–8* joined five operational Intelsat IV commercial comsats: three over the Atlantic—*F–2* launched 25 Jan 1971, *F–3* launched 19 Dec. 1971, and *F–7* launched 23 Aug. 1973; one over the Pacific—*F–4* launched 22 Jan. 1972; and one over the Indian Ocean—*F–5* launched 13 June 1972. *F–8* replaced *F–4*, which became a spare, orbiting over 178° E longitude. *F–8* was designed to operate seven years and could provide multiple access and simultaneous transmission. It had a capacity for 3000 circuits with transponders in earth mode and 9000 circuits with transponders in spot-beam-coverage mode, or 12 TV channels, or certain combinations. (NASA MORS, 22 Nov, 24 Dec 74; GSFC *Wkly SSR*, 21–26 Nov 74; NASA Release 74–305; INTELSAT Releases 74–62, 74–70; NASA Atlas-Centaur Mgr, interview, 30 July 75)

**21 November:** The U.S.S.R. launched *Molniya III–1* communications satellite from Plesetsk into orbit with a 40 683-km apogee, 628-km perigee, 12-hr 17.3-min period, and 62.8° inclination. The spacecraft was intended for long-distance telephone and cable radio communications in the Soviet Union, for transmission of color TV programs within the Orbita network, and for international cooperation. (Tass, FBIS–Sov, 22 Nov 74, U1; GSFC *Wkly SSR*, 21–26 Nov 74; LC Science Policy Research Div, interview, 9 Sept 75)

- In a Skylab awards ceremony at Marshall Space Flight Center, Dr. James C. Fletcher, NASA Administrator, presented Group Achievement Awards to the Skylab Program Offices at MSFC, Kennedy Space Center, Johnson Space Center, and NASA Hq. and to special Skylab teams. Dr. Robert A. Parker, NASA astronaut and Skylab program scientist at JSC, received an Outstanding Leadership Medal. Dr. Fletcher also presented 28 individual Exceptional Scientific Achievement Medals to principal investigators for Skylab experiments and to Comet Kohoutek scientific investigators. (MSFC Release 74–227; *Marshall Star*, 27 Nov 74)

*21–22 November:* New cost increases in the Viking program had raised the total cost of the mission to more than $1 billion, NASA officials told the House Committee on Science and Astronautics' Subcommittee on Space and Science and Applications. Dr. Rocco A. Petrone, NASA Associate Administrator, testified that the preliminary cost estimate of $364 million, presented to the Committee in 1969, had been nearly trebled by delay in the mission from 1973 to 1975, inflation, and technical problems in "developing the most advanced automated space equipment this country has ever attempted." The new total included a $930-million cost target, a $20-million contingency fund, and the cost of the Titan-Centaur launch vehicles. One orbiter and one lander spacecraft had been deleted from the program [see 10 Oct.], manpower had been reduced, and a number of tests had been eliminated. Dr. Petrone expressed "confidence in our present technical approach and our ability to meet cost goals." Dr. Edgar M. Cortright, Langley Research Center Director, attributed the "$120 million in unanticipated problems" experienced by mid-1973 to an underestimation of the complexity of the project at the outset: "We have been playing catch up ever since because of the low estimate to start with."

Questioned about a newspaper article saying the Viking lander might sink deep into Martian surface dust, Dr. Petrone said that tests mentioned in the article had been made some years earlier, not "two weeks ago" as reported. Dr. Noel W. Hinners, Associate Administrator for Space Science, testified that the lander's footpad area had been increased since the earlier tests and that radar observations had indicated the proposed landing sites for the first mission appeared "well within the capability" of the lander. Sites for the second mission were still being studied. (Transcript; O'Toole, *W Post*, 20 Nov 74, A3)

*22 November–6 January 1975:* NASA launched *Skynet IIB*, a U.K. military communications satellite, for the U.S. Air Force Space and Missile Systems Office as agents for the U.K. Ministry of Defense. The satellite was launched on a three-stage Thor-Delta launch vehicle from Eastern Test Range at 7:28 pm EST into an orbit with a 36 965-km apogee, 185-km perigee, and 24.6° inclination. The apogee kick motor, fired at 8:10 pm EST on 24 Nov., circularized the orbit with an apogee of 36 595 km, perigee of 35 896 km, period of 24 hrs 19.4 min, and inclination of 2.2°. The spacecraft then drifted to its geosynchronous station at 50° E longitude over the Indian Ocean. NASA adjudged the launch a success 6 Jan. 1975.

The first-generation *Skynet A* and *B* had been developed for the U.K. in the USAF's Initial Defense Communications Satellite Program (IDCSP) and launched by NASA 21 Nov. 1969 and 18 Aug. 1970. *Skynet IIA* failed to achieve a satisfactory orbit when launched 18 Jan. 1974 and reentered 25 Jan. The launch of *Skynet IIB* had been delayed from 19 Nov. by a defective component in the Delta 2nd-stage attitude control system. (NASA MORs, 7 Nov, 5 Dec 74; 6 Jan 75; GSFC *Wkly SSR*, 21–26 Nov 74; NASA Release 74–303; AP, *W Post*, 12 Nov 74, C5)

*22 November:* NASA and the National Science Foundation (NSF) had signed a Memorandum of Understanding for collaboration in research on terrestrial uses of solar energy, NASA announced. NSF, the lead Federal agency in solar energy research, chaired the 20-member Interagency Panel for Terrestrial Applications of Solar Energy. NASA's role in the National Solar Energy Program would be directed to projects in which NASA had special capabilities, although this participation was expected to touch the full range of solar energy research. (NASA Release 74–302)

*23–24 November:* President Ford and General Secretary Leonid I. Brezhnev of the Soviet Communist Party's Central Committee, meeting in Vladivostok, drew up the framework of a new, 10-yr strategic arms agreement, with a goal of completing the accord in 1975. The agreement would put a ceiling of 2400 for each country on the total of intercontinental ballistic missiles, submarine-launched missiles, and heavy bombers. Wthin that total, the number of missiles armed with multiple warheads would be limited to 1320. (*PD*, 2 Dec 74, 1489–97; 9 Dec 74, 1514–7)

*25 November:* Sen. Frank E. Moss (D–Utah) introduced an amendment to S. 2350 that would continue experimentation with earth resources remote-sensing satellite systems through 1 Jan. 1980 unless a permanent system was established before that date. The amendment was a substitute for S. 2350, the earth resources survey bill to develop an operational ERTS system and to establish an Office of Earth Resources Survey Systems within NASA and an Earth Resources Observation Administration within the Dept. of the Interior [see 18 Sept.]. Other provisions of the amendment would require the President, through NASA and DOI, to ensure continuity of satellite data during the experimental period and furnish Congress with any necessary budget amendments. (*CR*, 25 Nov 74, S19948)

*26 November:* Deferral of $72 million in NASA FY 1975 R&D funding was proposed by President Ford in a message to Congress. Aeronautical research and space support activities would be deferred by $36 million, affecting advanced systems for space exploration and aeronautics, analysis of data in the space science program, and maintenance and upgrading of equipment for the tracking and data-acquisition program. In manned space flight, $20 million in ASTP funds would be deferred. And in space science and applications, deferral of $16 million in no-year (not designated for a specific year) funds would delay the rate of buildup of Pioneer-Venus, SEASAT–A, Nimbus-G, Tiros-N, and the Heat Capacity Mapping Mission. These deferrals were parts of a $4.6-billion budget outlay reduction asked by the President. (*PD*, 2 Dec 74, 1500–1; *CR*, 26 Nov 74, S20096–7; *Federal Register*, 5 Dec 74, 42519–667)

- NASA and the National Science Foundation announced the award of two contracts of about $500 000 each to General Electric Co. Space Div. and Kaman Aerospace Corp. for preliminary design of very large wind systems for generating electricity. Windmills from 100 to 3000 kw would be examined by each company in a project managed by Lewis Research Center for NSF as part of the program for Research Applied to National Needs (RANN). A 100-kw system with vanes more than 37 m in dia, already under design and construction, would be erected at LeRC's Plum Brook test area. (LeRC Release 74–73)

*During November:* Marshall Space Flight Center awarded McDonnell Douglas Corp. a $299 454 contract for the Spacelab Payload Utilization and Planning Study for the space transportation system. (MSFC Release 74–243)

## December 1974

*2–8 December:* The U.S.S.R. launched *Soyuz 16*, carrying cosmonauts Col. Anatoly V. Filipchenko and Nikolay N. Rukavishnikov, from Baykonur Cosmodrome near Tyuratam at 2:40 pm local time (4:40 am EST) in preparation for the July 1975 U.S.–U.S.S.R. Apollo-Soyuz Test Project mission. The spacecraft, identical to the one that would fly during the ASTP mission, entered orbit with a 280-km apogee, 194-km perigee, 89.3-min period, and 51.8° inclination. Tass reported the purpose of the mission was to test the spacecraft's onboard systems, which had been "modernized to meet the demands of the joint flight"; make scientific and technical investigations; and observe and photograph sections of the earth's surface for the solution of economic problems.

In a prelaunch Tass interview, Filipchenko said that the mission was to last several days. "This would be enough for us to test carefully all the ship's systems, above all its docking gear." Using a passive simulator docking ring, *Soyuz 16* would make docking and separation maneuvers in almost all the ASTP modes. In addition, U.S.S.R. ASTP Technical Director Konstanin D. Bushuyev said in a launch-day Tass interview, the crew would check out the Soyuz life support system, which had been redesigned for better compatibility with Apollo. Soyuz normally had a cabin pressure of 10 newtons per sq cm (14.7 psi) and an atmosphere similar in oxygen content to terrestrial atmosphere. Apollo maintained its pure oxygen atmosphere at 3.5 newtons per sq cm (5 psi). To reduce the acclimation period for the crews' transition between the two atmospheres, the U.S.S.R. had agreed to reduce the pressure and increase the percentage of oxygen in the Soyuz atmosphere. The system's capacity also had been increased to accommodate four persons.

On 3 Dec. NASA, which had been tracking *Soyuz 16* since shortly after launch, began a 15-day joint tracking rehearsal with the U.S.S.R. After the mission, information gathered by nine U.S. tracking stations would be compared with data from Soviet tracking stations. Meanwhile the crew reduced the spacecraft's cabin pressure to 7.2 newtons per sq cm (10.4 psi) and doubled the oxygen content, from 20% to 40%, with no adverse effect. Color TV broadcasts from the spacecraft showed all was well. A trajectory correction 4 Dec. placed *Soyuz 16* in a circular orbit with a 225-km altitude, 88.9-min period, and 51.8° inclination, an orbit similar to ASTP requirements. Simulated docking and separation maneuvers "passed off without a hitch" 5–7 Dec., Tass reported. Medical, biological, earth resources, and astronomy experiments also were carried out during the six-day mission. At 1:04 pm Baykonur time (3:04 am EST) 8 Dec., *Soyuz 16* softlanded in the U.S.S.R. 300 km north of Dzhezkazgan in Kazakstan. Tass reported the cosmonauts were in good health.

Following the successful landing, U.S. ASTP Technical Director Glynn S. Lunney conveyed his congratulations "for the brilliant rehearsal before the Soviet-U.S. flight." In a Tass interview Bushuyev said,

"We have succeeded in checking in real space flight the whole package of the new system and units responsible for the success of the future linkup of the Soviet and U.S. spaceships." *Soyuz 16* was the third manned U.S.S.R. launch in 1974, following *Soyuz 14* (launched 3 July) and *Soyuz 15* (26 Aug.). (GSFC *SSR*, 31 Dec. 74; Tass, FBIS–Sov, 2–11 Dec 74; Wren, *NYT*, 3 Dec 74, 16; 4 Dec 74, 26; JSC Release 74–272)

*2, 12 December:* American astronauts and technicians, although denied access to Soviet spacecraft factories, would be permitted to inspect the Soyuz spacecraft for the joint Apollo-Soyuz Test Project at the Baykonur Cosmodrome launch site in May 1975, NASA officials said at a Johnson Space Center news conference. U.S. ASTP Technical Director Glynn S. Lunney said the arrangement would satisfy U.S. technical needs.

In a 12 Dec. press conference at Star City, Maj. Gen. Vladimir A. Shatalov, U.S.S.R. chief of cosmonaut training, said American crews would be permitted to view only facilities directly connected with the flight itself. "Everything that American astronauts need to be acquainted with on the Soyuz that will fly will be shown them beforehand." In the meantime astronauts would train with mockups and simulators. (JSC PIO, interview, 15 Sept 75; Keefer, *Today*, 5 Dec 74; Chriss, *LA Times*, 3 Dec 74; Toth, *LA Times*, 13 Dec 74; FBIS–Sov, 13 Dec 74, U1)

*2 December:* The surface of Jupiter's moon Io might be composed of evaporite salt deposits, rich in sodium and sulfur, NASA reported. A year-long Jet Propulsion Laboratory study had indicated that deposits had resulted from migration of salt-saturated aqueous solutions to Io's surface from a warm or hot interior followed by loss of the water to space. Salt-rich materials could have been produced by the leaching of carbonaceous meteorites. According to Dr. Fraser P. Fanale, JPL team leader, the deposits could account for Io's high reflective power, the absence of ice bands, and the dark reddish poles.

Sodium vapor emissions from Io had been first reported in February 1973 by Dr. Robert Brown of Harvard Univ. Subsequently, the cloud of sodium was observed to extend more than 320 000 km from Io.

In August Dr. Fanale with two other astronomers had proposed explanations for the sodium emissions. Proton bombardment experiments had indicated that sodium and other atoms had been blasted from Io's surface by the impact of Jupiter's magnetospheric protons, electrons, and ions and had formed a cloud around Io. Io's glow, according to intensive spectral studies, resulted from scattering of sunlight by the sodium cloud. (NASA Release 74–315; Fanale *et al.*, *Science*, 6 Dec 74, 922–5; Matson *et al.*, *Astro Journ*, 15 Aug 74, L43–6; JPL proj off, interview, 24 Sept 75, NSF Release 74–199)

- The first Apollo lunar surface experiment package (ALSEP), placed on the moon by *Apollo 12* astronauts in November 1969, had far exceeded its one-year life expectancy and was still functioning, along with instruments in ALSEP packages from *Apollo 14* through *17*, NASA announced. From ALSEP magnetometers, Dr. Palmer Dyal, Ames Research Center lunar investigator, had calculated that the moon had a magnetic field 1000 times weaker than the earth's, with an iron content about 9% by weight, in contrast to the earth's 30%. The heat flow measured by ALSEP instruments was half the earth's, indicating differentiation and upward concentrations of radioactivity early in the moon's history. Seismometers had uncovered a new class of moonquakes originating at depths of 100 km or less below the lunar surface. Dr. M. Nafi Toksoz of Massachusetts Institute of Technology earlier had reported that preliminary information indicated the moon had a small core less than 800 km across, lying inside a molten shell 600–1000 km down from the surface, where deep moonquakes had been detected. Seismic, magnetic, and heat-flow data had also suggested that the moon was further along than the earth in its evolutionary history. (NASA Release 74-313; NASA prog off, interview, 17 Nov 75; McElheny, *NYT*, 18 Nov 74)

- An agreement to establish an Aerosat (Aeronautical Satellite) Space Segment Program was signed in Washington by the European Space Research Organization, COMSAT General Corp., and the Government of Canada. Under the agreement—provided for in the Memorandum of Understanding signed 2 Aug.—ESRO, COMSAT General, and Canada would furnish the space segment capability for the intergovernmental program. COMSAT General would lease its share of the space segment capability to the Federal Aviation Administration. (ESRO Release, 3 Dec 74)

- The Dept. of Defense had officially approved development of the NAVSTAR Global Positioning System (GPS), the Air Force announced. The first phase of the three-phase joint armed services program called for five Rockwell International Corp. spacecraft to join a Navy Navigational Technology Satellite (NTS) in two-plane orbital constellations in 1977 to test the system. The five-meter, 635-kg spacecraft would be launched on Atlas F boosters by the Air Force Systems Command's Space Missile Test Center. Phase two would bring the number of satellites to nine in three orbital planes. The operational third phase—with 24 satellites orbiting at an 18 520-km altitude, providing three-dimensional location to air, sea, and surface vehicles in all kinds of weather—would be complete in the mid-1980s. (AFSC Release OIP 274.74)

- Inflation would force even more extensive cuts in NASA operations if funding did not increase in the next few fiscal years, *Aviation Week & Space Technology* reported in a survey of NASA belt-tightening. The magazine quoted NASA Comptroller William E. Lilly as saying cost contracts for major spacecraft programs had a 9–12% inflation rate. The overall effective impact of inflation was 7–7.5% at NASA, but these averages were much lower than for many specific areas. Although materials increases were averaging 15–20%, particular increases ranged from 0–100%. The 7% inflation rate in Center supplies was being absorbed by using inventories and by decreasing the use of supplies. Utility cost increases of 20–30% at NASA Centers had been countered by lower use—a 30% cut—but Centers were still paying 10–15% more. Overall wage rates could increase 20–30% over the next three years, Lilly had said, because of a trend-setting agreement between the Boeing Co. and the International Assn. of Machinists and Aerospace Workers. Support service contracts were showing a 7–9% yearly inflationary increase.

NASA Centers were seeking new solutions. At Goddard Space Flight Center, more fixed-price contracts were being issued, but predicting economic changes was problematic, according to William R. Mecca, GSFC Deputy Director for Administration and Management. GSFC had begun limiting data acquisition from spacecraft and drawing up contingency plans for many unmanned programs—like the International

Sun-Earth Explorer Spacecraft program and the International Ultraviolet Explorer program—to save costs if primary plans began to exceed spending limits.

Marshall Space Flight Center had reduced its work force by 850—about 15%—since June 1973 and had closed 47 structures. Lewis Research Center had cut its power needs 7%, using a transient data-recording system to reduce operating hours. Johnson Space Center, along with others, had been hiring young persons rather than high-salaried experienced persons. JSC electric costs were still increasing even though the Center had reduced its exterior lighting by 75% and its interior lighting by 33%. Kennedy Space Center was phasing construction for shuttle and other programs, outfitting facilities to meet minimum initial operational requirements rather than for long-range mission models. Consideration was being given to advance procurement for lower costs and for preventing slowdowns from shortages. (Covault, *Av Wk*, 2 Dec 74, 38–42; Lilly, interview, 19 Sept 75)

*3 December:* NASA's *Pioneer 11*, traveling toward Jupiter since 5 April 1973 launch, flew within 42 000 km of the planet's cloud tops and sent back the first polar images—as well as magnetospheric, temperature, and other measurements—before swinging on toward Saturn [see 2 Nov.–3 Jan. 1975]. (NASA Pioneer Mgr, interview, 12 Nov 75; Hall, Opp, *et al.*, *Science*, 2 May 75, 445–8)

• The European Space Research Organization announced the award of three contracts totaling nearly $66 million for design and development of two spacecraft and one satellite communications payload: $25.2 million to GEC–Marconi (United Kingdom) as the prime contractor for the MAROTS maritime communications satellite payload to be launched in 1977; $20.76 million plus launch and postlaunch support to Hawker Siddeley Dynamics Ltd. (U.K.) as prime contractor for the MESH consortium, for the MAROTS spacecraft platform and satellite integration and support; and $19.56 million to Dornier-System GMBh (Germany) as prime contractor for the STAR consortium, for the ISEE–B International Sun-Earth Explorer, to be launched with ISEE–A in 1977 in a cooperative NASA–ESRO program.

Norway, the first nonmember country to participate in an ESRO project, had agreed to contribute 1.5% to the MAROTS development budget of about $97 million and to share in fixed common costs. (ESRO Release, 3 Dec 74; *Av Wk*, 2 Dec 74, 38)

*4 December:* The NASA Space Shuttle Cost and Review Committee recommended program changes to Dr. James C. Fletcher, Administrator, to reduce rising costs. It also reported design changes made necessary by thermal weight constraints. The committee, headed by NASA Associate Administrator for Manned Space Flight John F. Yardley, recommended delaying the second shuttle orbiter flight, refurbishing the first test orbiter earlier than planned instead. Deferrals of three months to two years were recommended for components of the second shuttle launch complex at Kennedy Space Center, which had originally been scheduled for completion about the time of the first shuttle flights. Also recommended was a delay in the development of improved spacesuits and portable life support systems for extravehicular activities from the space shuttle. Instead, EVAs would use Apollo-style hardware. Emergency barriers on shuttle runways and a building at KSC for installing fairings and parachutes on solid-fueled rocket motors were judged unnecessary. Some tests were recommended for cancellation.

Design changes to accommodate thermal and weight constraints included payload doors of a graphite epoxy composite that expanded less when heated and weighed 400 kg less than the aluminum composite originally planned for use. Solid-fueled-rocket separation motors that would provide a higher thrust level for a shorter time when separating the large solid-propellant boosters from the orbiter after launch were found to cause no orbiter damage—solving a possible problem anticipated with the original separation motors. The committee also recommended switching to a new thermal coating material for the external liquid-hydrogen and liquid-oxygen tank to prevent tank overheating during launch. (NASA prog off, interview, 22 Sept 75; *Av Wk*, 9 Sept 74, 9–10; NASA Off of Admin, Daily Appointments Calendar)

- Dr. George M. Low, NASA Deputy Administrator, was presented the 1974 Rockefeller Public Service Award for Administration by President Ford at the Mayflower Hotel in Washington, D.C. As Deputy Associate Administrator for Manned Space Flight at NASA Hq., Deputy Director of Manned Space Flight Center, and then MSC Manager of the Apollo Spacecraft Program before becoming NASA Deputy Administrator in 1969, Dr. Low had been responsible for development of the Mercury and Gemini programs, an original planner of the Apollo program, and negotiator for the U.S.–U.S.S.R. Apollo-Soyuz Test Project. He had joined the National Advisory Committee for Aeronautics in 1949 and moved to NASA when it was established in 1958.

Dr. Robert M. White, Administrator of the National Oceanic and Atmospheric Administration, also received one of the five $10 000 awards, for physical resource development and protection. Established by John D. Rockefeller II in 1952 to honor excellence in the career Civil Service, the awards were administered by the Princeton Univ. Woodrow Wilson School of Public and International Affairs. (*PD*, 9 Dec 74, 1528–9; *Marshall Star*, 11 Dec 74; *NASA Activities*, 15 Dec 74, 8; NASA Biog)

*5 December:* Dr. James C. Fletcher, NASA Administrator, signed a Memorandum of Understanding on cooperation in energy research and development with the Electric Power Research Institute (EPRI). Dr. Chauncey Starr, President of EPRI, had signed the agreement 24 Nov. The agencies would cooperate in applying safety, reliability, and quality assurance techniques developed for the space program to electric power generating plants and in research and technology development for advanced energy systems. (NASA OAST PIO, interview, 3 Sept 75; NASA Release 74–325)

*6 December:* The International Telecommunications Satellite Organization announced it had contracted with Hughes Aircraft Co. for three Intelsat IVA satellites in addition to three ordered in 1973. The amendment to the 1973 contract totaled more than $40 million, including launch support services but not possible performance incentive payments. The first of the additional satellites, with twice the communications capability of the Intelsat IV series, was to be delivered in 1977. (INTELSAT Release 74–67)

*9 December:* Wind-tunnel tests at Ames Research Center had indicated that NASA's swing-wing aircraft configuration could provide a practical design for an advanced commercial jet transport, NASA announced. The aircraft's straight wing, mounted above the body, could be turned to various oblique angles for different flight speeds and was proposed for slow and fast design versions. Operating at speeds of 800 to 1400 km per hr, the needle-nosed fast version would maintain relatively economical fuel consumption—13 000 to 17 000 km per cu m per passenger—about twice the fuel economy of supersonic transports like the Concorde or the Russian Tu-144. For slower flight, the wing would be fixed at right angles to the fuselage, for landings and takeoffs with minimum power and noise. The design's flexibility would permit shorter flight times with little increase in overall fuel consumption. (NASA Release 74-318)

- A breathing system developed by Johnson Space Center using technology for solid-propellant rocket-motor cases was being evaluated in the field by Houston firemen. New air tanks were built of spiral-wound glass fiber over an aluminum liner to provide a lighter and more durable system. (NASA Release 74-320)

**10 December–16 January 1975:** NASA launched West German-built *Helios 1* solar probe at 2:11 am EST from Eastern Test Range on a Titan IIIE–Centaur vehicle, putting it on a path that would take it closer to the sun than any spacecraft had yet traveled. The spacecraft first entered a parking orbit and then an elliptical orbit of the sun with a 0.985–AU aphelion, 0.3095–AU perihelion, 190.15-day period, and 0.02° inclination to the ecliptic. All vehicle systems performed normally during launch.

The Titan-Centaur carried a Delta solid-fueled rocket-motor kick stage as a 4th stage and required seven interdependent ignition systems.

Following separation from the 1st stage, the Centaur, with 2300 kg of propellants remaining, began maneuvers to provide data for the 1977 Mariner Jupiter-Saturn mission, future synchronous orbit missions, and continued development of the launch vehicle. Maneuvers included a one-hour coast and restart, three-hour coast with thermal maneuvers, boost-pump operation experiment with igniter on, and zero-g boost-pump dead-head operation.

*Helios 1* was to investigate the fundamental solar processes and solar-terrestrial relations by studying the solar wind, magnetic and electric fields, cosmic rays, and cosmic dust from launch to the first perihelion, 0.31 AU from the sun. The 370-kg spool-shaped spacecraft carried 10 instruments—7 West German and 3 U.S.—to measure the solar wind, magnetic field, and related shock waves; study radio waves and oscillations in the electron plasma; measure the spatial gradient of solar and galactic cosmic rays; observe composition and dynamics of interplanetary dust; and monitor x-ray activity of the solar disc.

By 16 Jan. orientation and spin-rate maneuvers had been conducted and the high-gain antenna despun and pointed toward the earth. Control of the spacecraft had been passed to Germany. Instruments had been turned on and were operating normally with two exceptions: One of two antennas in the U.S. plasma-wave experiment had jammed during extension and unrolled inside the drive enclosure, producing an electrical ground and degrading all low-frequency data. Rewinding the antenna had not eliminated the ground. In addition, operation of the high-gain antenna for the West German plasma experiment produced a space charge near the spacecraft, preventing detection of electrons with energies less than 100 ev. The medium-gain antenna would be used during selected periods.

To prepare for its demanding mission, *Helios 1*, designed to withstand the highest temperatures ever demanded of a spacecraft, had been successfully subjected to six days at 644 K (700°F), the highest heat load expected at 0.3 AU, during spring 1974 tests at Jet Propulsion Laboratory. The spacecraft had also withstood radiation 11 times the solar intensity at the outer edge of the earth's atmosphere. Independent payload components, louver systems, optical surface reflectors and mirrors, and several layers of insulation dissipated spacecraft heat. In addition the spacecraft spun at one revolution per second to distribute evenly the heat coming from the sun. Temperatures within the experiment area were kept between 263 and 303 K (14° and 86°F).

A second Helios spacecraft would be launched in 12 to 18 mos and information from the two spacecraft would be received simultaneously from widely differing locations, for comparison with each other and with data from Pioneer- and IMP-class spacecraft.

Helios, the third cooperative U.S. and West German project, was agreed on during 1966 talks. Following technical discussions and mission definition, a Memorandum of Understanding was signed 10 June 1969. NASA provided three scientific instruments, spacecraft technical support, launch vehicle and operations, and initial tracking and data acquisition and flight operations. The German Federal Ministry for Research and Technology (BMFT) designed, built, and tested the spacecraft and provided seven scientific instruments, data acquisition and analysis, and spacecraft control.

Goddard Space Flight Center was responsible for NASA project management; Lewis Research Center had responsibility for the launch vehicle. (NASA MORs, 6 Dec 74, 16 Jan 75; NASA PAO, interview, 10 Dec 74; NASA Releases 74-314, 75-68; UPI, *NYT*, 9 Dec 74, 45)

*10 December:* A Memorandum of Understanding on space cooperation was signed at NASA Hq. by Dr. James C. Fletcher, NASA Administrator; Dr. Robert M. White, Administrator of the National Oceanic and Atmospheric Administration; and Prof. Maurice Levy, President of France's Centre National d'Études Spatiales. France agreed to design and build a data-collection system for Tiros-N meteorological satellite scheduled for launch in late 1977 and for the seven NOAA operational satellites to follow. (NASA Photo No. 74-H-1170)

- A United Air Lines, Inc., 747 airliner carrying instruments to monitor pollution in the upper atmosphere for NASA's Global Air Sampling Program completed flight testing and was certified as airworthy. Managed by Lewis Research Center, GASP would eventually use four 747 jets measuring dust particles and gases at altitudes of 6000 to 12 000 m to study the effects of emissions from jet aircraft and other pollution sources on the earth's atmosphere and weather. (LeRC Release 74-80; *Lewis News*, 27 Dec 74)

*11 December:* Edwin C. Kilgore, NASA Deputy Associate Administrator for Center Operations, had been named Director for Management Operations, Langley Research Center, effective 1 March 1975, NASA announced. Kilgore would assume the functions of LaRC Director for Administration Raymond G. Romatowski and most of the functions of Director for Center Development and External Affairs T. Melvin Butler. While at NASA Hq., Kilgore had served as Acting Associate Administrator for Aeronautics and Space Technology and also as Deputy Associate Administrator (Management) in the same office. In 1973 Kilgore had been awarded NASA's Outstanding Leadership Medal for developing an improved system for managing advanced technology programs.

Romatowski had accepted the position at NASA Hq. of Assistant Administrator for Personnel Programs. Butler would retire in December. (NASA Release 74-324A)

- The universe was 16 billion yrs old, 6 billion yrs older than earlier estimates, according to California Institute of Technology astronomer Dr. Allen R. Sandage, the *Washington Post* reported after an interview. After 10 yrs of observing 50 galaxies, Dr. Sandage had dated the galaxies from 14 billion yrs ago, meaning that 2 billion yrs were required for the primordial gas cloud to cool enough for stars to begin forming. Working with Swiss astronomer Gustav Tamman, Dr. Sandage had arrived at his new date when his observations had led him to recalculate the Hubble Constant, the rate of speed for the expansion of the universe. He had checked his figures by dating stars in globular clusters in the Milky Way and by measuring the amounts of metal still burning in their atmospheres. (O'Toole, *W Post*, 11 Dec 74, A1)

*13 December:* A golden age of planetary exploration was a realistic possibility, *Science* quoted Dr. James C. Fletcher, NASA Administrator, as saying in an interview. Dr. Fletcher believed space science was likely to be a major thrust of the space program in the 1980s, although NASA's budget was vulnerable to cuts in the present political climate. He ruled out manned programs other than the space shuttle, but NASA would not become an all-purpose technology development agency for energy, ground transportation, and similar nonspace systems. Applications and materials processing would be important. Dr. Fletcher believed NASA's program was acceptable to Congress and to the White House and that its future problem would not be wholesale cancellation of the space program but surviving annual budget cuts.

Increased competition for money and staff attention in congressional space-related committees could aggravate NASA's budgetary problems, according to *Science*. Observers had predicted that Congress would be faced with a choice between the shuttle and continuation of the space science and applications program. The decision could rest on the support of the shuttle program by organized labor, as it had in a 1972 Senate attempt to kill the shuttle. Some scientists had suggested that a $1-billion NASA budget cut would be better for science than a $100-million cut because the former would entail canceling the shuttle, in theory freeing money for space science. Dr. Fletcher had maintained, however, that a balance between programs would be kept. (Hammond, *Science*, 13 Dec 74, 1011-3)

- Three Kitt Peak National Observatory astronomers announced they had photographed large-scale structures, hot and cold regions, that were probably convection currents in the atmosphere of Alpha Orionis (the star Betelguese in the Constellation Orion), 500 light years from the earth. The announcement, made in Florida at the American Astronomical Society's annual meeting, was based on photos made 28 March using the four-meter Mayall Telescope. A new technique of interferometry analysis had permitted reconstruction of the star's image. In resolving the disc of Betelguese, the Kitt Peak astronomers became the first observers to distinguish actual forms on a star other than the sun. (NSF Release 74-229)

*15 December:* NASA's *Erts 1* Earth Resources Technology Satellite—routinely surveying crops, forests, watersheds, and mineral deposits on the earth since its 23 July 1972 launch—had transmitted photos of seven major Soviet antiballistic missile launch and radar installations at the Sary Shagan development site in southern Russia, the *San Diego Union* reported. The newspaper pointed out the ERTS satellites might be used to discover whether the Soviets were violating arms limitation agreements. (Macomber, *San Diego Union*, 15 Dec 75)

*16 December:* The first map of gravity forces on the moon's far side indicated that the strong gravitational tug of mascons (mass concentrations) on the lunar face was caused by lava flooding rather than by buried meteorites as previously thought, California Institute of Technology planetary scientist Dr. Alfred J. Ferrari reported in a press release based on a paper to be published in *Science*. The map was developed from measurements of the orbits of two small subsatellites left orbiting the moon by *Apollo 15* and *16* in July 1971 and April 1972 and of *Lunar Orbiter 5* (launched 1 Aug. 1967). Basins on the moon's far side, not lava-filled, exerted only weak gravitational forces on orbiting spacecraft, while lava-flooded basins on the front side exerted a strong pull. (Miles, *LA Times*, 17 Dec 74; Ferrari interview, 20 Nov 75; Ferrari, *Science*, 27 June 75, 1927–1300)

*17 December:* The U.S.S.R. launched *Meteor 20* meteorological satellite from Plesetsk into orbit with an 896-km apogee, 841-km perigee, 102.3-min period, and 81.2° inclination. The satellite carried instruments to photograph clouds and snow cover on the day and night sides of the earth and collect data on thermal energy reflected and emitted by the earth and its atmosphere, for day-to-day weather service. (GSFC *Wkly SSR*, 12–18 Dec 74; Tass, FBIS-Sov, 19 Dec 74, U1; *SBD*, 19 Dec 74, 270)

*17–19 December:* A major review of the first Viking lander spacecraft was held at Langley Research Center. The lander was approved for shipment to Kennedy Space Center in January 1975. The review followed successful completion of critical surface simulation testing [see 18 Sept.–21 Oct., 24 Nov.–14 Dec.]. (*Langley Researcher*, 20 Dec 74)

*18 December–6 January 1975:* NASA launched *Symphonie 1*, French and West German experimental communications satellite, at 9:39 pm EST from Eastern Test Range on a three-stage, thrust-augmented Thor-Delta launch vehicle. The 402-kg satellite entered a transfer orbit with a 38 071-km apogee, 402-km perigee, 688.2-min period, and 13.2° inclination.

Nine seconds after the Thor-Delta's 3rd stage separated from the 2nd stage, the inertial measurement unit (the altitude-sensing portion in the 2nd-stage guidance system) stopped functioning, but the failure did not affect the path of the already separated 3rd stage. In addition, the main engine cutoff was achieved by fuel depletion with 430 kg liquid oxygen remaining instead of with the planned depletion. The anomalies, which did not affect the mission, were under investigation.

Ground controllers in West Germany activated the apogee kick motor, the first using liquid propellant, at 6:00 pm EST 21 Dec., putting the spacecraft in circular geosynchronous orbit with 35 810-km apogee, 35 767-km perigee, 23-hr 56-min period, and 0.3° inclination. The spacecraft was despun and three-axis-stabilized and its large solar panels were successfully deployed 24 Dec. *Symphonie 1* drifted eastward to its final station at 115° W longitude over the west coast of Africa and began operations 2 Jan. 1975.

The NASA mission objective—to launch the spacecraft into a transfer orbit that would permit the onboard propulsion system to place it in an accurate synchronous orbit—was satisfied and the mission was adjudged successful 6 Jan. *Symphonie 1* would provide 1200 telephone circuits, 8 voice channels, and 2 color TV channels for experimental communications between Europe, Africa, and South America.

The first of two Symphonie satellites, provided for by a French-West Germany agreement signed June 1967, the satellite was developed by a French-German industrial consortium under the direction of the French Centre National d'Études Spatiales (CNES) and the West German Gesellschaft für Weltraumforschung (GFW). In October 1973 NASA agreed to provide the launch vehicle and services on a cost-reimbursable basis and in June 1974 signed a launch services contract with CNES–GFW. Goddard Space Flight Center in cooperation with French and West German ground stations was responsible for tracking the spacecraft during launch and transfer orbit operations. GSFC was also responsible for the launch vehicle. (NASA MORs, 13 Dec 74, 6 Jan 75; NASA prog off, interview, 29 Aug 75; NASA Release 74–316)

*18 December:* Post-storage checkout was proceeding at Marshall Space Flight Center on the first propulsive stage and instrument unit of the Saturn IB launch vehicle (SA–210) for the July 1975 Apollo-Soyuz Test Project mission. Fin-mounting brackets on the stage, taken out of storage 27 Nov., had been rod-peened (pounded with a small bundle of rods to place compressive stresses in the surface of the material) to avoid the stress-corrosion problems that had delayed the *Skylab 4* launch in November 1973. Bolt holes for the fins had been rounded to eliminate stress points. The complete Saturn vehicle was to be stacked in January 1975. (MSFC Release 74–241; *Spaceport News*, 12 Dec 74)

• Ames Research Center scientist Dr. Theodore Wydeven, Jr., had developed a new technique for coating salt lenses, NASA announced. Lenses made from large salt crystals had long been used in scientific applications because they were transparent to infrared light, but they were susceptible to fogging from even slight moisture. Using a method from his water purification experiments, Dr. Wydeven had prepared a Teflon-like membrane in an electric discharge to coat a salt lens, at a few cents a coating. The coating increased resistance to moisture and enhanced optical quality of the lens because of the film's antireflection properties. (NASA Release 74–319)

*18–20 December:* An experimental communications linkup between Mission Control Center at Johnson Space Center and the Soviet mission control near Moscow was made by U.S. and Soviet flight controllers rehearsing for the July 1975 Apollo-Soyuz Test Project mission. Procedures for interfacing voice, video, teletype, and telex facsimile between the two control centers through the tracking networks were checked out. A 10-hr procedural simulation of the launches of the Apollo and Soyuz spacecraft, with exchange of status reports and data on spacecraft position, was held at JSC the third day. The first JSC Mission Control Center simulation with Apollo crewmen was scheduled for mid-February 1975, with the first all-up simulation with both control centers and both Apollo and Soyuz crews in late March 1975. (JSC Release 74–275; Chriss, *LA Times*, 24 Dec 74, 8; JSC PIO, interview, 15 Sept 75)

*19 December:* Nelson A. Rockefeller was sworn in as Vice President of the United States after the House confirmed his nomination. The Senate had confirmed the nomination 10 Dec. (CR, 10 Dec 74, D1343; 19 Dec 74, D1413, D1416)

*20 December:* Johnson Space Center announced selection of Singer Co. Simulated Products Div. for an $8-million cost-plus-award fee contract to develop a space shuttle orbiter aeroflight simulator for crew training. Singer would design, manufacture, and test a simulator that would include a high-fidelity crew station, out-of-the-window color visual scene, motion base, and flight computer and simulator interface to simulate vehicle flight dynamics and motion during the atmospheric phase of missions. (JSC Release 74–276)

*21 December:* The U.S.S.R. launched *Molniya II–11* from Plesetsk into orbit with a 40 630-km apogee, 656-km perigee, 12-hr 16.8-min period, and 62.9° inclination. The comsat would help provide long-distance telephone and telegraph radio communications and transmit U.S.S.R. TV programs to stations of the Orbita network. (GSFC *Wkly SSR*, 19–25 Dec 74; Tass, FBIS–Sov, 24 Dec 74, U1; *SBD*, 27 Dec 74, 290)

*23 December:* Capt. Lee R. Scherer (USN, Ret.), Flight Research Center Director, had been named Director of Kennedy Space Center effective in mid-January 1975, NASA announced. He would succeed Dr. Kurt H. Debus, who retired 9 Oct. Before going to FRC in 1971, Capt. Scherer had been Lunar Exploration Director in the NASA Hq. Apollo Program Office, directing the scientific exploration of the first five manned landings on the moon. He had been Manager of the Lunar Orbiter Program 1963–1967. (NASA Ann, 23 Dec 74)

- Marshall Space Flight Center announced the award of a two-phase planning study contract for the space transportation system (STS) to McDonnell Douglas Astronautics Co. Phase 1, funded at $116 214 for six months, would define a master scheduling concept for effective use of the space shuttle, interim upper stage, space tug, and Spacelab. Phase 2, funded at $183 240 for an additional nine months, would validate a master planning technique and establish specifications for scheduling. (MSFC Release 74–242)

*25 December:* Tass announced that the flight of the U.S.S.R.'s *Salyut 3* orbiting space station, launched 25 June, would soon end, after completion of a six-month program that doubled the flight duration originally planned. During its six months operation the spacecraft had performed more than 400 scientific and technical experiments, received more than 8000 control commands from the earth, conducted more than 200 "dynamic operations," conducted 70 TV transmission and 2500 telemetric sessions, fired its thrusters more than 500 000 times, and generated 5000 kw of electricity with its solar panels. A two-man crew, launched 3 July on *Soyuz 14*, boarded the space station and spent 15 days living and working in space. A second crew, launched 26 Aug. on *Soyuz 15*, failed in several attempts to dock with the station and returned to the earth after 48 hrs in space. Since that time *Salyut 3* had operated in an unmanned automatic mode. The Western press later speculated that the mission was being ended to prevent any interference with the mission of Salyut 4 [see 26 Dec.–9 Feb. 1975]. The retrorockets were fired 24 Jan. 1975 and the station reentered the atmosphere over the

Pacific Ocean after 213 days in space. (TASS, FBIS–Sov, 26 Dec 74, U1–2; 24 Jan 75, U1; GSFC SSR, 28 Feb 75; *Soviet Aero*, 6 Jan 75, 1)

*26 December–9 February 1975:* The U.S.S.R. launched *Salyut 4* space station into earth orbit and 16 days later launched *Soyuz 17* with a two-man crew to join the Salyut and establish a manned orbital laboratory.

*Salyut 4* was launched 26 Dec. from Baykonur Cosmodrome near Tyuratam into orbit with a 252-km apogee, 214-km perigee, 89.2-min period, and 51.6° inclination. Tass announced the purpose of the mission was to test the design, onboard systems, and equipment and to conduct scientific studies and experiments in space. The news agency also reported that onboard systems and equipment were operating normally.

The Western press reported that major maneuvers 29 and 30 Dec. put the station at a record high altitude for a Salyut space station, reaching an orbit with a 350-km apogee and 337-km perigee by 31 Dec.

*Soyuz 17*, launched at 2:43 am 11 Jan. 1975 local time (4:43 pm EST 10 Jan.) from Baykonur Cosmodrome, carried cosmonauts Lt. Col. Aleksey A. Gubarev and Georgy M. Grechko into orbit with a 249-km apogee, 186-km perigee, 88.9-min period, and 51.6° inclination. Tass announced the mission would carry out joint experiments with *Salyut 4*, including a comprehensive checkup of the spacecraft's onboard systems in various flight conditions. In a launch-day interview with Tass, Maj. Gen. Georgy T. Beregovoy, head of the Soviet cosmonaut training center, said that *Soyuz 17* was in "no way connected with additional checkups of 'Soyuz' systems before its rendezvous with 'Apollo.'"

An orbital correction raised the spacecraft to an orbit with a 354-km apogee, 293-km perigee, 90.7-min period, and 51.6° inclination, permitting *Soyuz 17* to dock with *Salyut 4* on 12 Jan. The spacecraft was moved automatically to a distance of 10 m from the station and then the cosmonauts completed docking by manual control. The crew entered the space station, switched on the power and radio transmitters, and inspected the scientific equipment.

In a 14 Jan. interview with Tass, Dr. Konstantin P. Feoktistov, cosmonaut who had helped design the space station, said that *Salyut 4* improvements included three panels of solar batteries that were positioned like the tail of an aircraft and could be individually oriented toward the sun. The panels permitted the energy supply in the station to be replenished without using fuel for spacecraft orientation. Other improvements had been introduced in the flight control systems.

During a broadcast from the spacecraft over Moscow Radio 15 Jan., Gubarev and Grechko said they had set up the first Soviet space-based teletype communications station. The teletype could receive and copy communications from the earth while the cosmonauts were busy at other tasks. Tass reported 17 Jan. that the cosmonauts, neither of whom had been in space before, had become accustomed to weightlessness about the fifth or sixth day of flight.

During nearly 30 days aboard *Salyut 4*, Gubarev and Grechko used a newly developed high-resolution solar telescope to study solar phenomena, x-ray telescopes and sensors to study the flux of x-radiation from celestial bodies, and an infrared spectrometer to examine the earth's radiation. The cosmonauts also studied effects of weightlessness on the human vestibular function, cardiovascular system, and blood; made earth resources observations; and studied the earth's upper atmosphere. They resprayed two telescope mirrors dulled by exposure to space;

carried out biological "Oasis" experiments using insects, microorganisms, tissue cultures, and plants; and tried recycling water, condensed from the cabin atmosphere, for drinking and food preparation.

The crew began preparations for their return to earth 9 Feb., transferring equipment to *Soyuz 17*. The cosmonauts boarded *Soyuz 17* and undocked from the space station at 11:08 am Baykonur time (1:08 am EST). *Soyuz 17* softlanded in the U.S.S.R. 110 km northeast of Tselinograd, Kazakhstan, "in complex meteorological conditions" after 29 days 13 hrs 20 min in space. The *Soyuz 17* cosmonauts had broken the previous 23-day 18-hr 22-min Soviet record for time in space set by the *Soyuz 11* crew on *Salyut 1* (6–30 June 1971) before they were killed during reentry. On-the-spot and later medical checks showed that the *Soyuz 17* cosmonauts had "withstood well the long space flight."

*Salyut 4* was the fourth space station orbited by the U.S.S.R. *Salyut 1* had been launched 19 April 1971. A three-man crew, launched on *Soyuz 10*, docked with the station 24 April but returned to earth without entering. The *Soyuz 11* crew, launched 6 June 1971, boarded the station and conducted scientific experiments for 24 days before their reentry tragedy.

*Salyut 2* was launched into orbit 3 April 1973, but an explosion or wildly firing thrusters sent the station out of control, tearing off the solar panels and making the station uninhabitable. *Salyut 3* was orbited 25 June 1974 [see 25 Dec.]. (GSFC *Wkly SSR*, 19–25 Dec 74; GSFC *SSR*, 31 Dec 74; 28 Feb 75; Tass, FBIS–Sov, 31 Dec 74–13 Feb 75; *SF*, June 75, 235; April 75, 144–5; O'Toole, *W Post*, 31 Dec 74, A7; UPI, *NYT*, 16 Jan 75, 14; *SBD*, 14 Jan 75, 62; LC, S&T News Alert 2645)

*30 December:* Dr. James C. Fletcher, NASA Administrator, and Housing and Urban Development Secretary James T. Lynn signed and sent to the President and Congress their agencies' joint plan for implementing the Solar Heating and Cooling Demonstration Act of 1974. Under the Act, signed 3 Sept., NASA was responsible for development and procurement of solar equipment. HUD would coordinate the demonstration, including installation of solar equipment in residential structures, and collecting and disseminating information about solar energy. (MSFC Release 74-246)

- Dr. John F. Victory, former Executive Secretary of the National Advisory Committee for Aeronautics, died in Tucson, Ariz., at the age of 82. Dr. Victory, NACA's first employee when it was established in 1915, had been appointed Secretary of NACA in 1921 and in 1948 had been named its Executive Secretary, responsible for NACA's functioning as an independent research agency. When NASA succeeded NACA in 1958, he served as special assistant to Dr. T. Keith Glennan, NASA Administrator, and he remained with NASA until retirement in 1960. Dr. Victory, who had been known as "Mr. Aviation" in the industry, had received numerous awards, including a Presidential Medal for Merit, the first Air Foundation Certificate of Recognition, the Wright Brothers Memorial Trophy, and NACA's Distinguished Service Medal. (*NASA Activities*, Feb 75, 18)

*31 December:* A Soviet earth station to provide U.S.–U.S.S.R. Hot Line communications via Intelsat and Molniya satellites went into operation. The comsat station, near Moscow, was constructed under a bilateral agreement signed in Washington 30 Sept. 1971 to improve the Washington-Moscow direct communications link in an effort to reduce risk of war.

(ComSatCorp Off Internat'l Affairs, interview, 23 Jan 75; *A&A 1971, 1972*)
- President Ford signed S. 1283 into P.L. 93-577, the Federal Nonnuclear Energy Research and Development Act of 1974, to establish and conduct a national program of basic and applied research and development, including but not limited to demonstrations of practical applications, of all potentially beneficial energy sources and utilization technologies, within the Energy Research and Development Administration. (Text; *PD,* 6 Jan 75, 14)

*During December:* A total of 349 NASA Civil Service employees, including Associate Deputy Administrator Willis H. Shapley, retired under an "early out" retirement plan authorized by the Civil Service Commission. Under the plan employees could retire if they were at least 50 yrs of age and had 20 yrs of service or if they had 25 yrs of service, regardless of age. NASA had requested the retirement authority, which would extend through 28 Feb. 1975, to minimize effects of a reduction in force scheduled for February at Marshall Space Flight Center and of end of FY 1975 ceilings at Headquarters and elsewhere. NASA also announced a plan to reassign some Headquarters employees to Centers.

Retirees included 50 at Headquarters, 50 at Goddard Space Flight Center, 78 at Langley Research Center, 68 at MSFC, 28 at Lewis Research Center, 26 at Ames Research Center, 16 at Kennedy Space Flight Center, 14 at Johnson Space Flight Center, 10 at Flight Research Center, and 8 at Wallops Flight Center. By 31 Dec. NASA's permanent Civil Service strength totaled 24 619—303 over its end of FY 1975 ceiling. (NASA Gen Mgt Rev Rpt, 20 Jan 75, 11-12; NASA Ann, 12 Dec 74; NASA Deputy Admin memo to Hq employees, 23 Dec 74)
- Marshall Space Flight Center awarded more than $28.2 million in contracts of $25 000 or more. Among awards were $2 253 300 to Inscho's Mechanical Contractors, Inc., for construction of the X-Ray Telescope Facility at MSFC; also $695 000 to Lockheed Aircraft Corp. and $687 201 to Martin Marietta Corp., for Large Space Telescope support system module studies. (MSFC Release 75-18)

# 1974 in Summary

*Launches*

The U.S. orbited 28 payloads in 22 launches during the year. The U.S.S.R. orbited 95 payloads in 81 launches; Italy launched 2 payloads, using NASA launch vehicles; and Japan launched 1. U.S. launches included 16 payloads in 14 NASA launches and 12 spacecraft in 8 Dept. of Defense launches.

For the first year since 1967 the U.S. launched no men into space, although *Skylab 4* Astronauts Gerald P. Carr, Dr. Edward G. Gibson, and William R. Pogue completed the longest mission yet accomplished by man when they splashed down in the Pacific Ocean 8 Feb. after 84 days in space. The U.S.S.R. orbited three two-man crews during the year, including two crews to dock with the *Salyut 3* space station. The *Soyuz 14* crew docked and worked aboard the laboratory for nearly 16 days, but *Soyuz 15* failed to dock and it returned to earth after 48 hrs in space. *Soyuz 16* tested the Apollo-Soyuz Test Project configuration in December. The U.S.S.R. also launched the *Salyut 4* orbiting space station, to be joined by a two-man crew in January 1975.

The 16 NASA-launched payloads—11 paid for by domestic and foreign corporations or foreign governments—included 7 communications, 1 applications technology, 2 meteorological, and 6 scientific satellites. These were in addition to 2 Italian-launched cooperative scientific satellites. Comsats included two spacecraft launched for United Kingdom military communications—*Skynet IIA* into too low an orbit, followed later by the successful *Skynet IIB*. *Westar 1* and *2*, launched for Western Union Telegraph Co., were the first U.S. commercial domestic communications satellites. The most complex and versatile of the Applications Technology Satellites, *Ats 6*, went into orbit in May. The spacecraft was transmitting medical and educational telecommunications to small inexpensive ground stations in remote areas of the U.S. and would, in 1975, be positioned to broadcast educational programs to 5000 villages in India. *Symphonie 1* was launched for France and West Germany as the first of two experimental communications satellites. *Intelsat-IV F–8* was launched for Communications Satellite Corp. on behalf of the International Telecommunications Satellite Organization, and *Amsat Oscar 7* for Radio Amateur Satellite Corp. as a secondary payload on the *Noaa 4* launch.

The two meteorological satellites included Synchronous Meteorological Satellite *Sms 1*, launched for the National Oceanic and Atmospheric Administration to provide continuous day and night images of cloud cover over the U.S. and Atlantic Ocean for the first time. *Sms 1* was the first meteorological satellite in synchronous orbit. *Noaa 4* was launched as part of NOAA's global weather watch program.

All but one of NASA's six scientific payloads were cooperative efforts with foreign countries. *Miranda* (UK–X4) was launched for U.K. to measure the density of sun-reflecting particles near the spacecraft as well as test spacecraft systems. Spain's first satellite—*Intasat*, put into polar orbit as a secondary payload with *Noaa 4*—began studies of the ionosphere. Two spacecraft were launched in cooperation with West Germany: *Aeros 2* to measure aeronomic parameters of the upper atmosphere and solar ultraviolet radiation, *Helios 1* into solar orbit to investigate interstellar space closer to the sun than any previous spacecraft had and to gather extensive new data on the sun. NASA and The Netherlands joined to orbit The Netherlands Astronomical Satellite *Ans 1* to study stellar UV and x-ray sources. NASA's only all-U.S. scientific payload, *Hawkeye 1* (*Explorer 52*), was launched into polar orbit to study plasma properties of the magnetosphere over the north polar cap.

A Titan-Centaur proof-flight vehicle carrying a Viking spacecraft model and a SPHINX satellite failed to achieve orbit and was destroyed during launch operations. The new NASA launch vehicle later successfully launched *Helios 1*.

Italian launch crews, using NASA launch vehicles, launched two satellites into earth orbit from the San Marco launch platform in the Indian Ocean off the East African coast. *San Marco 4* in February carried two U.S. and one Italian experiment into orbit to measure diurnal variations in the equatorial neutral atmosphere. *Ariel 5* was launched in October for U.K., to study galactic and extragalactic x-ray sources, in the first Italian launch for a country other than the U.S. or Italy.

Japan launched its fifth satellite, *Tansei 2*—Japan's first on a guided booster.

In addition to the three manned Soyuz missions, the U.S.S.R. launched 2 Salyut space stations; 3 unmanned Soyuz spacecraft identified as *Cosmos 638, 656,* and *672*; 7 Molniya comsats; 5 Meteor meteorological satellites; 2 cooperative Intercosmos; 2 Luna lunar probes; and 71 spacecraft under the Cosmos designation.

*Space Science and Exploration*

Besides the new scientific satellites, NASA continued systematic exploration of the solar system as *Pioneer 11*, launched in April 1973, sped to within 42 000 km of Jupiter 3 Dec. The spacecraft photographed the giant planet and its moons, transmitting new information on the planet's weather patterns, atmosphere, and radiation belts. Data from *Pioneer 11* and its predecessor *Pioneer 10*, which had encountered Jupiter in December 1973 revealed that Jupiter was hotter than previously estimated and composed largely of liquid hydrogen. Cyclones and anticyclones stretched completely around the planet with rising grey-white cloud ridges extending 20 km above Jupiter's distinctive orange-brown belts. Jupiter's great red spot was probably the vortex of a great storm that had raged along a 400 000-km front for centuries. At year's end *Pioneer 11* was speeding toward rendezvous with Saturn to give man his closest look at that planet while *Pioneer 10* raced to cross the orbits of Saturn, Uranus, Neptune, and Pluto and become the first spacecraft to leave the solar system.

Meanwhile *Mariner 10*—after passing within 5800 km of Venus and sending back new information on particle environment, mass, and density of that planet—made two successful flybys of Mercury in March and September. The spacecraft transmitted data that gave scientists mass measurements 100 times better than previous ones, described a detached, well-defined bow shock wave, and showed temperatures varying from 90 K on the night side to 700 K on the day side ($-300°F$ to $800°F$). During the two encounters the spacecraft photographed 45% of the planet's surface. By the end of the year *Mariner 10* was headed for a third Mercury encounter in March 1975.

Preparations for the two 1975 Viking missions to Mars continued with the qualification-testing of Viking science instruments, engineering subsystems, and entire orbiter and lander spacecraft. During the last quarter of the year the lander successfully completed critical surface simulation tests that included subjecting the spacecraft to temperatures approximating Martian conditions and dropping it from 0.6 m to determine the effects of landing shocks on the integrated Viking equipment.

A detailed spacecraft design for the 1977 Mariner mission to Jupiter and Saturn was completed while a major effort to find a satisfactory way to protect the spacecraft from Jupiter's strong radiation continued.

A major new space science start, approved in NASA's FY 1975 budget, was the 1978 dual Pioneer-Venus mission to obtain detailed measurements of the Venus atmosphere with an orbiter and four atmospheric entry probes. Hughes Aircraft Co. was chosen in February for spacecraft design, with an option for spacecraft development, and experimenters and scientific instrumentation were selected in June.

NASA continued preparations to place the Large Space Telescope in earth orbit to make detailed astronomical observations 10 times deeper into space than possible before. During 1974, preliminary system definitions and advanced technological development were carried out on selected elements of the LST.

Besides preparing for future missions, scientists continued analysis of data accumulated from previous missions. A Lunar Data Analysis and Synthesis program began to correlate existing data and data still being collected by the five Apollo lunar surface-experiment packages (ALSEP) left on the moon during the lunar landing program. With the conclusion of the Skylab flight program, large volumes of new data on the sun and the earth and the most comprehensive inflight data yet on man's adaptation to space flight became available.

During the year sounding rockets, balloons, telescopes, and instrumented aircraft continued to increase knowledge of the universe. NASA launched nearly 80 sounding rockets from sites around the world. A significant accomplishment was the launch of 54 sounding rockets within 24 hrs to support the Joint NASA-Air Force Atmospheric Layering and Density Distribution of Ions and Neutrals (ALADDIN) program.

During 1974 scientists in programs outside NASA as well as in, using data from both orbital and suborbital programs, discovered a 13th moon of Jupiter, observed a pulsar that was a member of a double star system, obtained new data on the size and shape of the Crab Nebula, discovered a galaxy six times larger than any previously known in the universe, and made the greatest advance yet in cometary research with the study of the Comet Kohoutek, including first-time identification of water molecules in a comet.

The U.S.S.R. continued a program of space exploration with the launch of lunar probes *Luna 22* in May and *Luna 23* in October. *Luna 22* studied the lunar surface and atmosphere from lunar orbit. *Luna 23* was damaged while landing on the moon's surface and its work was discontinued after three days of equipment testing.

*Space Transportation and Technology*

The space shuttle program accelerated, with funding increasing from $475 million in FY 1974 to $800 million in FY 1975. But this total reflected an $89-million cut by the Office of Management and Budget from planned space shuttle funding, delaying the program four to six months. The final major contract was awarded and design work neared completion;

by year's end the program moved into hardware test and fabrication stages. Construction of the shuttle runway at Kennedy Space Center began in April and throughout the year contracts were awarded for additional space shuttle facilities. The European Space Research Organization selected VFW-Fokker/ERNO Raumfahrttechnik GmbH as prime contractor to design and build Spacelab for missions on the shuttle. Primary efforts were directed toward planning early shuttle missions and making analyses for a NASA mission model.

In solid propulsion technology, the final motor—the first with a restart capability—in a series of high-efficiency solid-fueled motors was successfully static-fired. The October firing marked the first time a spacecraft solid-fueled motor had been fired, quenched, reignited, and requenched. Another significant step was the static-firing of a thermally sterilized solid-fueled motor. Electric propulsion—attractive as auxiliary propulsion for long-life station-keeping and attitude control—took a stride forward with the tests of a 4.5-millinewton (0.001-lb-thrust) ion engine aboard *Ats 6*. In spacecraft energy system technology, solar cells were made from a silicon ribbon grown directly from the liquid state, a critical first step in low-cost manufacture of solar cells. Other advances were made in laser, nuclear, atmosphere-entry, material-and-structures, guidance, and control-and-information technology.

*Aeronautics*

The X–24B lifting body completed 15 flights in the NASA and Air Force program to develop a safe piloted vehicle for reentry from space flight and to develop hypersonic technology. Following the first supersonic flight in March, the vehicle's performance and handling qualities were evaluated in the speed range between mach 1.6 and landing.

NASA's aeronautics research and development program to reduce fuel consumption, reduce undesirable environmental effects, improve safety, and advance technology continued during the year. Flight tests of an F–111 equipped with the NASA-developed supercritical wing demonstrated a 15% reduction in fuel consumption. Other tests using flexible aircraft skin concepts, composite materials, active-control technology, and alternative fuels also demonstrated reduced drag and potential fuel savings.

In NASA's refan program—an effort to reduce noise of the JT8D engine that powered most of the Nation's narrow-body commercial fleet—three refanned engines demonstrated in ground tests that the ground area affected by objectionable noise from commercial aircraft could be reduced by 75%. In another technique, a DC–8 using a NASA-developed two-segment landing approach reduced the ground area affected by excessive noise 53%.

Efforts to measure and reduce aircraft engine exhaust in the atmosphere continued with the Global Air Sampling Program. A United Air Lines, Inc., Boeing 747 equipped with an atmospheric sampling instrument began measuring aircraft pollution on commercial air routes, while the clean combuster technology program was expanded to reduce engine emissions.

Programs to improve aircraft safety continued to make progress. Techniques to reduce hazardous trailing vortices were flight-tested. A NASA-developed prototype system to track aircraft wakes near airports, allowing traffic controllers to adjust aircraft spacing more closely, was being operationally evaluated at John F. Kennedy Airport in New York City.

The Dept..of Defense A–10 close-air-support program moved into production stage, with the first of six aircraft in final assembly and checkout. Production of an initial increment of 22 aircraft was approved in December, with a full production decision expected late in 1975.

DOD's B–1 advanced strategic bomber moved through engine preliminary flight-readiness tests in April, rollout in October, and first flight in December. Following flight testing, a production decision would be made in 1976. The first operational F–15 advanced tactical fighter was delivered in November. At year's end 20 test aircraft had accumulated more than 300 flight hrs.

*International Cooperation*

Aerospace activities continued to become more international. Of the 16 NASA-launched payloads, 10 were cooperative efforts with foreign organizations. In addition, the 2 Italian-launched payloads carried U.S. experiments. During the year negotiations were completed on an agreement for NASA to furnish reimbursable launches for three future synchronous-orbit Japanese satellites. NASA–ESRO Spacelab working groups met throughout the year on experimental objectives and user requirements. Investigators from 42 countries and 5 international organizations were selected for Earth Resources Technology Satellite follow-on investigations, and Italy and Iran signed agreements with NASA to build earth stations to receive ERTS data directly.

The 1975 U.S.–U.S.S.R. Apollo-Soyuz Test Project neared flight readiness. Joint compatibility tests of communications and docking systems were carried out and intensive joint flight-crew and flight-controller training was begun. Apollo hardware was completed. The U.S.S.R launched two unmanned and one manned Soyuz spacecraft (*Cosmos 638* and *672* and *Soyuz 16*) to check out new systems and equipment for the joint mission. NASA and the Soviet Academy of Sciences approved public information plans and plans for prelaunch testing at the U.S. and U.S.S.R. launch sites.

U.S. and Soviet specialists exchanged information about the moon, planets, environmental problems, and biomedical results from U.S. and U.S.S.R. space flights. U.S. and Soviet scientists preparing a joint experiment using *Ats 6* data met in the U.S. in October for technical coordination. And plans were formulated in October for a U.S. life science experiment to fly on the next available Soviet biological satellite.

For 101 days, June–September, the U.S. and 69 other countries participating in the Global Atmospheric Research Program's Atlantic Tropical Experiment monitored nearly every known meteorological factor along a 52-million-sq-km tropical area of land and sea from the eastern Pacific Ocean across Latin America, the Atlantic Ocean, and Africa to the western Indian Ocean. More than 4000 persons using 41 ships, 40 instrumented buoys, 12 aircraft, and 6 satellites probed from 1.6 km below the sea surface to the top of the atmosphere to improve weather predictions, assess pollution, determine the feasibility of large-scale weather modification, and establish new bonds of international cooperation.

*Applications and Energy*

During 1974 space technology continued to make significant contributions to the quality of life. In addition to 1974-launched *Ats 6, Sms 1, Westar 1* and *2, Intelstat-IV F–8*, and *Noaa 4* to improve weather predictions and communications on the earth—*Erts 1*, launched in July 1972, completed 29 mos in operation. *Erts 1* had transmitted more than 100 000 photos covering three fourths of the world's land masses and coastal areas. ERTS data—used to monitor urban and agricultural development, locate air and water pollution, update maps, and find new water and mineral resources—had been provided to more than 300 U.S. and foreign investigators in government, universities, and industry. ERTS–B development continued on schedule.

NASA's pollution monitoring program—to develop and demonstrate technology to monitor atmospheric and water pollution—completed its second year. Procurement for design and construction of the Nimbus-G Oceanographic and Air Pollution Observing Satellite, planned for 1978 launch, was under way. The Lower Atmospheric Composition and Temperature Experiment (LACATE) was demonstrated in May and a test in the advanced applications flight experiment program—using balloons, sounding rockets, and aircraft as well as ground-based experiments—successfully gathered atmospheric data over a five-hour period.

DOD expanded and improved its Defense Satellite Communications System and let contracts for Phase I NAVSTAR global positioning satellites. The Defense Meteorological Satellite program continued to furnish weather data for military and civilian use.

Space technology continued to be applied to medicine. A digital computing cardiotachometer to monitor heart rates, an automated blood pressure monitoring system, and a new method of attaching artificial limbs were among the NASA-developed contributions demonstrated in 1974.

An Office of Energy Programs was established in NASA to focus the application of its aerospace technology to energy needs. In addition to programs to reduce aircraft air pollution and energy consumption and develop new fuels, NASA, under the Solar Heating and Cooling Demonstration Act of 1974 and in cooperation with the Dept. of Housing and Urban Development and other Federal agencies, was investigating the use of solar energy for heating and cooling buildings. An experimental solar house at Marshall Space Flight Center would ultimately derive up to 85% of its heating and cooling energy from the sun. Solar heating and cooling were being demonstrated in school.

NASA continued research on wind energy systems, with the construction of a 100-kw wind-powered generator at Lewis Research Center; low-pollution, low-fuel-consumption automotive engines; aerodynamically designed vehicles to increase fuel economy; energy conversion systems.

NASA distributed some 300 new Tech Briefs and 30 special compilations of new technology items to 15 000 subscribers in business and industry. Use of Regional Dissemination Centers for space technology applicable to other uses increased 34%, to 4000 clients.

*Tracking and Data Acquisition*

The Spaceflight Tracking and Data Network (STDN) supported some 40 flight projects during 1974, including all NASA's earth-orbital missions as well as space projects of foreign governments and private industry. Congress authorized funds for procurement activities for a Tracking and Data Relay Satellite System (TDRSS)—two earth-orbiting satellites leased by NASA to relay data between low-altitude spacecraft and a single ground station. The authority to proceed further was subject to additional congressional review of a NASA-owned versus a leased system. (Pres Rpt 74; NASA Release 74–330; *NASA Activities*, 15 Dec 74; *A&A 74*)

# Appendix A

## SATELLITES, SPACE PROBES, AND MANNED SPACE FLIGHTS, 1974

The following table includes all payloads that have (a) orbited; (b) as probes, ascended to at least the 6500-kilometer altitude that traditionally has distinguished probes from sounding rockets; or (c) conveyed one or more human beings into space whether orbit was attained or not. Date of launch is referenced to local time at the launch site. An asterisk follows dates that are one day later by Greenwich Mean Time. A double asterisk follows dates one day earlier by GMT.

The data were compiled from open sources including the United Nations Public Registry; the *Satellite Situation Report*, compiled by the Operations Control Center at Goddard Space Flight Center; and public information releases of NASA, Department of Defense, National Oceanic and Atmospheric Administration, and other agencies, as well as Communications Satellite Corporation. Soviet data are from the U.N. Public Registry, the *Satellite Situation Report*, translations from the Tass News Agency, statements in the Soviet press, and international news service reports. New sources of information for 1974 were speeches and press briefings given by Soviet officials in the United States to prepare for the U.S.–U.S.S.R. Apollo-Soyuz Test Project's 1975 mission. Data on satellites of other foreign nations are from the U.N. Public Registry, the *Satellite Situation Report*, governmental announcements, and international news service reports.

World space activity in 1974 decreased slightly. Total launches, 106, were down from the 109 launches in 1973, as were total payloads orbited—126 in 1974, down from 133 in 1973. The 5 multiple-payload launches—2, 8-satellite launches for the U.S.S.R.; 2, 3-satellite launches for DOD, and a 3-satellite launch for NASA—account for the difference between launches and payloads.

Of the 1974 world total, the United States launched 22 boosters carrying 28 payloads, a slight decline from 1973, when 23 U.S. boosters carried 26 payloads into space. Of these 1974 totals, DOD was responsible for 8 launches and 12 payloads. Of the 16 NASA-launched payloads, 11 were launched for domestic and foreign corporations or foreign governments: *Skynet IIA* and *IIB* and *Miranda* (UK–X4) for the United Kingdom, *Intasat* for Spain, *Symphonie 1* for France and West Germany, *Aeros 2* for West Germany, *Intelsat–IV F–8* for ComSatCorp on behalf of the International Telecommunications Satellite Organization, *Westar 1* and *2* for Western Union Telegraph Company, *Noaa 4* for NOAA, and *Amsat Oscar 7* for the Radio Amateur Satellite Corporation. Two satellites were cooperative programs with foreign governments: *Ans 1* with The Netherlands and *Helios 1* with West Germany. Three NASA-sponsored payloads—*Ats 6*, *Sms 1*, and *Hawkeye 1* (*Explorer 52*)—were launched for applications and scientific missions.

The Soviet Union once again far exceeded U.S. space totals, launching 95 payloads in 81 launches, with activity decreasing slightly from 1973's 107 payloads in 86 launches.

For the United States the year was the first since 1967 marked by an absence of manned launches. The last of three Skylab crews, already orbiting as the year began, reentered 8 Feb. on *Skylab 4* after 84 days in space. NASA launches included 8 communications, 2 meteorological, and 6 scientific satellites. Highlights of the year came with the launches of *Ats 6*, the most complex and powerful of the Applications Technology Satellites, to transmit medical and educational telecommunications to small inexpensive ground stations in remote areas of the United States and of *Helios 1* to investigate interstellar space closer to the sun than any previous spacecraft. The 12 DOD payloads included *Nts 1* Navigation Technology Satellite to perfect technology for use in the NAVSTAR program and *S3–1* to measure air density and upper atmospheric composition and to provide data on atmospheric efforts on spacecraft orbits.

The Soviet Union launched 2 space stations and 3 two-man crews during 1974, besides unmanned scientific, lunar, and applications spacecraft. Following the launch of the *Salyut 3* space station, a crew in *Soyuz 14* docked with the station and worked on board for 16 days before returning to the earth. A second crew, in *Soyuz 15*, failed to dock after several attempts and returned to earth after two days in space. A second station, *Salyut 4*, was launched in December to be joined in January 1975 by a crew aboard *Soyuz 17*. *Soyuz 16*, carrying two cosmonauts, was launched to check out docking and life support systems in preparation for the 1975 ASTP mission. In addition, the U.S.S.R. launched 3 unmanned Soyuz spacecraft, 7 Molniya comsats, 5 Meteor meteorological satellites, 2 cooperative Intercosmos, 2 Luna lunar probes, and 71 Cosmos spacecraft.

Italian crews launched 2 satellites on NASA launch vehicles: *San Marco 4* carried one Italian and two U.S. experiments into orbit to collect atmospheric data. The U.K.-built *Ariel 5* carried one NASA and five U.K experiments to study x-ray sources.

Japan launched its fifth satellite, *Tansei 2*, to test spacecraft control devices.

As we have cautioned in previous years, the "Remarks" column of these appendixes is never complete, because of printing time and the inescapable lag between each flight and the analysis and interpretation of results. Also, some missions are still continuing. The information in this table was compiled in February 1976.

## SATELLITES, SPACE PROBES, AND MANNED SPACE FLIGHTS, 1974

| Launch Date | Spacecraft, Country, Int'l Designation, Vehicle, Launch Site | Payload Data | Apogee (km) | Perigee (km) | Period (min) | Inclination (degrees) | Remarks |
|---|---|---|---|---|---|---|---|
| 17 Jan. | *Cosmos 628* U.S.S.R. 1974-1A Unavailable Plesetsk | Total weight: Unavailable. Objective: "Investigation of the upper atmosphere and outer space." Description: Unavailable. | 1 014 | 959 | 104.8 | 82.9 | Still in orbit. |
| 18 Jan.* | *Skynet IIA* U.K.–U.S. 1974-2A Thor-Delta ETR | Total weight: 435 kg. Objective: Place spacecraft in transfer orbit of sufficient accuracy to permit apogee motor boost into geosynchronous orbit for X-band military communications for U.K. Description: Drum-shaped fiberglass and aluminum spacecraft 2.1 m high, 1.9 m in dia; 17-w transponder and horn antenna provided communications. Tracking, telemetry, and command subsystem. Monopropellant-hydrazine reaction control system. Fiberglass substrate thermal shields and aluminumized H-film insulation blankets. Six solar panels, two backup batteries, and power control unit. Thiokol TE-M-604-1 solid-fueled apogee boost motor. | 3 406 | 95 | 121.6 | 37.6 | U.K.-built communications satellite launched by NASA under agreement with USAF. Placed in unplanned eccentric orbit by 3-stage thrust-augmented Thor-Delta when regulator board of DC-to-DC converter in 2nd-stage electronics package short-circuited. Engine took "hard over" position when 2nd-stage restart command was given 24 Jan., causing stage to tumble. Spacecraft reentered 25 Jan. 1974. |
| 24 Jan. | *Cosmos 629* U.S.S.R. 1974-3A Unavailable Plesetsk | Total weight: Unavailable. Objective: "Investigation of upper atmosphere and outer space." Description: Unavailable. | 285 | 196 | 89.3 | 62.8 | Reentered 5 Feb. 1974. |
| 30 Jan. | *Cosmos 630* U.S.S.R. 1974-4A Unavailable Plesetsk | Total weight: Unavailable. Objective: "Investigation of upper atmosphere and outer space." Description: Unavailable. | 332 | 203 | 90.0 | 72.9 | Reentered 13 Feb. 1974. |
| 6 Feb. | *Cosmos 631* U.S.S.R. | Total weight: Unavailable. Objective: "Investigation of upper atmos- | 545 | 519 | 95.2 | 74.0 | Still in orbit. |

| Launch Date | Spacecraft, Country, Int'l Designation, Vehicle, Launch Site | Payload Data | Apogee (km) | Perigee (km) | Period (min) | Inclination (degrees) | Remarks |
|---|---|---|---|---|---|---|---|
| | 1974-5A Unavailable Plesetsk | phere and outer space." Description: Unavailable. | | | | | |
| 12 Feb. | *Cosmos 632* U.S.S.R. 1974-6A Unavailable Baykonur-Tyuratam | Total weight: Unavailable. Objective: "Investigation of upper atmosphere and outer space." Description: Unavailable. | 306 | 175 | 89.3 | 65.0 | Reentered 26 Feb. 1974. |
| 13 Feb. | DOD spacecraft United States 1974-7A Titan IIIB-Agena WTR | Total weight: Unavailable. Objective: Develop space flight techniques and technology. Description: Unavailable. | 404 | 128 | 89.8 | 110.4 | Launched by USAF. Press reported spacecraft supported Navy's ocean surveillance program. Reentered 17 March 1974. |
| 16 Feb. | *Tansei 2* Japan 1974-8A MU-3C Kagoshima | Total weight: 56 kg. Objective: Test spacecraft control devices. Description: Octagonal prism-shaped satellite 45 cm high, 75 cm wide. Yo-yo despinner and magnetic attitude control devices. Housekeeping, measuring, and radio telemetry systems. | 3 229 | 283 | 121.7 | 31.2 | Fifth Japanese satellite and first launched by guided booster rocket with thrust-vector-control system for stabilization. Still in orbit. |
| 18 Feb. | *San Marco 4* Italy-U.S. 1974-9A Scout San Marco platform | Total weight: 170 kg. Objective: Measure diurnal variations of equatorial neutral atmosphere's density, composition, and temperature; data to be correlated with *Explorer 51* data for study of thermosphere. Description: Spherical spacecraft 70 cm in dia with 4, canted, 48.5-cm monopole antennas for telemetry and command. Series of mica windows on equator of shell structure. Two nickel-cadmium batteries, mercuric oxide battery pack, and voltage conversion and control electronics completed power system. Passive thermal control system. Triaxial magnetometer and digital-sun-sensor attitude control | 930.5 | 234.5 | 96.0 | 2.9 | Fourth satellite in joint Italian-NASA cooperative space program. Launched by Italian crew using NASA launch vehicle. Vehicle performance placed Italian-built Explorer spacecraft in higher than planned orbit. Both U.S. experiments performed normally but Italian air-drag experiment malfunctioned shortly after launch. After 23 mos in orbit by 30 Jan. 1976, *San Marco 4* had obtained more than 17 diurnal cycles of data on equatorial thermosphere. Still acquiring data. |

## ASTRONAUTICS AND AERONAUTICS, 1974

| Date | Name | Weight | | | | Remarks |
|---|---|---|---|---|---|---|
| 27 Feb. | *Cosmos 633*<br>U.S.S.R.<br>1974-10A<br>Unavailable<br>Plesetsk | system. Three experiments: neutral mass spectrometer (U.S.), omegatron mass spectrometer (U.S.), and atmospheric-drag density accelerometer (Italy), using spacecraft structure as integral part of air density balance.<br>Total weight: Unavailable.<br>Objective: "Investigation of upper atmosphere and outer space."<br>Description: Unavailable. | 486 | 273 | 92.2 | 71.0 | Reentered 4 Oct. 1974. |
| 5 Mar. | *Meteor 16*<br>U.S.S.R.<br>1974-11A<br>Unavailable<br>Plesetsk | Total weight: Unavailable.<br>Objectives: Collect meteorological information for weather forecasts. Photograph clouds and snow cover on day and night sides of globe and obtain data on heat energy reflected or emitted by earth and its atmosphere.<br>Description: Carried cameras, power supply system with independent orientation of solar array toward sun, radio telemetry, and data-collection systems. | 892 | 830 | 102.2 | 81.2 | Still in orbit. |
| 5 Mar. | *Cosmos 634*<br>U.S.S.R.<br>1974-12A<br>Unavailable<br>Plesetsk | Total weight: Unavailable.<br>Objective: "Investigation of upper atmosphere and outer space."<br>Description: Unavailable. | 489 | 269 | 92.1 | 70.1 | Reentered 9 Oct. 1974. |
| 8 Mar.* | *Miranda* (UK-X4)<br>U.K.-U.S.<br>1974-13A<br>Scout D<br>WTR | Total weight: 92 kg.<br>Objective: Launch satellite into earth orbit of sufficient accuracy to permit evaluating performance of accurate 3-axis gas-jet attitude control system, demonstrating in-orbit performance of infrared sensor, high-reliability Canopus sensor, and digital, albedo horizon sensor; measuring density of sun-reflecting particles near spacecraft; and determining in-orbit degradation of silicon solar cells.<br>Description: Rectangular satellite of aluminum honeycomb-sandwich construction, 83.5 cm high, 66.5 cm wide. Two deployable kapton solar array panels containing 1800 solar cells extended spacecraft width to 250 cm. Three-ampere-hour nickel-cadmium battery for launch and high power demands. Four 46.5-cm- | 926.8 | 727.7 | 101.1 | 97.8 | U.K.-built UK-X4 experimental satellite launched into sunlit polar orbit by NASA on 4-stage, solid-propellant Scout D booster. Malfunctioning regulator in propane gas control system was corrected by reorienting spacecraft, permitting successful completion of mission within planned 6 mos. Still in orbit; mission completed. |

| Launch Date | Spacecraft, Country, Int'l Designation, Vehicle, Launch Site | Payload Data | Apogee (km) | Perigee (km) | Period (min) | Inclination (degrees) | Remarks |
|---|---|---|---|---|---|---|---|
| | | long antennas folded out at corners of side facing earth. | | | | | |
| 14 Mar. | Cosmos 635<br>U.S.S.R.<br>1974-14A<br>Unavailable<br>Plesetsk | Total weight: Unavailable.<br>Objective: "Investigation of upper atmosphere and outer space."<br>Description: Unavailable. | 325 | 201 | 89.8 | 72.8 | Reentered 26 March 1974. |
| 16 Mar. | DOD spacecraft<br>United States<br>1974-15A<br>Thor-Burner II<br>WTR | Total weight: Unavailable.<br>Objective: Practical applications of space-based technology.<br>Description: Unavailable. | 878 | 781 | 101.4 | 98.9 | Reported by press as part of Defense Meteorological Satellite Program (DMSP). Still in orbit. |
| 20 Mar. | Cosmos 636<br>U.S.S.R.<br>1974-16A<br>Unavailable<br>Baykonur-Tyuratam | Total weight: Unavailable.<br>Objective: "Investigation of upper atmosphere and outer space."<br>Description: Unavailable. | 385 | 164 | 90.0 | 65.0 | Reentered 3 April 1974. |
| 26 Mar. | Cosmos 637<br>U.S.S.R.<br>1974-17A<br>Unavailable<br>Baykonur-Tyuratam | Total weight: Unavailable.<br>Objective: "Investigation of upper atmosphere and outer space."<br>Description: Unavailable. | 35 604 | 35 595 | 1426.6 | 0.2 | First Soviet satellite in geosynchronous orbit; possible test of stationary comsat. Still in orbit. |
| 3 Apr. | Cosmos 638<br>U.S.S.R.<br>1974-18A<br>Unavailable<br>Baykonur-Tyuratam | Total weight: Unavailable.<br>Objective: Unmanned test of docking systems for 1975 U.S.–U.S.S.R. Apollo-Soyuz Test Project mission.<br>Description: Unavailable. | 308 | 189 | 89.5 | 51.8 | Identified by Soviet official as ASTP docking test. Reentered 13 April 1974. |
| 4 Apr. | Cosmos 639<br>U.S.S.R.<br>1974-19A<br>Unavailable<br>Plesetsk | Total weight: Unavailable.<br>Objective: "Investigation of upper atmosphere and outer space."<br>Description: Unavailable. | 224 | 205 | 88.8 | 81.3 | Reentered 15 April 1974. |

## ASTRONAUTICS AND AERONAUTICS, 1974

| Date | Satellite | | | | | Remarks |
|---|---|---|---|---|---|---|
| 10 Apr. | DOD spacecraft<br>United States<br>1974-20A<br>Titan IIID<br>WTR | Total weight: Unavailable.<br>Objective: Develop space flight techniques and technology.<br>Description: Unavailable. | 288 | 152 | 88.0 | 94.0 | Three satellites launched on single USAF booster; 20A reentered 28 July 1974. |
| | and<br>DOD spacecraft<br>1974-20B | Total weight: Unavailable.<br>Objective: Develop space flight techniques and technology.<br>Description: Unavailable. | 829 | 783 | 101.0 | 94.0 | Still in orbit. |
| | and<br>DOD spacecraft<br>1974-20C | Total weight: Unavailable.<br>Objective: Develop space flight techniques and technology.<br>Description: Unavailable. | 528 | 502 | 94.0 | 94.0 | Still in orbit. |
| 11 Apr. | Cosmos 640<br>U.S.S.R.<br>1974-21A<br>Unavailable<br>Plesetsk | Total weight: Unavailable.<br>Objective: "Investigation of upper atmosphere and outer space."<br>Description: Unavailable. | 225 | 198 | 88.7 | 88.7 | Reentered 23 April 1974. |
| 13 Apr. | Westar 1<br>United States<br>1974-22A<br>Thor-Delta<br>ETR | Total weight: 572 kg at launch; 300 kg in orbit.<br>Objective: Place spacecraft in transfer orbit of sufficient accuracy for onboard propulsion system to place it in stationary synchronous orbit while retaining sufficient fuel for mission lifetime requirements. Transmit TV, voice, and data throughout U.S.<br>Description: Cylindrical satellite 1.8 m dia and 3.7 m high including 1.5-m-dia parabolic antenna with despin and pointing mechanism. Spin-stabilized. Powered by 20 500 solar cells plus onboard batteries. Redundant hydrazine reaction control systems. Each of 12 independent fixed-gain transponders could relay 1 color TV and 1200 voice channels. | 36 238<br>After apogee motor firing,<br>35 592 | 230<br><br>35 166 | 642.0<br><br>1415.3 | 24.7<br><br>0.6 | First U.S. commercial domestic communications satellite. Launched by NASA for Western Union Telegraph Co. on 3-stage thrust-augmented Thor-Delta with 9 strap-on rockets into transfer orbit. Apogee kick motor, fired 16 April, placed spacecraft in stationary equatorial synchronous orbit at 99° W. longitude by 9 May. First of series of 3. Commercial operations began 16 July. First mail delivery via satellite made 5 Sept. First commercial TV broadcast over Westar system made 21 July 1975. |
| 20 Apr. | Molniya 1-27<br>U.S.S.R.<br>1974-23A<br>Unavailable | Total weight: Unavailable.<br>Objective: Transmit long-range telecommunications and TV programs to stations in Orbita network in extreme north and | 40 705 | 623 | 737.6 | 62.9 | Still in orbit. |

## ASTRONAUTICS AND AERONAUTICS, 1974

| Launch Date | Spacecraft, Country, Int'l Designation, Vehicle, Launch Site | Payload Data | Apogee (km) | Perigee (km) | Period (min) | Inclination (degrees) | Remarks |
|---|---|---|---|---|---|---|---|
| 23 Apr. | Pleetsk | far east of U.S.S.R. Description: Unavailable. | | | | | |
| | *Cosmos 641* U.S.S.R. 1974-24A Unavailable Pleetsk | Total weight: Unavailable. Objective: "Investigation of upper atmosphere and outer space." Description: Unavailable. | 1 483 | 1 387 | 114.6 | 74.0 | Eight satellites launched on a single booster. Reported by press as part of Soviet navy communications system. Still in orbit. |
| | and | | | | | | |
| | *Cosmos 642* 1974-24B | Total weight: Unavailable. Objective: "Investigation of upper atmosphere and outer space." Description: Unavailable. | 1 481 | 1 319 | 113.8 | 74.0 | Still in orbit. |
| | and | | | | | | |
| | *Cosmos 643* 1974-24C | Total weight: Unavailable. Objective: "Investigation of upper atmosphere and outer space." Description: Unavailable. | 1 482 | 1 353 | 114.2 | 74.0 | Still in orbit. |
| | and | | | | | | |
| | *Cosmos 644* 1974-24D | Total weight: Unavailable. Objective: "Investigation of upper atmosphere and outer space." Description: Unavailable. | 1 483 | 1 335 | 114.0 | 74.0 | Still in orbit. |
| | and | | | | | | |
| | *Cosmos 645* 1974-24E | Total weight: Unavailable. Objective: "Investigation of upper atmosphere and outer space." Description: Unavailable. | 1 483 | 1 369 | 114.9 | 74.0 | Still in orbit. |
| | and | | | | | | |
| | *Cosmos 646* 1974-24F | Total weight: Unavailable. Objective: "Investigation of upper atmosphere and outer space." Description: Unavailable. | 1 486 | 1 403 | 114.8 | 74.0 | Still in orbit. |
| | and | | | | | | |
| | *Cosmos 647* 1974-24G | Total weight: Unavailable. Objective: "Investigation of upper atmos- | 1 484 | 1 423 | 115.0 | 74.0 | Still in orbit. |

| Date | Name / ID / Location | Objective / Description | | | | | Remarks |
|---|---|---|---|---|---|---|---|
| | | ...phere and outer space." Description: Unavailable. | 1 480 | 1 439 | 115.2 | 74.0 | Still in orbit. |
| | Cosmos 648<br>1974-24H | | | | | | |
| 24 Apr. | Meteor 17<br>U.S.S.R.<br>1974-25A<br>Unavailable<br>Plesetsk | Total weight: Unavailable. Objective: Collect meteorological information for weather forecasts. Photograph cloud and snow cover on day and night sides of globe and obtain data on thermal energy reflected and emitted by earth and its atmosphere. Description: Carried cameras, power supply system with independent orientation of solar array toward sun, radio telemetry, and data-collection systems. | 892 | 863 | 102.5 | 81.2 | Still in orbit. |
| 26 Apr. | Molniya II-9<br>U.S.S.R.<br>1974-26A<br>Unavailable<br>Plesetsk | Total weight: Unavailable. Objective: Transmit long-range telecommunications and TV programs to stations in Orbita network. Description: Unavailable. | 40 700 | 599 | 737.0 | 62.9 | Still in orbit. |
| 29 Apr. | Cosmos 649<br>U.S.S.R.<br>1974-27A<br>Unavailable<br>Plesetsk | Total weight: Unavailable. Objective: "Investigation of upper atmosphere and outer space." Description: Unavailable. | 294 | 179 | 89.2 | 62.8 | Reentered 11 May 1974. |
| 29 Apr. | Cosmos 650<br>U.S.S.R.<br>1974-28A<br>Unavailable<br>Plesetsk | Total weight: Unavailable. Objective: "Investigation of upper atmosphere and outer space." Description: Unavailable. | 1 401 | 1 368 | 113.5 | 74.0 | Still in orbit. |
| 15 May | Cosmos 651<br>U.S.S.R.<br>1974-29A<br>Unavailable<br>Baykonur-Tyuratam | Total weight: Unavailable. Objective: "Investigation of upper atmosphere and outer space." Description: Unavailable. | 262 | 249 | 89.6 | 65.0 | Still in orbit. |

| Launch Date | Spacecraft, Country, Int'l Designation, Vehicle, Launch Site | Payload Data | Apogee (km) | Perigee (km) | Period (min) | Inclination (degrees) | Remarks |
|---|---|---|---|---|---|---|---|
| 15 May | Cosmos 652 U.S.S.R. 1974-30A Unavailable Baykonur-Tyuratam | Total weight: Unavailable. Objective: "Investigation of upper atmosphere and outer space." Description: Unavailable. | 321 | 184 | 89.6 | 51.8 | Reentered 23 May 1974. |
| 15 May | Cosmos 653 U.S.S.R. 1974-31A Unavailable Plesetsk | Total weight: Unavailable. Objective: "Investigation of upper atmosphere and outer space." Description: Unavailable. | 296 | 186 | 89.3 | 63.5 | Reentered 27 May 1974. |
| 17 May | Cosmos 654 U.S.S.R. 1974-32A Unavailable Baykonur-Tyuratam | Total weight: Unavailable. Objective: "Investigation of upper atmosphere and outer space." Description: Unavailable | 263 | 250 | 89.6 | 65.0 | Still in orbit. |
| 17 May | Sms 1 United States 1974-33A Thor-Delta ETR | Total weight: 628 kg at launch; 272 kg in orbit. Objective: Launch spacecraft into synchronous orbit of sufficient accuracy for mission requirements, evaluate and check out spacecraft in orbit, and turn operational control over to NOAA/NESS for regular daytime and nighttime meteorological observations in support of National Operational Meteorological Satellite System. Description: Cylindrical spacecraft 1.91 m in dia, 3.44 m long from top of magnetometer to bottom of apogee boost motor. Scanning mirror looked out through opening in solar array covering outer walls of spacecraft. Redundant sampling unit and processor command system. Dual telemetry unit. Attitude determination and control unit using earth, sun, and nutation sensors and auxiliary propulsion systems. Solar cell power system supplemented by | 32 895 After apogee motor firing, 32 828.3 After APS firings, 35 519 | 180 30 747 35 455 | 576.5 1239.2 1420.8 | 24.5 2.0 2.0 | Synchronous Meteorological Satellite was first weather satellite in synchronous orbit. Launched by NASA as operational prototype for NOAA's GOES system on 3-stage thrust-augmented Thor-Delta with 9 strap-on rockets. Provided continuous day and night images of cloud cover over U.S. and Atlantic Ocean for first time. Although spacecraft entered lower than normal transfer orbit, apogee boost motor raised it to near-synchronous and APS to synchronous orbit. Spacecraft drifted to station at 45° W longitude over east coast of Brazil to support Global Atmospheric Research Program GATE experiment. Sms 1 was moved to 75° W longitude by mid-Nov. 1974 to become part of NOAA system; operational control turned over to NOAA 10 Dec. 1974. Still able to provide full operational data, satellite was moved to stand-by status at 105° W longitude over east |

## ASTRONAUTICS AND AERONAUTICS, 1974

| Date | Name | Description | Orbit | Apogee | Period | Incl. | Remarks |
|---|---|---|---|---|---|---|---|
| | | 2 nickel cadmium batteries. Visible infrared spin-scan radiometer and space environmental monitoring system. | | | | | Pacific Ocean after successor, *Goes 1* became operational in Jan. 1976. |
| 17 May | *Intercosmos 11* U.S.S.R. 1974-34A Unavailable Baykonur-Tyuratam | Total weight: Unavailable. Objective: Study ultraviolet and x-ray radiation of sun and upper atmosphere. Description: Unavailable. | 511 | 483 | 94.6 | 50.6 | Cooperative U.S.S.R., East German, and Czechoslovakian satellite. Still in orbit. |
| 21 May | *Cosmos 655* U.S.S.R. 1974-35A Unavailable Plesetsk | Total weight: Unavailable. Objective: "Investigation of upper atmosphere and outer space." Description: Unavailable. | 540 | 522 | 95.3 | 74.1 | Still in orbit. |
| 27 May | *Cosmos 656* U.S.S.R. 1974-36A Unavailable Baykonur-Tyuratam | Total weight: Unavailable. Objective: "Investigation of upper atmosphere and outer space." Description: Unavailable. | 322 | 185 | 89.6 | 51.8 | Reported by press as probable test of Soyuz spacecraft. Reentered 29 May 1974. |
| 29 May | *Luna 22* U.S.S.R. 1974-37A Unavailable Baykonur-Tyuratam | Total weight: Unavailable. Objective: Scientific investigation of moon and circumlunar space. Description: Unavailable. | 226 Lunar orbit, 220 | 176 220 | 88.5 130.0 | 51.6 19.6 | *Luna 22* entered lunar orbit 2 June 1974, began photographing lunar surface, measuring characteristic features of moon's surface, and analyzing chemical composition of lunar rock, with orbit change bringing spacecraft within 25 km of surface 9 June. During flight to moon, spacecraft measured density of meteor substance, intensity and energy spectra of solar rays, concentration of luna plasma, and intensity of magnetic fields. Initial program of TV photography ended 13 June but other scientific investigation and photography continued later as additional orbit changes brought *Luna 22* again as close as 30 km to moon's surface. *Luna 22* was still transmitting in Sept. 1975. Still in orbit. |
| 30 May | *Cosmos 657* U.S.S.R. 1974-38A Unavailable Plesetsk | Total weight: Unavailable. Objective: "Investigation of upper atmosphere and outer space." Description: Unavailable. | 283 | 174 | 89.1 | 62.7 | Reentered 13 June 1974. |

| Launch Date | Spacecraft, Country, Int'l Designation, Vehicle, Launch Site | Payload Data | Apogee (km) | Perigee (km) | Period (min) | Inclination (degrees) | Remarks |
|---|---|---|---|---|---|---|---|
| 30 May | *Ats 6* United States 1974-39A Titan IIIC ETR | Total weight: 1403 kg. Objective: Launch spacecraft in near-geostationary orbit; erect and evaluate in orbit 9.14-m antenna structure for good quality TV signals to small, inexpensive ground receivers; stabilize spacecraft using 3-axis control system with slewing capability in roll and pitch, permitting antenna pointing and accuracy commensurate with antenna; support and demonstrate user-oriented applications experiments; demonstrate new technology for aircraft and maritime traffic control and spacecraft-to-spacecraft relay. Description: Structure included rectangular earth-viewing module (EVM) 1.37 m long and wide, 1.66 m high, containing most experiments and all earth-viewing sensors; electrical, telemetry, and command attitude control subsystems; and communications. A 4-m graphite-fiber-reinforced plastic truss supported 9-m-dia deployable parabolic reflector. Overall length from magnetometer on top of reflector to bottom of EVM, 8.5 m. Two structural arms, each extending 8 m from hub supporting reflector, held semicylindrical solar panels containing 21 600 solar cells. Environmental measurements experiment (EME) package was mounted on top of hub. Two 19-cell nickel-cadmium batteries supplemented solar power. Communications system consisted of multifrequency transponder, redundant transmitting-receiving device with 6 receivers and 9 transmitters. Hydrazine monopropellant propulsion subsystem and control sensors, torquers, gyros, and onboard computers provided 3-axis stabilization and precision pointing. | Parking orbit, 626 After 2nd transtage burn, 35 845.6 | 152 35 774.1 | 1437.28 | 1.76 | Most complex, versatile, and powerful of NASA's Applications Technology Satellites. Placed in synchronous orbit at 94° W longitude over Galapagos Island. *Ats 6* transmitted education and medical programs to remote areas of U.S. In May 1975 spacecraft was moved over Lake Victoria in Central Africa to broadcast educational programs to 5000 villages in India, one half by direct broadcast. Arrived on station by July; began program in cooperation with Indian Space Research Organization 1 Aug. 1975. Also relayed communications for U.S.-U.S.S.R. ASTP docking mission in July 1975. After 1-yr Indian program, was to be repositioned over west coast of South America for 3rd year of experiments. |
| 3 June | *Hawkeye 1* United States 1974-40A Scout E WTR | Total weight: 26.6 kg Objective: Study plasma properties of magnetosphere in vicinity of magnetic neutral point over earth's north polar cap. Description: Eight-sided truncated cone | 124 477 | 469 | 2996 | 89.8 | *Explorer 52*, also named *Hawkeye 1*, was 6th in Univ. of Iowa's Injun series. Launched by NASA on special-configuration 5-stage Scout E incorporating BE-3A solid-fueled rocket motor in 5th stage and |

## ASTRONAUTICS AND AERONAUTICS, 1974

| Date | Spacecraft | Description | Weight (kg) | Perigee/Apogee | Period (min) | Inclination (°) | Remarks |
|---|---|---|---|---|---|---|---|
| 6 June | Cosmos 658<br>U.S.S.R.<br>1974-41A<br>Unavailable<br>Baykonur-Tyuratam | 0.75 m at base, 0.25 m at top, and 0.75 m high. Solar cells mounted on sides and bottom of outer shell and silver-cadmium battery provided power for spin-stabilized spacecraft. Extendable appendages included 2, 23-m antennas, 1.6-m flux gate magnetometer, and 1.6-m search coil magnetic antenna. Science instrumentation also included low-energy proton-electron differential energy analyzer and ELF-VLF receivers. | 273 | 209 | 89.3 | 65.0 | attitude-correction system to orient vehicle before 5th-stage ignition. All booms, antennas, and experiments were turned on and operating satisfactorily by 20 June. Spacecraft making systematic observations of polar magnetosphere.<br>Reentered 18 June 1974. |
| 6 June | DOD spacecraft<br>United States<br>1974-42A<br>Titan IIIB<br>WTR | Total weight: Unavailable.<br>Objective: Develop spaceflight techniques and technology.<br>Description: Unavailable. | 410 | 120 | 89.8 | 110.4 | Launched by USAF.<br>Reentered 23 July 1974. |
| 13 June | Cosmos 659<br>U.S.S.R.<br>1974-43A<br>Unavailable<br>Plesetsk | Total weight: Unavailable.<br>Objective: "Investigation of upper atmosphere and outer space."<br>Description: Unavailable. | 374 | 161 | 89.9 | 62.9 | Reentered 26 June 1974. |
| 18 June | Cosmos 660<br>U.S.S.R.<br>1974-44A<br>Unavailable<br>Plesetsk | Total weight: Unavailable.<br>Objective: "Investigation of upper atmosphere and outer space."<br>Description: Unavailable. | 1 969 | 395 | 109.0 | 83.0 | Still in orbit. |
| 21 June | Cosmos 661<br>U.S.S.R.<br>1974-45A<br>Unavailable<br>Plesetsk | Total weight: Unavailable.<br>Objective: "Investigation of upper atmosphere and outer space."<br>Description: Unavailable. | 546 | 509 | 95.2 | 74.0 | Still in orbit. |
| 25 June** | Salyut 3<br>U.S.S.R.<br>1974-46A<br>Unavailable | Total weight: 18 500 kg?<br>Objectives: Improve station's design, onboard systems, and equipment, and perform scientific research in space. | 252<br>After circularization, | 213 | 89.2 | 51.6 | Salyut 3 orbiting laboratory joined 5 July 1974 by 2-man crew of Pavel R. Popovich and Yuri P. Artyukhin for 15 days of scientific and technical studies and medical |

239

| Launch Date | Spacecraft, Country, Int'l Designation, Vehicle, Launch Site | Payload Data | Apogee (km) | Perigee (km) | Period (min) | Inclination (degrees) | Remarks |
|---|---|---|---|---|---|---|---|
| | Baykonur-Tyuratam | Description: Two cylinders and connecting conical structure, 21 m long, contained 3 isolated rooms with total volume of 99 cu m. Three pivoting winglike solar panels extended from structures. Attitude control system with reaction control thrusters and fine-control gas jet for spacecraft orientation. Probe-drogue docking hardware with airlock hatch closure for extravehicular capability. Orion telescope and earth observation equipment with portholes for observation and photography. Exterior survey system permitted inspection of outboard systems. Dosimeters measured radiation exposure. Special gymnastic equipment; food preparation, life support, and water regeneration equipment; recreation area. | 268 | 265 | 89.2 | 51.6 | and biological experiments. Crew undocked and returned to earth 19 July 1974. Second crew of Lt. Col. Gennady V. Sarafanov and Col. Lev S. Demin was launched on *Soyuz 15* on 26 Aug. 1974 but failed, after several attempts, to dock with station; returned to earth 29 Aug. 1974. *Salyut 3*'s "main program of work" ended 23 Sept. 1974 after module containing research materials and experiments was separated from space station. Module was landed and recovered in U.S.S.R. *Salyut 3* reentered 24 Jan. 1975. |
| 26 June | *Cosmos 662* U.S.S.R. 1974-47A Unavailable Plesetsk | Total weight: Unavailable. Objective: "Investigation of upper atmosphere and outer space." Description: Unavailable. | 833 | 289 | 95.5 | 70.9 | Still in orbit. |
| 27 June | *Cosmos 663* U.S.S.R. 1974-48A Unavailable Plesetsk | Total weight: Unavailable. Objective: "Investigation of upper atmosphere and outer space." Description: Unavailable. | 1 003 | 968 | 104.8 | 82.9 | Still in orbit. |
| 29 June | *Cosmos 664* U.S.S.R. 1974-49A Unavailable Plesetsk | Total weight: Unavailable. Objective: "Investigation of upper atmosphere and outer space." Description: Unavailable. | 337 | 203 | 89.9 | 72.8 | Reentered 11 July 1974. |
| 29 June | *Cosmos 665* U.S.S.R. 1974-50A | Total weight: Unavailable. Objective: "Investigation of upper atmosphere and outer space." | 39 379 | 626 | 710.7 | 62.8 | Western press reported very high orbit of *Cosmos 665* would permit full global surveillance; quoted Western observers as |

## ASTRONAUTICS AND AERONAUTICS, 1974

| Date | Name / Designation / Site | Description / Objective | | | | Remarks |
|---|---|---|---|---|---|---|
| | Unavailable Plesetsk | Description: Unavailable. | 338 | 89.9 | 51.5 | linking launch with talks between President Nixon and Soviet leaders on nuclear arms limitations. Still in orbit. |
| 3 July | *Soyuz 14* U.S.S.R. 1974-51A Unavailable Baykonur-Tyuratam | Total weight: 6570 kg? Objective: Test methods and systems of controlling spacecraft, improve life-sustaining complex, explore new possibilities of using both separate craft and groups of space objects to accomplish scientific and applied tasks in near-earth orbit. Description: Three-unit spacecraft 9.5 m long with 2 crew cabins totaling 8.9 cu m interior space. | | | | *Soyuz 14* carried Cosmonauts Pavel R. Popovich and Yuri P. Artyukhin to dock 5 July with *Salyut 3* orbiting space station (launched 25 June). Crew studied geological morphological formations on earth's surface, atmospheric phenomena, physical characteristics of outer space, manufacturing techniques, and influence of space on human organism. After 15 days 17 hrs 30 min in space, crew boarded *Soyuz 14* and softlanded 19 July 1974 near Dzhezkazgan. |
| 9 July | *Meteor 18* U.S.S.R. 1974-52A Unavailable Plesetsk | Total weight: Unavailable. Objective: Collect meteorological information for rapid data supply. Description: Carried equipment for photographing clouds and underlying surface, studying distribution of precipitation and ice zones, and studying influences of corpuscular flows on upper layers of atmosphere. Also carried attitude control system, solar-power array, and radio telemetric systems. | 891 | 102.5 | 81.2 | Still in orbit |
| 12 July | *Cosmos 666* U.S.S.R. 1974-53A Unavailable Plesetsk | Total weight: Unavailable. Objective: "Investigation of upper atmosphere and outer space." Description: Unavailable. | 471 | 91.6 | 62.7 | Reentered 25 July 1974. |
| 14 July | *Nts 1* United States 1974-54A Atlas F WTR | Total weight: Unavailable. Objective: Test techniques for DOD's NAVSTAR 'Global Positioning System. Description: Carried receivers and transmitters to broadcast spacecraft's exact position continuously and atomic clock to transmit ultraprecise time signals. | 13 762 | 468.4 | 125.1 | Navigation Technology Satellite launched by USAF, originally known as Timation 3A. Built by NRL. Still in orbit. |
| 16 July | *Aeros 2* West Germany-U.S. 1974-55A Scout D WTR | Total weight: 127 kg. Objective: Place satellite in earth orbit that would permit measuring main aeronomic parameters of upper atmosphere and solar uv radiation in wavelength of main absorption. | 879.0 | 97.44 | 95.65 | West German-built satellite launched by NASA into polar orbit. GSFC experiment (NATE) and 4 West German experiments would provide model of ionosphere from which upper atmospheric conditions could be predicted. Although 2 tape recorders |

241

| Launch Date | Spacecraft, Country, Int'l Designation, Vehicle, Launch Site | Payload Data | Apogee (km) | Perigee (km) | Period (min) | Inclination (degrees) | Remarks |
|---|---|---|---|---|---|---|---|
| 23 July | Molniya II-10 U.S.S.R. 1974-56A Unavailable Plesetsk | Description: Circular cylindrical shell welded to bottom conical shell to form structural unit 9.16 m in dia, 7.10 m high. Power provided by flat honeycomb solar cell array attached to spacecraft cylinder top and supplemented by 1 silver-zinc and 1 nickel-cadmium battery. Attitude measurement and control system, monopropellant-hydrazine vernier apogee-boost motor, uplink and downlink telecommunications, 4 telemetry antennas, and spacecraft command system. Spin-stabilized. Carried 5 scientific instruments: mass spectrometer, retarding potential analyzer, impedance probe, EUV spectrometer, and neutral-atmosphere temperature experiment (NATE). | 40 724 | 603 | 737.6 | 62.9 | Still in orbit failed, additional U.S. and French tracking collected 70%-80% of real-time data, sufficient for scientific objectives. Planned operational lifetime of 6 mos was exceeded. Spacecraft reentered 25 Sept. 1975. |
| 25 July | Cosmos 667 U.S.S.R. 1974-57A Unavailable Baykomur-Tyuratam | Total weight: Unavailable. Objective: "Investigation of upper atmosphere and outer space." Description: Unavailable. | 311 | 183 | 89.5 | 65.1 | Reentered 7 Aug. 1974. |
| 25 July | Cosmos 668 U.S.S.R. 1974-58A Unavailable Plesetsk | Total weight: Unavailable. Objective: "Investigation of upper atmosphere and outer space." Description: Unavailable. | 490 | 269 | 92.1 | 71.0 | Reentered 21 Feb. 1975. |
| 26 July | Cosmos 669 U.S.S.R. 1974-59A Unavailable Plesetsk | Total weight: Unavailable. Objective: "Investigation of upper atmosphere and outer space." Description: Unavailable. | 230 | 212 | 88.9 | 81.3 | Reentered 8 Aug. 1974. |

## ASTRONAUTICS AND AERONAUTICS, 1974

| Date | Spacecraft | Details | | | | Remarks |
|---|---|---|---|---|---|---|
| 29 July | *Molniya 1-S* U.S.S.R. 1974-60A Unavailable Baykonur-Tyuratam | Total weight: Unavailable. Objective: Further perfection of satellite communications systems. Description: Carried equipment for experimental TV broadcasting, long-range radio communications, and systems for spacecraft orientation, orbit correction, thermo regulation, and power supply. | 35 790 | 35 787 | 1436.2 | 0.0 | First operational U.S.S.R. synchronous communications satellite. Still in orbit. |
| 6 Aug. | *Cosmos 670* U.S.S.R. 1974-61A Unavailable Baykonur-Tyuratam | Total weight: Unavailable. Objective: "Investigation of upper atmosphere and outer space." Description: Unavailable. | 291 | 209 | 89.5 | 50.6 | Reported by press as reconnaissance satellite launched to observe war on Cyprus between Turkish and Greek Cypriots. Reentered 9 Aug. 1974. |
| 7 Aug. | *Cosmos 671* U.S.S.R. 1974-62A Unavailable Plesetsk | Total weight: Unavailable. Objective: "Investigation of upper atmosphere and outer space." Description: Unavailable. | 343 | 180 | 89.7 | 62.8 | Reentered 20 Aug. 1974. |
| 8 Aug.* | DOD spacecraft United States 1974-63A Thor-Burner II WTR | Total weight: Unavailable. Objective: Develop space flight techniques and technology. Description: Unavailable. | 875 | 805 | 101.6 | 98.8 | Launched by USAF. Reported by press as Defense Meteorological Satellite. Still in orbit. |
| 12 Aug. | *Cosmos 672* U.S.S.R. 1974-64A Unavailable Baykonur-Tyuratam | Total weight: Unavailable. Objective: "Investigation of upper atmosphere and outer space." Description: Unavailable. | 277 | 142 | 88.7 | 51.8 | Unmanned Soyuz spacecraft tested docking systems for July 1975 U.S.-U.S.S.R. ASTP mission. Reentered 18 Aug. 1974. |
| 14 Aug. | DOD spacecraft United States 1974-65A Titan IIIB-Agena WTR | Total weight: Unavailable. Objective: Develop space flight techniques and technology. Description: Unavailable. | 407 | 123 | 89.8 | 110.4 | Launched by USAF. Reported by press as Navy ocean surveillance satellite. Reentered 29 Sept. 1974. |
| 16 Aug. | *Cosmos 673* U.S.S.R. 1974-66A Unavailable Baykonur-Tyuratam | Total weight: Unavailable. Objective: "Investigation of upper atmosphere and outer space." Description: Unavailable. | 634 | 605 | 97.1 | 81.2 | Still in orbit. |

| Launch Date | Spacecraft, Country, Int'l Designation, Vehicle, Launch Site | Payload Data | Apogee (km) | Perigee (km) | Period (min) | Inclination (degrees) | Remarks |
|---|---|---|---|---|---|---|---|
| 26 Aug. | *Soyuz 15* U.S.S.R. 1974-67A Unavailable Baykonur-Tyuratam | Total weight: 6570 kg? Objective: Continue scientific research and experiments in space begun 3 July 1974 during flight of *Soyuz 14* and *Salyut 3*. Description: Three-unit spacecraft 9.5 m long with 2 crew cabins totaling 8.9 cu m of interior space. | 239 | 175 | 88.6 | 51.7 | *Soyuz 15*, carrying crew members Lt. Col. Gennady V. Sarafanov and Col. Lev S. Demin, failed in several attempts to dock with *Salyut 3* orbiting space station. *Soyuz 15* returned to earth 29 Aug. (28 Aug. EDT) after 48 hrs 12 min in space. Made first Soviet night landing. |
| 29 Aug. | *Cosmos 674* U.S.S.R. 1974-68A Unavailable Baykonur-Tyuratam | Total weight: Unavailable. Objective: "Investigation of upper atmosphere and outer space." Description: Unavailable. | 311 | 174 | 89.4 | 65.0 | Reentered 7 Sept. 1974. |
| 29 Aug. | *Cosmos 675* U.S.S.R. 1974-69A Unavailable Plesetsk | Total weight: Unavailable. Objective: "Investigation of upper atmosphere and outer space." Description: Unavailable. | 1 425 | 1 363 | 113.7 | 74.1 | Still in orbit. |
| 29 Aug. | *Ans 1* Netherlands-U.S. 1974-70A Scout WTR | Total weight: 129.8 kg. Objective: Increase scientific knowledge of stellar ultraviolet and x-ray sources. Description: Rectangular double-bar configuration, 1.23 m high, 0.73 m in depth, and 1.44 m wide with 2 winglike solar panels deployed. Honeycombed aluminum spacecraft carried 4 canted turnstile antennas. Power provided by 2050 solar cells and 1 nickel-cadmium battery. Spacecraft subsystems included passive thermal control, 3-axis attitude control and stabilization, and onboard computer and telemetry command. Carried 3 experiments: cassegrain telescope with spectrometer and 5 uv-sensitive photomultipliers, 2-detector soft x-ray instrument with 0.2- to 5-kev spectral range, and 2-detector hard x-ray instrument with 2- to 15-kev spectral range and Bragg crystal spectrometer. | 1 167 | 254 | 99.0 | 98.1 | Cooperative Netherlands Astronomical Satellite launched by NASA into polar orbit. Spacecraft entered highly elliptical orbit, rather than planned circular orbit, because of malfunction in Scout guidance package. Changes in spacecraft and experiment operations accommodated to actual orbit. All experiments turned on by 8 Sept. During 1975 *Ans 1* returned new data on changing intensity and spectrum of x-radiation from Cygnus X-1 and measured high-count soft x-radiation from Sirius. Still in orbit, transmitting on command. |

# ASTRONAUTICS AND AERONAUTICS, 1974

| Date | Name | Details | | | | Remarks |
|---|---|---|---|---|---|---|
| 11 Sept. | Cosmos 676<br>U.S.S.R.<br>1974-71A<br>Unavailable<br>Plesetsk | Total weight: Unavailable.<br>Objective: "Investigation of upper atmosphere and outer space."<br>Description: Unavailable. | 814 | 795 | 101.0 | 74.0 | Still in orbit. |
| 19 Sept. | Cosmos 677<br>U.S.S.R.<br>1974-72A<br>Unavailable<br>Plesetsk<br>and | Total weight: Unavailable.<br>Objective: "Investigation of upper atmosphere and outer space."<br>Description: Unavailable. | 1 466 | 1 398 | 114.5 | 74.0 | Eight satellites launched on single booster. Reported by U.S. press as probably part of worldwide naval communications system. Still in orbit. |
| | Cosmos 678<br>1974-72B<br>and | Total weight: Unavailable.<br>Objective: "Investigation of upper atmosphere and outer space."<br>Description: Unavailable. | 1 533 | 1 466 | 116.0 | 74.0 | Still in orbit. |
| | Cosmos 679<br>1974-72C<br>and | Total weight: Unavailable.<br>Objective: "Investigation of upper atmosphere and outer space."<br>Description: Unavailable. | 1 511 | 1 466 | 115.8 | 74.0 | Still in orbit. |
| | Cosmos 680<br>1974-72D<br>and | Total weight: Unavailable.<br>Objective: "Investigation of upper atmosphere and outer space."<br>Description: Unavailable. | 1 492 | 1 466 | 115.5 | 74.0 | Still in orbit. |
| | Cosmos 681<br>1974-72E<br>and | Total weight: Unavailable.<br>Objective: "Investigation of upper atmosphere and outer space."<br>Description: Unavailable. | 1 472 | 1 466 | 115.3 | 74.0 | Still in orbit. |
| | Cosmos 682<br>1974-72F<br>and | Total weight: Unavailable.<br>Objective: "Investigation of upper atmosphere and outer space."<br>Description: Unavailable. | 1 466 | 1 453 | 115.1 | 74.0 | Still in orbit. |
| | Cosmos 683<br>1974-72G<br>and | Total weight: Unavailable.<br>Objective: "Investigation of upper atmosphere and outer space."<br>Description: Unavailable. | 1 467 | 1 434 | 114.9 | 74.0 | Still in orbit. |

| Launch Date | Spacecraft, Country, Int'l Designation, Vehicle, Launch Site | Payload Data | Apogee (km) | Perigee (km) | Period (min) | Inclination (degrees) | Remarks |
|---|---|---|---|---|---|---|---|
| | *Cosmos 684* 1974-72H | Total weight: Unavailable. Objective: "Investigation of upper atmosphere and outer space." Description: Unavailable. | 1 467 | 1 416 | 114.7 | 74.0 | Still in orbit. |
| 20 Sept. | *Cosmos 685* U.S.S.R. 1974-73A Unavailable Baykonur-Tyuratam | Total weight: Unavailable. Objective: "Investigation of upper atmosphere and outer space." Description: Unavailable. | 279 | 201 | 89.3 | 65.0 | Reentered 2 Oct. 1974. |
| 26 Sept. | *Cosmos 686* U.S.S.R. 1974-74A Unavailable Plesetsk | Total weight: Unavailable. Objective: "Investigation of upper atmosphere and outer space." Description: Unavailable. | 486 | 270 | 92.1 | 71.0 | Reentered 1 May 1975. |
| 10 Oct. | *Westar 2* United States 1974-75A Thor-Delta ETR | Total weight: 572 kg at launch; 300 kg in orbit. Objective: Place satellite in transfer orbit of sufficient accuracy to permit onboard propulsion system to place spacecraft in geosynchronous orbit while retaining sufficient fuel for mission lifetime requirements to provide TV, voice, and data transmission throughout U.S. Description: Spacecraft similar to *Westar 1*, launched 13 April 1974. | 36 107 After apogee motor firing, 35 766 | 231 35 244 | 641.2 1422.0 | 24.76 0.5 | Domestic communications satellite launched by NASA for Western Union Telegraph Co. as backup for *Westar 1*. Launched into transfer orbit on 3-stage thrust-augmented Thor-Delta with 9 strap-on rockets. Apogee kick motor, fired 13 Oct., placed spacecraft in geosynchronous orbit. *Westar 2* drifted eastward, arriving on station at 90° W longitude over equator southeast of Hawaii 21 Nov. 1974. Onboard motor was fired again 7 Feb. 1975, moving spacecraft to final position at 123° W longitude, over Los Angeles, 27 Feb. 1975. Westar system made first commercial TV broadcast 21 July 1975. |
| 11 Oct. | *Cosmos 687* U.S.S.R. 1974-76A Unavailable Plesetsk | Total weight: Unavailable. Objective: "Investigation of upper atmosphere and outer space." Description: Unavailable. | 697 | 286 | 94.5 | 74.0 | Still in orbit. |

| Date | Name/Designation | Description | Col4 | Col5 | Col6 | Col7 | Remarks |
|---|---|---|---|---|---|---|---|
| 15 Oct. | *Ariel 5* U.K.–Italy–U.S. 1974-77A Scout San Marco platform | Total weight: 132.3 kg. Objective: Investigate galactic and extragalactic x-ray sources. Description: Sixteen-sided polyhedron 0.96 m in dia, 0.86 m high, consisting of assembly of aluminum honeycomb panels for mounting experiments and control instruments. Central box, 0.36 m square, accommodated x-ray experiment. Spacecraft powered by 4536 silicon solar cells mounted on outside surface and supplemented by nickel-cadmium battery. Attitude sensing and control of spin-stabilized spacecraft provided by sun and earth sensors and propane gas jet system. Data-handling, telemetry, and command subsystem included 2-core store computer with 8000 8-bit word memory, redundant receivers and transmitters, and common 4-element antenna. *Ariel 5* carried 6 x-ray experiments, 4 designed for pointing at specific x-ray sources and 2 providing scanning information. | 569.5 | 502.0 | 95.4 | 2.86 | Fifth in series of U.K.–U.S. cooperative satellites, built by U.K. and carrying 1 U.S. and 5 U.K. experiments, was launched by Italian launch crew on NASA vehicle. By 30 Oct. 1974 all instruments were turned on and operating normally. In August 1975, recorded x-ray source in Orion 4 times brighter than any previously observed; source identified as recurrent nova. *Ariel 5* still in orbit, transmitting on command. |
| 18 Oct. | *Cosmos 688* U.S.S.R. 1974-78A Unavailable Plesetsk | Total weight: Unavailable. Objective: "Investigation of upper atmosphere and outer space." Description: Unavailable. | 351 | 176 | 89.8 | 62.8 | Reentered 30 Oct. 1974. |
| 18 Oct. | *Cosmos 689* U.S.S.R. 1974-79A Unavailable Plesetsk | Total weight: Unavailable. Objective: "Investigation of upper atmosphere and outer space." Description: Unavailable. | 1 018 | 976 | 105.1 | 82.9 | Still in orbit. |
| 22 Oct. | *Cosmos 690* U.S.S.R. 1974-80A Unavailable Plesetsk | Total weight: Unavailable. Objective: Investigate effects of space on living organisms and test life support systems. Description: Cesium 137 gamma ray source on board. | 366 | 216 | 90.3 | 62.8 | Carried experimental systems with laboratory animals, other biological organisms, and equipment to evaluate effects of weightlessness and radiation. Returned to earth with animals 12 Nov. 1974. |
| 24 Oct. | *Molniya 1-28* U.S.S.R. 1974-81A Unavailable Plesetsk | Total weight: Unavailable. Objective: Operation of long-range telecommunications, and transmission of TV programs to stations in Orbita network. Description: Unavailable. | 39 707 | 645 | 717.7 | 62.8 | Comsat to relay TV programs and establish remote multichannel radio communications. Still in orbit. |

ASTRONAUTICS AND AERONAUTICS, 1974

| Launch Date | Spacecraft, Country, Int'l Designation, Vehicle, Launch Site | Payload Data | Apogee (km) | Perigee (km) | Period (min) | Inclination (degrees) | Remarks |
|---|---|---|---|---|---|---|---|
| 25 Oct. | Cosmos 691 U.S.S.R. 1974-82A Unavailable Plesetsk | Total weight: Unavailable. Objective: "Investigation of upper atmosphere and outer space." Description: Unavailable. | 318 | 170 | 89.4 | 65.0 | Reentered 6 Nov. 1974. |
| 28 Oct. | Meteor 19 U.S.S.R. 1974-83A Unavailable Plesetsk | Total weight: Unavailable. Objective: Collect meteorological data for weather service. Description: Carried instruments for photographing cloud formations and snow cover on lighted and unlighted sides of globe and obtain data on thermal energy reflected and radiated by earth and its atmosphere. Spacecraft systems included solar power array, attitude control system, and radio telemetry. | 904 | 842 | 102.4 | 81.2 | Still in orbit. |
| 28 Oct. | Luna 23 U.S.S.R. 1974-84A Unavailable Baykonur-Tyuratam | Total weight: Unavailable. Objective: Continue scientific investigation of moon and circumlunar space. Description: Unavailable. | 246 Lunar orbit, 104 | 183 94 | 88.7 117.0 | 51.5 138.0 | Luna 23 entered lunar orbit 2 Nov. 1974. Orbital corrections made 4 and 5 Nov. placed spacecraft in a 105- by 17-km orbit, where automatic station was separated and landed in southern part of moon's Sea of Crisis 6 Nov. 1974. Because of rugged terrain, device designed to sample lunar surface was damaged and planned sampling program abandoned. Modified program to test lander's equipment continued until 9 Nov. 1974. |
| 29 Oct. | DOD spacecraft United States 1974-85A Titan IIID WTR and | Total weight: Unavailable. Objective: Develop space flight techniques and technology. Description: Unavailable. | 279 | 152 | 88.8 | 96.7 | Three satellites launched by USAF on single booster included S3-1 to measure air density, upper-atmospheric composition data, and provide information on atmospheric effects on spacecraft orbits. Satellite 85A reentered 19 March 1975. |
| | DOD spacecraft 1974-85B | Total weight: Unavailable. Objective: Develop space flight techniques | 542 | 540 | 95.3 | 96.1 | Still in orbit. |

248

| Date | Spacecraft | Description | | | | |
|---|---|---|---|---|---|---|
| | DOD spacecraft 1974-85C | and technology. Description: Unavailable. Total weight: Unavailable. Objective: Develop space flight techniques and technology. Description: Unavailable. | 3 711 | 147 | 125.4 | 97.0 | Reentered 26 May 1975. |
| 31 Oct. | Intercosmos 12 U.S.S.R. 1974-86A Unavailable Plesetsk | Total weight: Unavailable. Objective: Investigate earth's atmosphere, ionosphere, and micrometeorite flux. Description: Equipment included mass spectrometer and instruments for studying micrometeorites. | 697 | 241 | 94.0 | 74.0 | Cooperative program of U.S.S.R., Bulgaria, Czechoslovakia, East Germany, Hungary, Poland, and Romania. Spacecraft reentered 11 July 1975. |
| 1 Nov. | Cosmos 692 U.S.S.R. 1974-87A Unavailable Plesetsk | Total weight: Unavailable. Objective: "Investigation of upper atmosphere and outer space." Description: Unavailable. | 288 | 199 | 89.4 | 63.0 | Reentered 13 Nov. 1974. |
| 4 Nov. | Cosmos 693 U.S.S.R. 1974-88A Unavailable Plesetsk | Total weight: Unavailable. Objective: "Investigation of upper atmosphere and outer space." Description: Unavailable. | 234 | 212 | 89.0 | 81.4 | Reentered 16 Nov. 1974. |
| 15 Nov. | Noaa 4 United States 1974-89A Thor-Delta WTR | Total weight: 345 kg. Objective: Place spacecraft in sun-synchronous orbit with local equator-crossing time of 8:40 am. to permit regular day and nighttime observation of global cloud cover and collection of global measurements of earth's atmospheric structure; evaluate and check out spacecraft in orbit; and turn operational control over to NOAA/NESS for NOMSS. Description: Rectangular box-shaped spacecraft, 1.02 m square at base and 1.22 m high; carried 4 antennas. Spacecraft power provided by deployable 4.46-sq-m solar array and supplementary battery. Dynamic control devices and attitude control system provided continuous earth orientation for 3-axis stabilized spacecraft. Active and passive thermal control maintained temperatures of 280–310 K (40°–95°F). Instruments included 2 | 1 458 | 1 444 | 115.0 | 101.74 | Three satellites launched by NASA on single 2-stage thrust-augmented Thor-Delta with 3 strap-on rockets. Noaa 4 (ITOS-G), launched for NOAA's global weather watch, was 4th operational satellite in series of 2nd-generation meteorological satellites. Spacecraft was checked out and turned over to NOAA/NESS control 4 Dec. Was still providing global coverage in Jan. 1976. |

| Launch Date | Spacecraft, Country, Int'l Designation, Vehicle, Launch Site | Payload Data | Apogee (km) | Perigee (km) | Period (min) | Inclination (degrees) | Remarks |
|---|---|---|---|---|---|---|---|
| | and | scanning radiometers, 2 very-high-resolution radiometers, 2 improved vertical-temperature-profile radiometers, and solar proton monitor. | | | | | |
| | *Amsat Oscar 7* AMSAT—U.S. 1974-89B | Total weight: 29.5 kg. Objective: Provide noncommercial public service and educational use for amateur radio community and U.S. schools and conduct experimental program of multiple-access communication techniques using low-powered earth terminal. Description: Eight-sided polyhedron 0.42 m in dia and 0.43 m high. Eight solar cell arrays on spacecraft surface with supplemental nickel-cadmium battery supplied spacecraft power. Spacecraft systems included 2- to 10-m repeater with input frequency of 145.85–145.95 mhz and output frequency of 29.40–29.50 mhz, 70-cm to 2-m repeater with input frequency of 432.125–432.175 mhz and output frequency of 145.975–145.925 mhz, 2 beacon transmitters, morse code telemetry encoder, teletype telemetry encoder, and codestore unit. | 1 458 | 1 454 | 115.0 | 101.7 | *Amsat Oscar 7*, Orbiting Satellite Carrying Amateur Radio, was built by U.S., Australian, Canadian, and West German amateur radio groups, working through Radio Amateur Satellite Corp., using surplus satellite hardware and equipment donated by aerospace organizations. Satellite was operating normally after pickaback launch with *Noaa 4*. Still transmitting. |
| | and | | | | | | |
| | *Intasat* Spain—U.S. 1974-89C | Total weight: 20 kg. Objective: Place satellite in high-inclination near-circular orbit to measure ionospheric total electron content and ionospheric irregularities and scintillations. Description: Twelve-sided truncated conical body 0.44 m in dia and 0.41 m high, carried four 0.49-m monopole telemetry antennas mounted on spacecraft bottom, with permanent magnet and damping bar attached to top. Two-element dipole beacon antenna, each element 1.75 m long, was oriented along spacecraft X axis. Spacecraft power provided by 660 solar cells mounted on side panels and supple- | 1 468 | 1 468 | 115.0 | 101.7 | First Spanish satellite was operating normally after pickaback launch with *Noaa 4*. Expected operational lifetime was 2 yrs. Still transmitting. |

| Date | Satellite | Description | Perigee/Apogee | | Period | Incl. | Remarks |
|---|---|---|---|---|---|---|---|
| 16 Nov. | Cosmos 694<br>U.S.S.R.<br>1974-90A<br>Unavailable<br>Plesetsk | mented by nickel-cadmium battery. Attitude control system maintained Z-axis orientation parallel to earth's magnetic field. Spacecraft magnetically stabilized with spin rate of 2 revolutions per orbit. Experiment consisted of ionospheric beacon transmitter to transmit data to 25-30 experimenters worldwide. | 319 | 201 | 89.7 | 72.8 | Reentered 29 Nov. 1974. |
| 20 Nov. | Cosmos 695<br>U.S.S.R.<br>1974-91A<br>Unavailable<br>Plesetsk | Total weight: Unavailable.<br>Objective: "Investigation of upper atmosphere and outer space."<br>Description: Unavailable. | 466 | 272 | 91.9 | 71.0 | Reentered 15 July 1975. |
| 21 Nov. | Molniya III-1<br>U.S.S.R.<br>1974-92A<br>Unavailable<br>Plesetsk | Total weight: Unavailable.<br>Objective: Operation of long-range telecommunications and transmission of color TV programs to stations in Orbita network.<br>Description: Carried relay equipment designed to operate in 1-cm waveband. | 40 683 | 628 | 737.3 | 62.8 | First of new model of Molniya communications satellites. Still in orbit. |
| 21 Nov. | Intelsat-IV F-8<br>United States<br>1974-93A<br>Atlas-Centaur<br>ETR | Total weight: 1387 kg at launch; 700 kg in orbit.<br>Objective: Conduct design, performance, and flight readiness reviews for PCC; ensure compatability of Intelsat with NASA launch vehicles and launch environmental conditions; and launch satellite into transfer orbit of sufficient accuracy for onboard apogee motor to place spacecraft in geosynchronous orbit to provide 3000 to 9000 telephone circuits simultaneously or 12 color TV channels, or combination of telephone, TV, and other forms of communication.<br>Description: Cylindrical spacecraft 2.38 m in dia, 5.28 m high. Spin-stabilized spacecraft carried 45 012 solar cells, 12 transponders, 2 global transmit antennas, 2 global receive antennas, and 2 steerable spot-beam transmit antennas. | 35 899.2<br>After apogee motor firing, 35 780 | 536.5<br><br>35 630 | 639.8<br><br>1431.9 | 26.0<br><br>1.8 | Launched by NASA for ComSatCorp, manager for INTELSAT. Was 6th in Intelsat IV series. Apogee kick motor, fired 23 Nov., placed spacecraft in geosynchronous orbit drifting eastward to arrive on station over Pacific Ocean at 174° E longitude 12 Dec. Commercial operations began 18 Dec. Still operating as primary comsat over Pacific in 1976. |

ASTRONAUTICS AND AERONAUTICS, 1974

| Launch Date | Spacecraft, Country, Int'l Designation, Vehicle, Launch Site | Payload Data | Apogee (km) | Perigee (km) | Period (min) | Inclination (degrees) | Remarks |
|---|---|---|---|---|---|---|---|
| 22 Nov. | Skynet IIB U.K.–U.S. 1974–94A Thor-Delta ETR | Total weight: 435 kg. Objective: Launch spacecraft into transfer orbit permitting boost into synchronous orbit of sufficient accuracy to provide in-orbit X-band military communications for U.K. Description: Similar to Skynet IIA, launched 18 Jan. 1974. | 39 965 After apogee motor firing, 36 595 | 185 35 896 | 1495.4 | 24.6 2.2 | Second of 2 U.K. second-generation operational comsats launched by NASA under agreement with USAF. Launched into transfer orbit on 3-stage thrust-augmented Thor-Delta with 3 strap-on rockets. Apogee kick motor, fired 24 Nov. 1974, placed spacecraft in geosynchronous orbit, drifting to final position at 50° E longitude, over Indian Ocean. Still in orbit. |
| 27 Nov. | Cosmos 696 U.S.S.R. 1974–95A Unavailable Plesetsk | Total weight: Unavailable. Objective: "Investigation of upper atmosphere and outer space." Description: Unavailable. | 320 | 203 | 89.7 | 72.9 | Reentered 9 Dec. 1974. |
| 2 Dec. | Soyuz 16 U.S.S.R. 1974–96A Unavailable Baykonur-Tyuratam | Total weight: 6570 kg? Objective: Test spacecraft onboard systems, modernized to meet demands of 1975 U.S.–U.S.S.R. ASTP mission; make scientific and technological investigations; and observe and photograph sections of earth's surface. Description: Soyuz 16 announced to be used for ASTP; included descent vehicle, orbital module, and instrument-assembly module, totaling 7.5-m length without passive docking test ring. Total volume of habitable modules, 10 cu m. Carried life support system, newly designed for better compatibility with Apollo, with cabin pressure of 1000 newtons per sq cm (14.7 psi) and atmosphere similar in oxygen content to terrestrial. New equipment also included compatible docking equipment, with donut-shaped metal flange and latches, for use as passive docking test ring; updated radio communications; and locating devices, observation window, power supply system, and attitude control system. | 280 | 194 | 89.3 | 51.8 | Soyuz 16 carrying ASTP backup crew, Cosmonauts Anatoly V. Filipchenko and Nikolay N. Rukavishnikov, orbited earth in test of modified Soyuz spacecraft to be used in 1975 U.S.–U.S.S.R. ASTP mission. In addition to checking out spacecraft systems, crew performed medical, biological, earth resources, and astronomy experiments. Soyuz 16 softlanded in U.S.S.R. 300 km north of Dzhezkazgan 8 Dec. 1974 with crew in good health. |

## ASTRONAUTICS AND AERONAUTICS, 1974

| Date | Spacecraft | | | | | | |
|---|---|---|---|---|---|---|---|
| 10 Dec. | *Helios 1*<br>West Germany-<br>U.S.<br>1974-97A<br>Titan IIIE-<br>Centaur<br>ETR | Total weight: 370 kg.<br>Objective: Investigate fundamental solar processes and solar-terrestrial relations by studying solar wind, magnetic and electron fields, cosmic rays, and cosmic dust in region between earth orbit and 0.3 AU from sun.<br>Description: Sixteen-sided polyhedron 1.75 m in dia, 0.55 m high, with conical solar arrays attached at both ends. Largest diameter of solar arrays, 2.77 m. Height of spacecraft, including telecommunications antenna mounted above central body and protruding double-hinged rigid booms attached to central body, 4.2 m. Two deployable booms, used as antenna for radiowave experiment, 3.2 m tip to tip, mounted perpendicularly to rigid booms. Power provided by solar panels, silver-zinc batteries, and regulator. Active and passive thermal control including louvers, optical surface reflectors and mirrors, multiple-layer insulation blankets, and heaters—maintained temperatures of 200–300 K (15°–85° F). Attitude control system—including sensors, propulsion system, nutation damper, and DC stepper motor—maintained 60 rpm spin rate and attitude of spacecraft, controlled despin and pointing of high-gain antenna, and provided sun-reference pulses for experiments. Three antennas, high-power amplifier, data encoder, and core storage unit were used for communications and telemetry. | Heliocentric orbit, 0.985 AU | 166.7<br>0.3095 AU | 166.7 | 88.0<br>190.15 days | 30.3<br>0.02 | West German-built cooperative spacecraft launched by NASA into elliptical solar orbit in first operational use of Titan-Centaur launch vehicle. *Helios 1* flew within 45 million km of sun, closer than any previous spacecraft, 15 March 1975 with all instruments functioning normally. Spacecraft performance continued excellent during 2nd perihelion 21 Sept. 1975, returning new data on structure of solar fields and particles close to sun. |
| 13 Dec. | *Cosmos 697*<br>U.S.S.R.<br>1974-98A<br>Unavailable<br>Plesetsk | Total weight: Unavailable.<br>Objective: "Investigation of upper atmosphere and outer space."<br>Description: Unavailable. | | 394 | 174 | 90.2 | 62.8 | Reentered 25 Dec. 1974. |
| 17 Dec. | *Meteor 20*<br>U.S.S.R.<br>1974-99A<br>Unavailable<br>Plesetsk | Total weight: Unavailable.<br>Objective: Collect meteorological information for operational weather service.<br>Description: Carried equipment for photographing clouds and snow cover on day and night sides of globe and collecting data on thermal energy reflected and emitted by earth and its atmosphere. | | 896 | 841 | 102.3 | 81.2 | Still in orbit. |

| Launch Date | Spacecraft, Country, Int'l Designation, Vehicle, Launch Site | Payload Data | Apogee (km) | Perigee (km) | Period (min) | Inclination (degrees) | Remarks |
|---|---|---|---|---|---|---|---|
| 18 Dec. | Cosmos 698 U.S.S.R. 1974-100A Unavailable Plesetsk | Total weight: Unavailable. Objective: "Investigation of upper atmosphere and outer space." Description: Unavailable. | 548 | 515 | 95.3 | 74.0 | Still in orbit. |
| 18 Dec. | Symphonie 1 France-West Germany-U.S. 1974-101A Thor-Delta ETR | Total weight: 402 kg. Objective: Launch spacecraft into transfer orbit permitting boost into synchronous equatorial orbit to test and demonstrate communications equipment for TV, radio, telephone, telegraph, and data transmission and provide equivalent of 2 color TV channels and 8 voice channels, or 1200 telephone circuits between Europe, Africa, and South America. Description: Flat 6-sided structure 1.85 m in dia, 0.5 m high with 3 deployable solar wings mounted 120° apart and extended 2.5 m. Satellite was spin-stabilized at 120 rpm during transfer phase; then 3-axis-stabilized by nitrogen cold gas attitude control system in final orbit. Thermal control provided by heat shield and thermal paint. Liquid-fueled apogee kick motor, superhigh-frequency antenna feed and reflectors, reception horn, and solar sensors were mounted on spacecraft top. | 38 071 After apogee motor firing, 35 810 | 402 35 767 | 688.21 1436.2 | 13.2 0.3 | First of 2 experimental comsats to be developed by France and West Germany and launched by NASA. Launched on 3-stage thrust-augmented Thor-Delta with 9 strap-on rockets. Apogee kick motor, fired 21 Dec. 1974, placed satellite in geosynchronous orbit, drifting 5° per day to final station at 115° W longitude over west coast of Africa. Operations began 2 Jan. 1975. Still operating. |
| 21 Dec. | Molniya 11-11 U.S.S.R. 1974-102A Unavailable Plesetsk | Total weight: Unavailable. Objective: Operate long-range telecommunications and transmit TV programs to stations in Orbita network. Description: Unavailable. | 40 630 | 656 | 736.8 | 62.9 | Still in orbit. |
| 24 Dec. | Cosmos 699 U.S.S.R. 1974-103A Unavailable Baykonur- | Total weight: Unavailable. Objective: "Investigation of upper atmosphere and outer space." Description: Unavailable. | 442 | 430 | 93.3 | 65.0 | Still in orbit. |

## ASTRONAUTICS AND AERONAUTICS, 1974

| Date | Satellite | Description | | | Inclination |
|---|---|---|---|---|---|
| 26 Dec. | Salyut 4<br>U.S.S.R.<br>1974-104A<br>Unavailable<br>Baykonur-Tyuratam | Total weight: 18 500 kg?<br>Objective: Test station design, onboard systems, and equipment; conduct scientific and technical experiments in outer space.<br>Description: Similar to Salyut 3 launched 25 June 1974, but with modified scientific equipment and systems. Instruments included set of spectrometers—code-named "Emission"—mounted in rear part of station to scan earth's horizon in spots where electron activity was strongest, newly developed photometer to observe atomic oxygen in ionosphere, and new high-resolution solar telescope. Three panels of solar batteries automatically and individually oriented themselves toward sun. | 252 | 214 | 89.2 | 51.6 |

After launch, Salyut 4 orbiting space station was maneuvered to record high altitude for Salyut, reaching 350-km apogee and 337-km perigee by 31 Dec. Station was joined by Aleksey A. Gubarev and Georgy M. Grechko, launched 11 Jan. 1975 aboard Soyuz 17, for nearly 30 days of scientific experiments and testing of Salyut 4's onboard systems. Crew returned to earth 9 Feb. 1975. Second crew, Vasily G. Lazarev and Oleg G. Makarov, were launched 5 April 1975 but mission was aborted when booster 3rd stage malfunctioned. Cosmonauts were safely returned to earth in emergency landing in Siberian mountains. Third crew, Pyotr I. Klimuk and Vitaly I. Sevastyanov launched on Soyuz 18 24 May 1975, docked with Salyut 4 on 26 May. Cosmonauts worked aboard orbiting laboratory for Soviet manned space flight record of 63 days, landing 26 July 1975 in central Kazakhstan. On 19 Nov. 1975 unmanned Soyuz 20, launched 17 Nov. 1975, docked with Salyut 4 to test technology for permanent space station. Soyuz 20 carried biological test systems to provide data for design of new cosmonaut life support systems. Soyuz 20 separated and returned to softland in Kazakhstan 16 Feb. 1976 after 3 mos docked in orbit.

| 26 Dec. | Cosmos 700<br>U.S.S.R.<br>1974-105A<br>Unavailable<br>Plesetsk | Total weight: Unavailable.<br>Objective: "Investigation of upper atmosphere and outer space."<br>Description: Unavailable. | 998 | 965 | 104.7 | 82.9 |

Still in orbit.

| 27 Dec. | Cosmos 701<br>U.S.S.R.<br>1974-106A<br>Unavailable<br>Plesetsk | Total weight: Unavailable.<br>Objective: "Investigation of upper atmosphere and outer space."<br>Description: Unavailable. | 335 | 170 | 89.6 | 71.4 |

Reentered 9 Jan. 1975.

---

\* Time at launch site; 1 day later by Greenwich Mean Time.
\*\* Time at launch site; 1 day earlier by Greenwich Mean Time.

# Appendix B

## MAJOR NASA LAUNCHES, 1974

The chronological table of major NASA launches in 1974 is intended to provide an accurate historical record, compiling information gathered from open sources and scattered elsewhere in this volume. It includes payloads carried by all rocket vehicles larger than sounding rockets launched by NASA or under NASA direction.

In 1974 NASA successfully launched 3 payloads for NASA programs out of a total of 16 payloads launched: *Sms 1* operational prototype meteorological satellite, *Ats 6* experimental communications satellite, and *Hawkeye 1* (*Explorer 52*) scientific satellite. Of the rest, 11—*Skynet IIA* and *IIB*, *Miranda* (UK-X4), *Intasat, Symphonie 1, Aeros 2, Intelsat-IV F-8, Westar 1* and *2, Noaa 4,* and *Amsat Oscar 7*—were launched for other organizations or governments. *Ans 1* and *Helios 1* were launched as cooperative scientific programs with foreign governments. A Titan IIIE-Centaur proof flight was aborted by NASA range safety officers when the second-stage engine failed to ignite.

In addition to the 16 NASA-launched payloads, Italian crews launched two payloads on NASA launch vehicles: *San Marco 4*, carrying U.S. and Italian experiments, and *Ariel 5*, a cooperative United States–United Kingdom scientific satellite.

An attempt has been made to classify vehicle and payload performance, using three categories for evaluation: successful (S), partially successful (P), and unsuccessful (U). A fourth category, unknown (Unk), has been added for payloads when vehicle failures did not give the payload a chance to operate. These classifications are unofficial and necessarily arbitrary; many of the results cannot be neatly categorized. Some ignore the fact that much may be learned from missions that have been classified as unsuccessful. Others do not account for payloads with a design lifetime of long duration which might later fail to meet that design requirement and be officially adjudged unsuccessful after this publication goes to press.

Launch dates are referenced to local time at the launch site. Open sources were used, verified when in doubt with program offices in NASA Headquarters and with NASA Centers. For further information on each item, see Appendix A of this volume and the entries in the main chronology as referenced in the index. Information in the table is as of February 1976.

## MAJOR NASA LAUNCHES, 1974

| Date | Name (NASA Code) | General Mission | Launch Vehicle (Site) | Performance Vehicle | Performance Payload | Remarks |
|---|---|---|---|---|---|---|
| 18 Jan.* | Skynet IIA | Operational military communications satellite | Thor-Delta 2313 (ETR) | U | Unk | Placed in extremely eccentric orbit when electronics package in booster 2nd stage failed. Engine took "hard over" position when 2nd-stage restart command was given and stage tumbled. Spacecraft reentered 25 Jan. 1974. |
| 11 Feb. | Titan IIIE-Centaur (TC-1) and SPHINX | Vehicle proof flight / Space Plasma High Voltage Interaction Experiment | Titan IIIE-Centaur (ETR) | U | Unk / Unk | Vehicle, carrying Viking spacecraft model and SPHINX satellite, was destroyed after Centaur engine failed to start. Inflight data indicated liquid-oxygen boost pump had failed. |
| 18 Feb. | San Marco 4 | Scientific satellite, atmospheric research | Scout (San Marco platform) | S | P | Italian-built satellite launched on NASA launch vehicle by Italian crew. Carried 2 U.S. experiments, still acquiring data on upper atmosphere after 2 yrs in orbit. Italian air-drag balance experiment failed shortly after launch. |
| 8 Mar.* | Miranda (UK-X4) | Satellite technology | Scout D (WTR) | S | S | U.K.-built satellite launched into sunlit polar orbit by NASA. Malfunctioning regulator in propane gas control system was corrected, permitting successful completion of mission within planned 6 mos. |
| 13 Apr. | Westar 1 (Westar-A) | Commercial domestic communications satellite | Thor-Delta 2914 (ETR) | S | S | First of 2 U.S. commercial domestic comsats, launched by NASA into transfer orbit and then into equatorial geosynchronous orbit over Pacific Ocean for Western Union Telegraph Co. Commercial operations began 16 July 1974; first commercial TV broadcast was made 15 July 1975. |
| 17 May | Sms 1 (SMS-A) | Operational prototype meteorological satellite | Thor-Delta 2914 (ETR) | P | S | First meteorological satellite in synchronous orbit. Provided continuous day and night images of cloud cover over U.S. and Atlantic Ocean for first time. Placed in lower than planned transfer orbit but series of motor firings raised orbit to geosynchronous. Operational control turned over to NOAA 10 Dec. 1974. Moved to standby over Pacific after Goes 1 became operational in Jan. 1976. |
| 30 May | Ats 6 (ATS-F) | Applications technology satellite | Titan IIIC (ETR) | S | S | Most complex and versatile of NASA's Applications Technology Satellites. Transmitted educational and medical programs to remote areas in U.S. Moved over Kenya July 1975 to transmit educational programs to villages in India for 1 yr. Relayed communications for July 1975 ASTP mission. |

| Date | Name | Description | Vehicle | P/S | Notes |
|---|---|---|---|---|---|
| 3 June | Hawkeye 1 (Explorer 52) | Scientific satellite, magnetospheric physics | Scout E (WTR) | S | Sixth in Univ. of Iowa's "Injun" series, launched by NASA into planned highly elliptical orbit. Satellite began making systematic observations of polar magnetosphere 20 June. |
| 16 July | Aeros 2 (Aeros-B) | Scientific satellite, aeronomy | Scout D (WTR) | S | West-German-built satellite launched by NASA into polar orbit. Failures of 2 tape recorders caused spacecraft to operate in real-time mode only but did not jeopardize mission objectives. The 6-mo design lifetime had been exceeded when spacecraft reentered in September 1975. |
| 30 Aug. | Ans 1 | Scientific satellite, x-ray astronomy | Scout D (WTR) | P | First Netherlands-U.S. cooperative satellite, launched by NASA. Anomaly in Scout guidance package placed spacecraft in elliptical rather than planned circular orbit. Changes in spacecraft and experiment operations were made to accommodate to actual orbit. During 1975 Ans I returned new data on changing intensity and spectrum of x-rays from Cygnus X–1 and measured high-count soft x-rays from Sirius. |
| 10 Oct. | Westar 2 (Westar-B) | Commercial domestic communications satellite | Thor-Delta 2914 (ETR) | S | Second of 2 domestic comsats launched by NASA for Western Union Telegraph Co. Apogee kick motor, fired 13 Oct., placed satellite in geosynchronous orbit. Arrived on temporary station 21 Nov. 1974 and on final station, over Los Angeles, 27 Feb. 1975. |
| 15 Oct. | Ariel 5 (UK–5) | Scientific satellite, x-ray astronomy | Scout (San Marco platform) | S | U.K.-built satellite launched on NASA launch vehicle by Italian crew. One U.S. and 4 U.K. experiments collected data on x-ray sources inside and outside galaxy. Still transmitting data. |
| 15 Nov. | Noaa 4 (ITOS-G) and | Operational meteorological satellite | Thor-Delta 2310 (WTR) | S | Three satellites launched on a single booster by NASA. Noaa 4 launched into sun-synchronous orbit. Provided day and night meteorological observations for NOAA's global weather watch program. Control turned over to NOAA/NESS 4 Dec. 1974. |
|  | Amsat Oscar 7 and | Radio transmitter |  |  | Launched by NASA as secondary payload. Built by U.S., Australian, Canadian, and West German amateur radio organizations working through Radio Amateur Satellite Corp. Transmitted radio communications for amateur radio operators worldwide. |
|  | Intasat | Scientific satellite, ionospheric physics |  | S | Launched by NASA for Spain as secondary payload. Spain's first satellite. Measured ionospheric electron content and ionospheric irregularities and scintillations. |
| 21 Nov. | Intelsat-IV F–8 | Operational commercial communications satellite | Atlas-Centaur (ETR) | S | Launched into elliptical transfer orbit for ComSatCorp. Apogee kick motor, fired 23 Nov. by ComSatCorp., placed spacecraft in geosynchronous orbit drifting to final position over Pacific Ocean. Commercial operations began 18 Dec. 1975, continuing in 1976. |

| Date | Name (NASA Code) | General Mission | Launch Vehicle (Site) | Performance Vehicle | Performance Payload | Remarks |
|---|---|---|---|---|---|---|
| 22 Nov. | Skynet IIB | Operational military communications satellite | Thor-Delta 2313 (ETR) | S | S | Second of 2 U.K. 2nd-generation operational comsats, launched by NASA into transfer orbit. Apogee kick motor, fired 24 Nov., placed spacecraft in geosynchronous orbit drifting to final position over Indian Ocean. Providing in-orbit X-band military communications. |
| 10 Dec. | Helios 1 (Helios-A) | Scientific probe, solar physics | Titan IIIE-Centaur (ETR) | S | S | West-German-built cooperative probe launched by NASA into elliptical solar orbit. Helios 1 flew within 45 million km of sun, closer than any other spacecraft, 15 March 1975, with all instruments functioning normally. Performance continued excellent during 2nd perihelion 21 Sept. 1975, returning new data on structure of solar fields and particles close to sun. |
| 18 Dec. | Symphonie 1 (Symphonie-A) | Experimental communications satellite | Thor-Delta 2914 (ETR) | S | S | First of 2 experimental comsats, developed by France and West Germany and launched by NASA into transfer orbit. Apogee kick motor, fired 21 Dec. 1974, placed satellite in geosynchronous orbit drifting to its final station over west coast of Africa. Operations began 2 Jan. 1975. |

*Time at launch site; 1 day later by Greenwich time.

# Appendix C

## MANNED SPACE FLIGHTS, 1974

This chronology contains basic information on all manned space flights begun during 1974 and—taken with Appendix C to the 1965–1973 volumes of this publication—provides a summary record of manned exploration of the space environment through 1974.

For the first year since 1967 the United States launched no men into space, although the three-man *Skylab 4* crew ended its 84-day mission during the year, splashing down in the Pacific Ocean 8 Feb. The United States has no manned flights scheduled until the July 1975 Apollo-Soyuz Test Project with the U.S.S.R.

The Soviet Union launched three two-man missions into space in 1974. *Soyuz 14* was launched in July to rendezvous with the *Salyut 3* space station, orbiting since 25 June. The crew rendezvoused and docked and then performed scientific experiments aboard the station for nearly 16 days before softlanding in the U.S.S.R. 19 July. The August *Soyuz 15* mission failed to dock with *Salyut 3* after several attempts. It landed at night after 48 hours in space. The *Soyuz 16* mission, in December, orbited for five days, successfully testing spacecraft and life support systems redesigned for ASTP.

By the end of 1974 the United States had conducted a total of 30 manned space flights—2 suborbital, 19 in earth orbit, 3 in lunar orbit, and 6 lunar landing—with a total of 41 different crewmen. Of the 41 American astronauts, 10 had participated in two flights each, 4 had flown three times, and 3 had flown four times. The Soviet Union had made 23 manned flights, all in earth orbit, with 32 cosmonauts. Six had participated in two flights each and two had flown three times. Cumulative totals for manned spacecraft hours in flight were 7643 hours 42 minutes for the United States and 2901 hours 33 minutes for the Soviet Union. Cumulative total man-hours in space were 21 851 hours 15 minutes for the United States and 6011 hours 46 minutes for the U.S.S.R.

This chronology was compiled in February 1976 using the latest data available within NASA at that date. The information may be subject to modification as data are refined.

## MANNED SPACE FLIGHTS, 1974

| Date Launched | Date Recovered | Designation, Crew | Weight (kg) | Duration, Revolutions | Remarks |
|---|---|---|---|---|---|
| 3 July | 19 July | Soyuz 14<br>Pavel R. Popovich<br>Yuri P. Artyukhin | 6570? | 15 days 17 hrs 30 min<br>236 revs | Launched from Baykonur Cosmodrome near Tyuratam at 11:51 pm local time (2:51 pm EDT) to rendezvous with *Salyut 3* orbiting space station (launched 25 June 1974) during 17th orbit 5 July. Docking with station successfully demonstrated techniques Soviets would use during 1975 Apollo-Soyuz Test Project. Crew studied geological-morphological objects on earth's surface, atmospheric formations and phenomena, physical characteristics of outer space, and manufacturing techniques. Crew also tested *Salyut 3*'s improved design and studied effects of space on human organism. *Soyuz 14*, undocked 19 July and softlanded 140 km southeast of Dzheskazgan at 5:21 pm Baykonur time (8:21 am EDT). |
| 26 Aug. | 29 Aug. | Soyuz 15<br>Gennady V. Sarafanov<br>Lev S. Demin | 6570? | 48 hrs 12 min<br>30 revs | Launched from Baykonur Cosmodrome near Tyuratam at 12:58 pm local time (3:58 pm EDT), *Soyuz 15* failed in several attempts to dock with *Salyut 3* space station. Spacecraft and crew made first Soviet night reentry, softlanding 48 km southwest of Tselinograd at 1:10 am 29 Aug. Baykonur time (4:10 pm EDT 28 Aug.). |
| 2 Dec. | 8 Dec. | Soyuz 16<br>Anatoly V. Filipchenko<br>Nikolay N. Rukavishnikov | 6570? | 5 days 22 hrs 24 min<br>90 revs | Launched from Baykonur Cosmodrome near Tyuratam at 2:40 pm local time (4:40 am EST) to test spacecraft systems modified for the 1975 ASTP mission. Using passive simulator docking ring, ASTP backup crew demonstrated docking and separation maneuvers in almost all ASTP modes. Crew also tested life support system redesigned for compatibility with Apollo. Crew reduced cabin pressure to 7.2 newtons per sq cm (10.4 psi) and doubled oxygen content, from 20% to 40%, with no adverse effect. Orbit was changed 4 Dec. to place *Soyuz 16* in circular orbit similar to ASTP requirements. Medical, biological, earth resources, and astronomy experiments were also carried out. Spacecraft softlanded 300 km north of Dzheskazgan at 1:04 pm Baykonur time (3:04 am EST) 8 Dec. |

# Appendix D

## NASA SOUNDING ROCKET LAUNCHES, 1974

Following is a chronological table of the 77 sounding rockets of the Arcas class and above launched by NASA in 1974. Rockets were launched from sites in six foreign countries as well as the United States and carried payloads for 3 government agencies including NASA, 18 U.S. universities, 3 U.S. companies, and 4 European organizations. Onboard experiments probed near space, providing scientists with new data in aeronomy; galactic, radio, and x-ray astronomy; and solar, magnetospheric, and ionospheric physics. Five flight tests continued development of new sounding rocket capability.

Principal sources of information for the table were the "Quick Look Sounding Rocket Data" reports issued after launches and the twice yearly "NASA Sounding Rocket Compendium," both compiled by the Sounding Rocket Division's Office of Operations at Goddard Space Flight Center, and the NASA "Report of Sounding Rocket Launching" series.

Missions listed were successful unless otherwise noted in the Remarks. The date of the launch is referenced to local time at the launch site, with an asterisk following dates that are one day later by Greenwich Mean Time.

## NASA-LAUNCHED SOUNDING ROCKETS, 1974

| Launch Date | Rocket, NASA Designation, Launch Site | Apogee (km) | Remarks |
|---|---|---|---|
| 5 Jan. | Aerobee 200<br>26.023UA/DP<br>White Sands, N. M. | 232.6 | Johns Hopkins Univ. aeronomy experiment to study Comet Kohoutek. Also carried DOD experiment. |
| 7 Jan.* | Aerobee 200<br>26.022DG/DP<br>White Sands, N. M. | 193.1 | Naval Research Laboratory galactic astronomy experiment to study Comet Kohoutek. Photographed 5-million-km hydrogen cloud enveloping comet. |
| 12 Jan.* | Aerobee 200<br>26.024UG/UP<br>White Sands, N. M. | 82.2 | Univ. of Colorado galactic astronomy experiment to study Comet Kohoutek. Flight terminated by range safety officer before data could be collected. |
| 15 Jan. | Black Brant VC<br>21.014DS/NP<br>White Sands, N. M. | 234.2 | Naval Research Laboratory solar physics experiment to support *Skylab 4* mission. |
| 16 Jan. | Nike-Apache<br>14.511UI<br>Wallops Island, Va. | 205.0 | Univ. of Illinois experiment to measure ionospheric properties in 70- to 80-km region on a day of high electron density. |
| 16 Jan. | Super Arcas<br>15.105UI<br>Wallops Island, Va. | 103.0 | Pennsylvania State Univ. experiment to measure positive and negative ion conductivities to 65 km. Launched in conjunction with Nike-Apache 14.511UI. Payload ejected, descended by parachute, providing data 45 min. |
| 16 Jan. | Aerobee 200A<br>26.014CS<br>White Sands, N. M. | 210.7 | American Science and Engineering, Inc., solar physics experiments to support Apollo Telescope Mount on *Skylab 4* mission. Experiment data were unsatisfactory. |
| 16 Jan.* | Nike-Tomahawk<br>18.160GE<br>Poker Flats, Alaska | 297.7 | GSFC magnetospheric physics experiment. Released 1 trimethylaluminum and 4 barium clouds. |
| 20 Jan.* | Nike-Tomahawk<br>18.161GE<br>Poker Flats, Alaska | 297.7 | GSFC magnetospheric physics experiment. Released 1 trimethylaluminum and 4 barium clouds. |

| Date | Vehicle / Location | | Value | Description |
|---|---|---|---|---|
| 21 Jan.* | Nike-Tomahawk 18.162GE | Poker Flats, Alaska | 297.7 | GSFC magnetospheric physics experiment. Released 1 trimethylaluminum and 4 barium clouds. |
| 22 Jan. | Aerobee 200A 26.007CS | White Sands, N.M. | 238.3 | Aerospace Corp. solar physics experiment. |
| 27 Jan. | Nike-Tomahawk 18.139IE | Andoeya, Norway | 231.7 | Royal Norwegian Council for Scientific and Industrial Research magnetospheric physics experiment; 5 out of 7 instruments collected satisfactory data. |
| 31 Jan. | Arcas 15.051GM | Antigua, West Indies | 54.7 | GSFC meteorology experiment to measure vertical profiles of ozone in solar UV irradiance as part of climatic impact assessment program. Decreased data were received because of recovery system anomaly and unusually high fall rate. |
| 8 Feb.* | Aerobee 200A 26.021UG | White Sands, N.M. | 185.6 | Columbia Univ. galactic astronomy experiment. |
| 11 Feb.* | Aerobee 200A 26.031UG | White Sands, N.M. | 227.4 | Univ. of Colorado galactic astronomy experiment. |
| 20 Feb. | Nike-Tomahawk 18.133IE | Kiruna, Sweden | 230.1 | Swedish Space Corp. magnetospheric physics experiment. |
| 20 Feb. | Nike-Tomahawk 18.134IE | Kiruna, Sweden | 230.1 | Swedish Space Corp. magnetospheric physics experiment. |
| 22 Feb. | Nike-Apache 14.526UA | Fort Churchill, Canada | 136.7 | Univ. of Pittsburgh aeronomy experiment to study enhancement of nitric oxide in auroral arcs. |
| 26 Feb. | Black Brant VC 21.018US | White Sands, N.M. | 97.8 | Univ. of Colorado solar physics experiment. Mission was unsuccessful because of premature payload separation. |

| Launch Date | Rocket, NASA Designation, Launch Site | Apogee (km) | Remarks |
|---|---|---|---|
| 28 Feb.* | Nike-Apache 14.527UA Fort Churchill, Canada | | Univ. of Pittsburgh aeronomy experiment to study extreme uv spectrum in auroral arc. No data collected because of vehicle breakup at 3.2 sec after launch. |
| 1 Mar.* | Nike-Apache 14.528UA Fort Churchill, Canada | 135.0 | Univ. of Pittsburgh aeronomy experiment to study enhancement of nitric oxide in auroral arc. Payload was constructed from recovered payload flown on 14.526. |
| 4 Mar.* | Nike-Apache 14.518GA Fort Churchill, Canada | 154.0 | GSFC aeronomy experiment to obtain optical spectrum of aurora in extreme uv region. |
| 10 Mar. | Nike-Tomahawk 18.158UE Poker Flats, Alaska | 187.0 | Rice Univ. magnetospheric physics experiment to study energetic fields and particles in auroral area. |
| 15 Mar.* | Aerobee 200A 26.011GG White Sands, N. M. | 199.4 | GSFC optical astronomy experiment. One camera failed but second camera obtained adequate data. |
| 20 Mar.* | Aerobee 200 26.029UG Fort Churchill, Canada | 251.1 | Univ. of Michigan magnetospheric physics experiment launched in conjunction with good aurora and *Explorer 51* overpass. |
| 21 Mar. | Nike-Cajun 10.420GM Kourou, French Guiana | 88.7 | GSFC meteorological experiment to measure diurnal variation of ozone in stratosphere and lower mesosphere, measure nitrogen dioxide in stratosphere, and compare chemiluminescent and optical ozone measurements. Good data obtained on ascent but parachute malfunction and resulting high fall rate caused 50% data loss on descent. |
| 21 Mar. | Nike-Cajun 10.411GM Kourou, French Guiana | 82.8 | GSFC meteorological experiment to measure diurnal variation of ozone in stratosphere and lower mesosphere. |
| 21 Mar. | Super Arcas 15.123GM Kourou, French Guiana | | GSFC meteorological experiment to measure photometry of uv sunlight in support of 10.411 and 10.420. No data were collected, because of vehicle failure. |
| 22 Mar. | Nike-Cajun 10.421GM Kourou, French Guiana | 119.0 | GSFC meteorological experiment to measure temperature, pressure, density, and winds in upper atmosphere Explosions of all 31 grenades were heard through sound-ranging array. |
| 22 Mar. | Nike-Cajun 10.422GM Kourou, French Guiana | 121.0 | GSFC meteorological experiment to measure temperature, pressure, density, and winds in upper atmosphere. Explosions of all 31 grenades were heard. |

| Date | Vehicle | Altitude (km) | Description |
|---|---|---|---|
| 22 Mar. | Super Arcas 15.124GM Kourou, French Guiana | 89.0 | GSFC meteorological experiment to measure photometry of UV sunlight in support of 10.411 and 10.420. Telemetry signals were erratic and no useful data were collected. |
| 5 Apr.* | Aerobee 350 17.012CG White Sands, N.M. | 236.6 | Lockheed Missiles & Space Co., Inc., x-ray astronomy experiment. |
| 17 Apr. | Black Brant VC 21.026UE Poker Flats, Alaska | 270.0 | Univ. of Minnesota magnetospheric physics experiment. |
| 19 Apr. | Nike-Apache 14.520UI Wallops Island, Va. | 198.7 | Univ. of Illinois ionospheric physics experiment. |
| 8 May | Astrobee F 12.021GT White Sands, N.M. | 155.1 | GSFC performance flight test. Test unsuccessful, apparently because of propulsion failure. |
| 14 May | Nike-Malemute 28.001GT Wallops Island, Va. | 500 | GSFC performance flight test. Trajectory lower and longer than predicted. |
| 15 May | Nike-Apache 14.482UI Wallops Island, Va. | 138.2 | Pennsylvania State Univ. experiment to measure ion composition in D and lower E regions of ionosphere. |
| 15 May | Aerobee 200 26.018US White Sands, N.M. | 281.3 | Harvard College Observatory solar physics experiment. |
| 29 May | Hawk T1-7132WT Wallops Island, Va. | 46.0 | WFC single-stage Hawk performance flight test. Vehicle performed as planned. |
| 15 June* | Black Brant VC 21.017UG White Sands, N.M. | 188.3 | Univ. of California galactic astronomy experiment. |
| 19 June | Aerobee 200A 26.036US White Sands, N.M. | 200.2 | Univ. of Hawaii solar physics experiment. Successful data collection despite propulsion cutoff 10 sec earlier than planned. |

| Launch Date | Rocket, NASA Designation, Launch Site | Apogee (km) | Remarks |
|---|---|---|---|
| 22 June | Aerobee 200A 26.026GG White Sands, N.M. | 206.5 | GSFC galactic astronomy experiment. |
| 27 June | Aerobee 200A 26.015CS White Sands, N.M. | 211.1 | American Science and Engineering, Inc., solar physics experiment. |
| 29 June | Black Brant VC 21.028IA Wallops Island, Va. | 234.0 | Univ. of Bonn, West Germany, aeronomy experiment in support of Atmospheric Layering and Density Distribution of Ions and Neutrals (ALADDIN) program. |
| 29 June | Super Arcas 15.104UI Wallops Island, Va. | 89.3 | Pennsylvania State Univ. ionospheric physics experiment. |
| 29 June* | Super Arcas 15.125UE Wallops Island, Va. | 81.4 | Pennsylvania State Univ. magnetospheric physics experiment. |
| 29 June* | Nike-Apache 14.529UA Wallops Island, Va. | 133.0 | Univ. of Pittsburgh aeronomy experiment. |
| 29 June* | Nike-Apache 14.521UI Wallops Island, Va. | 195.1 | Univ. of Illinois ionospheric physics experiment. |
| 29 June* | Super Arcas 15.126UE Wallops Island, Va. | 68.9 | Pennsylvania State Univ. magnetospheric physics experiment. |
| 30 June | Nike-Apache 14.522UI Wallops Island, Va. | 190.5 | Univ. of Illinois ionospheric physics experiment. Telemetry signal weak and noisy. All data lost after 2 min. |

# ASTRONAUTICS AND AERONAUTICS, 1974

| Date | Vehicle | Altitude | Experiment |
|---|---|---|---|
| 30 June | Nike-Apache 14.508UA Wallops Island, Va. | 204.2 | Univ. of Michigan aeronomy experiment. |
| 30 June | Nike-Apache 14.523UI Wallops Island, Va. | 198.1 | Univ. of Illinois ionospheric physics experiment. Good data despite telemetry anomalies. |
| 2 July | Nike-Tomahawk 18.156IE/UE Soendre Stroemfjord, Greenland | Not available | Joint Univ. of Texas and Danish Meteorological Institute magnetospheric physics experiment. No tracking available. |
| 4 July | Nike-Apache 14.488GE Fort Churchill, Canada | 139.6 | GSFC magnetospheric physics experiment to record solar particle tracks to determine abundance and energy spectral distribution. |
| 5 July | Nike-Apache 14.488GE Fort Churchill, Canada | 140.0 | GSFC magnetospheric physics experiment to record solar particle tracks to determine abundance and energy spectral distribution. |
| 8 July | Nike-Tomahawk 18.157IE/UE Soendre Stroemfjord, Greenland | Not available | Joint Univ. of Texas and Danish Meteorological Institute magnetospheric physics experiment. No tracking available. |
| 20 July | Aerobee 170 13.084UG White Sands, N.M. | 1955. | Univ. of Wisconsin galactic astronomy experiment to map low-energy cosmic x-ray background within 40° of galactic plane. |
| 25 July | Super Arcas 15.106UI White Sands, N.M. | 74.7 | Univ. of Houston ionospheric physics experiment. |
| 14 Aug. | Nike-Apache 14.417UA Wallops Island, Va. | 131.7 | Dudley Observatory aeronomy experiment. |
| 6 Sept. | Aerobee 200 26.025DG White Sands, N.M. | 206.8 | Naval Research Laboratory galactic astronomy experiment. |
| 18 Sept. | Aerobee 170 13.035US White Sands, N.M. | 252.5 | Harvard College Observatory solar physics experiment. |

ASTRONAUTICS AND AERONAUTICS, 1974

| Launch Date | Rocket, NASA Designation, Launch Site | Apogee (km) | Remarks |
|---|---|---|---|
| 24 Sept. | Nike-Malemute 28.002GT Wallops Island, Va. | 465.1 | GSFC performance flight test. Vehicle apogee lower than predicted. |
| 2 Oct.* | Aerobee 200 26.037GG White Sands, N.M. | 165.9 | GSFC x-ray astronomy experiment. |
| 4 Oct. | Black Brant VC 21.015GP White Sands, N.M. | 226.9 | GSFC heat pipe experiment. |
| 10 Oct. | Astrobee D 23.005UE White Sands, N.M. | 74.4 | Pennsylvania State Univ. magnetospheric physics experiment. Experiment was unsuccessful. |
| 19 Oct. | Aerobee 200A 26.016AS White Sands, N.M. | 201.3 | Los Alamos Scientific Laboratory solar physics experiment. No data were collected because of attitude control system failure. |
| 3 Nov. | Aerobee 170 13.107UG Barking Sands, Hawaii | 169.3 | Columbia Univ. galactic astronomy experiment to make x-ray survey of Crab Nebula. |
| 8 Nov. | Hawk 30.002WT Wallops Island, Va. | 73.0 | WFC performance flight test of single-stage Hawk. Vehicle performed as planned. |
| 16 Nov. | Aerobee 170A-HE 13.092DG White Sands, N.M. | 181.4 | Naval Research Laboratory galactic astronomy experiment to study far uv spectra of Orion and Barnards Loop Nebula. Experiment was unsuccessful. |
| 16 Nov. | Aerobee 170A 13.034UG White Sands, N.M. | 166.6 | Massachusetts Institute of Technology galactic astronomy experiment to make soft and ultrasoft x-ray survey using proportional counters. |

| Date | Vehicle | | Weight | Description |
|---|---|---|---|---|
| 25 Nov. | Nike-Tomahawk 18.154UA | White Sands, N.M. | 132.6 | Univ. of Colorado aeronomy experiment. |
| 25 Nov. | Black Brant VC 21.031UC | White Sands, N.M. | 183.8 | Univ. of California at Berkeley galactic astronomy experiment. |
| 25 Nov. | Aerobee 200A 26.042UG | White Sands, N.M. | 207.3 | Univ. of Wisconsin galactic astronomy experiment. Mission was unsuccessful. |
| 27 Nov. | Aerobee 200 26.041US | White Sands, N.M. | 184.6 | Univ. of Southern California solar physics experiment. |
| 6 Dec. | Nike-Tomahawk 18.167UA | Wallops Island, Va. | 305.8 | Univ. of Colorado aeronomy experiment to measure distribution of nitric oxide molecules between 100 and 300 km, in conjunction with overpass of *Explorer 51*. |
| 13 Dec. | Nike-Tomahawk 18.140IE | Andoeya, Norway | 235.0 | Royal Norwegian Council for Scientific and Industrial Research magnetospheric physics (auroral) experiment. |
| 27 Dec.* | Aerobee 200A 26.030UC | White Sands, N.M. | 196.7 | Columbia Univ. galactic astronomy experiment to study x-ray occultation of Crab Nebula. |

*Time at launch site; 1 day later by Greenwich Mean Time.

# Appendix E

## X-24B LIFTING-BODY FLIGHTS, 1974

Lifting bodies—wingless vehicles that receive aerodynamic lift from their body shape alone—were being flown to develop concepts for a maneuverable manned vehicle that could reenter the atmosphere from space, as well as to provide technology for possible future hypersonic cruise aircraft. The program began in the early 1950s when NASA scientists and engineers at the Ames and Langley Research Centers studied the concept of the lifting body. Since 1963, Flight Research Center has flight-tested three versions of the ARC-designed M2 configuration, LaRC's HL-10, and, in joint programs with the Air Force, the X-24A and X-24B lifting bodies.

Following a joint program flight-testing the Air Force X-24A, NASA and the Air Force in 1971 signed a memorandum of understanding to incorporate test programs of the new double-delta-shaped X-24B, with its improved lift-to-drag ratio, and NASA's HL-10 and M2-F3. The HL-10, which had completed its first flight series in 1970, was held in storage. The M2-F3 made its final flight in December 1972. Beginning with a captive flight 19 July 1973, NASA and Air Force pilots tested X-24B performance at low supersonic, transonic, and subsonic speeds, with emphasis on landing. NASA was responsible for vehicle maintenance, instrumentation, and ground support. The Air Force supplied the X-24B vehicle, base services, fuel, B-52 operations, and XLR-11 engine and maintenance. NASA and the Air Force shared pilot, mission planning, and data analysis responsibilities.

The X-24B was launched from a B-52 aircraft, usually at 13 700-m altitude. In captive flights, the vehicle remained attached to the B-52. Unpowered free flights glided through maneuvers to landings on a dry lake bed. Powered flights ignited the rocket engine, to reach higher altitudes and speeds before landing.

The following table includes all X-24B missions leaving the ground in 1974, even those eventually aborted because of weather or mechanical problems. The flight numbering system included vehicle designation ("B" for X-24B vehicles)—free-flight number—B-52 carry number. For example, "B-9-16" was the 9th free flight of the X-24B and the 16th time the vehicle was carried by the B-52. A "C" or an "A" replacing the free-flight number indicated a captive or an aborted flight.

Information for the tables was compiled from "Post Launch Flight Reports" issued at FRC, "Flash Reports" issued by the NASA Headquarters Office of Aeronautics and Space Technology, and the "Manned Lifting Body Flight Log," a *NASA Facts* issuance released in March 1976 by Dryden Flight Research Center (named Flight Research Center until January 1976).

## X-24B LIFTING-BODY FLIGHTS, 1974

| Flight Date and Number | Pilot, Maximum Altitude, Maximum Speed | Remarks |
|---|---|---|
| 15 Feb. 74 (B-8-15) | Maj. Michael V. Love 13 700 m Mach 0.68 | Glide flight to investigate longitudinal stability, make tuft studies, investigate fin pressure, and train pilot. |
| 5 Mar. 74 (B-9-16) | John A. Manke 18 400 m Mach 1.09 | First supersonic flight. Objectives were to fly at mach 1.1, obtain stability and control data at mach 1.1 for power-on and power-off conditions, obtain performance and longitudinal trim data at mach 0.8 and 0.9 and investigate fin pressure to identify asymmetric flow. |
| 19 Mar. 74 (B-A-17) | Maj. Michael V. Love | Flight aborted because of poor weather conditions. |
| 22 Apr. 74 (B-A-18) | Maj. Michael V. Love | Flight aborted when fire erupted in B-52's TV monitor system. |
| 23 Apr. 74 (B-A-19) | Maj. Michael V. Love | Flight aborted when liquid-oxygen feed lines in B-52 failed. |
| 25 Apr. 74 (B-C-20) | Maj. Michael V. Love Not applicable Not applicable | Captive flight to determine if left-aileron dead-band problem encountered in previous powered flights had been eliminated. Problem solved. |
| 30 Apr. 74 (B-10-21) | Maj. Michael V. Love 15 900 m Mach 0.88 | First powered flight for pilot; training mission. |
| 24 May 74 (B-11-22) | John A. Manke 17 100 m Mach 1.14 | Fifth powered flight. Primary objectives were to fly at mach 1.25, evaluate stability and control at speed greater than mach 1, investigate buffet at mach 0.65 with 20° upper flap, check out engine at 207 newtons per sq cm chamber pressure, investigate fin and rudder pressure, and study boundary-layer noise and vibration. Because only 3 of 4 engines ignited, vehicle reached only mach 1.14. All other objectives were achieved. |
| 14 June 74 (B-12-23) | Maj. Michael V. Love 20 000 m Mach 1.23 | Objectives—to fly at mach 1.25; evaluate stability and control at speed over mach 1; perform boundary layer noise and vibration experiment; and survey fin, rudder, and flap pressure—were successfully completed. |
| 28 June 74 (B-13-24) | John A. Manke 20 800 m Mach 1.39 | Objectives—to fly at mach 1.4; evaluate stability and control at speed over mach 1; perform boundary-layer noise and vibration experiment; and survey fin, rudder, and flap pressure—were successfully completed. |
| 8 Aug. 74 (B-14-25) | Maj. Michael V. Love 22 400 m Mach 1.54 | Objectives—to fly at mach 1.55; evaluate stability and control at speed over mach 1 with 5° and 8° angles of attack; survey fin, rudder, and flap pressure; and performed boundary-layer noise and vibration experiment—were successfully completed. |
| 29 Aug. 74 (B-15-26) | John A. Manke 22 100 m Mach 1.10 | Objectives—to evaluate stability and control at speed over mach 1 and to make body pressure survey—were completed. Objective of flying at mach 1.68 was not achieved because of premature engine shutdown at 108 sec into flight. |
| 25 Oct. 74 (B-16-27) | Maj. Michael V. Love 22 000 m Mach 1.76 | Fastest X-24B flight to date. Objectives—to fly at mach 1.72; evaluate stability and control at mach 1 with 5°, 8°, and 12° angles of attack; evaluate supersonic performance and longitudinal trim; make body pressure survey; and perform boundary-layer noise and vibration experiment—were successfully completed. |
| 15 Nov. 74 (B-17-28) | John A. Manke 22 000 m Mach 1.62 | Objectives—to fly at mach 1.66, evaluate stability and control at mach 1 with 8° and 12° angles of attack, make approach and landing tests with upper flaps at −24°, make body pressure survey, and perform boundary-layer noise and vibration experiment—were successfully completed. |
| 17 Dec. 74 (B-18-29) | Maj. Michael V. Love 21 000 m Mach 1.59 | Objectives—to evaluate longitudinal stability and control; approach and land with upper flap at −24°; and survey fin, rudder, and flap pressure—were successfully completed. |

# Appendix F

## ABBREVIATIONS OF REFERENCES

Listed here are the abbreviations used for citing sources in the text. Not all the sources are listed, only those that are abbreviated.

| | |
|---|---|
| *AAAS Bull* | American Association for the Advancement of Science's *AAAS Bulletin* |
| *A&A* | American Institute of Aeronautics and Astronautics' magazine, *Astronautics & Aeronautics* |
| *A&A 1974* | NASA's *Astronautics and Aeronautics, 1974: A Chronology* (this publication) |
| ABC | American Broadcasting Company |
| AEC Release | Atomic Energy Commission News Release |
| *Aero Daily* | *Aerospace Daily* newsletter |
| *Aero Med* | *Aerospace Medicine* magazine |
| *AF Mag* | Air Force Association's *Air Force Magazine* |
| *AFHF Newsletter* | *Air Force Historical Foundation Newsletter* |
| *AFJ* | *Armed Forces Journal* magazine |
| AFSC *Newsreview* | Air Force Systems Command's *Newsreview* |
| AFSC Release | Air Force Systems Command News Release |
| AIA Release | Aerospace Industries Association of America News Release |
| AIAA *Facts* | American Institute of Aeronautics and Astronautics' *Facts* |
| AIAA Release | American Institute of Aeronautics and Astronautics News Release |
| AIP *Newsletter* | American Institute of Physics *Newsletter* |
| AP | Associated Press news service |
| ARC *Astrogram* | NASA Ames Research Center's *Astrogram* |
| *Astro Journ* | American Astronomical Society's *Astrophysical Journal* |
| *Atlanta JC* | *Atlanta Journal Constitution* newspaper |
| *Av Wk* | *Aviation Week & Space Technology* magazine |
| *B News* | *Birmingham News* newspaper |
| *B Sun* | *Baltimore Sun* newspaper |
| *Bull Atom Sci* | Education Foundation for Nuclear Science's *Bulletin of the Atomic Scientists* |
| *Bus Wk* | *Business Week* magazine |
| *C Daily News* | *Chicago Daily News* newspaper |
| *C Trib* | *Chicago Tribune* newspaper |
| Can Press | Canadian Press news service |
| CBS | Columbia Broadcasting System |
| *C&E News* | *Chemical & Engineering News* magazine |
| *Cl PD* | Cleveland *Plain Dealer* newspaper |
| *Cl Press* | *Cleveland Press* newspaper |
| *Columbia J Rev* | *Columbia Journalism Review* magazine |
| ComSatCorp Release | Communications Satellite Corporation News Release |
| *CQ* | *Congressional Quarterly* |
| *CR* | *Congressional Record* |
| *CSM* | *Christian Science Monitor* newspaper |
| CTNS | Chicago Tribune News Service |
| *D News* | *Detroit News* newspaper |
| *D Post* | *Denver Post* newspaper |

| | |
|---|---|
| DASA Release | Defense Atomic Support Agency News Release |
| DFRC | See FRC. |
| DJ | Dow Jones news service |
| DOC PIO | Department of Commerce Public Information Office |
| DOD Release | Department of Defense News Release |
| DOT Release | Department of Transportation News Release |
| EOP Release | Executive Office of the President News Release |
| FAA Release | Federal Aviation Administration News Release |
| FBIS–Sov | Foreign Broadcast Information Service, Soviet number |
| FonF | Facts on File |
| FRC Release | Flight Research Center News Release (after 8 Jan. 1976, became Dryden Flight Research Center News Release) |
| FRC X–Press | NASA Flight Research Center's X–Press |
| GE Forum | General Electric Forum magazine |
| Goddard News | NASA Goddard Space Flight Center's Goddard News |
| GSFC Release | NASA Goddard Space Flight Center News Release |
| GSFC SSR | NASA Goddard Space Flight Center's Satellite Situation Report |
| GT&E Release | General Telephone & Electronics News Release |
| H Chron | Houston Chronicle newspaper |
| H Post | Houston Post newspaper |
| JA | Journal of Aircraft magazine |
| JPL Lab-Oratory | Jet Propulsion Laboratory's Lab-Oratory |
| JPL Release | Jet Propulsion Laboratory News Release |
| JPRS | Department of Commerce Joint Publications Research Service |
| JSC Release | NASA Lyndon B. Johnson Space Center (Manned Spacecraft Center until 17 Feb. 1973) News Release |
| JSC Roundup | NASA Lyndon B. Johnson Space Center's Space News Roundup |
| JSR | American Institute of Aeronautics and Astronautics' Journal of Spacecraft and Rockets magazine |
| KC Star | Kansas City Star newspaper |
| KC Times | Kansas City Times newspaper |
| KSC Release | NASA John F. Kennedy Space Center News Release |
| LA Her-Exam | Los Angeles Herald-Examiner newspaper |
| LA Times | Los Angeles Times newspaper |
| Langley Researcher | NASA Langley Research Center's Langley Researcher |
| LARC Release | NASA Langley Research Center News Release |
| LATNS | Los Angeles Times News Service |
| LERC Release | NASA Lewis Research Center News Release |
| Lewis News | NASA Lewis Research Center's Lewis News |
| M Her | Miami Herald newspaper |
| M News | Miami News newspaper |
| M Trib | Minneapolis Tribune newspaper |
| Marshall Star | NASA George C. Marshall Space Flight Center's Marshall Star |
| MJ | Milwaukee Journal newspaper |
| MSFC Release | NASA George C. Marshall Space Flight Center News Release |
| N Hav Reg | New Haven Register newspaper |
| N News | Newark News newspaper |
| N Va Sun | Northern Virginia Sun newspaper |
| NAA News | National Aeronautic Association News |
| NAA Record Book | National Aeronautic Association's World and U.S.A. National World Aviation–Space Records |
| NAC Release | National Aviation Club News Release |
| NAE Release | National Academy of Engineering News Release |
| NANA | North American Newspaper Alliance |
| NAS Release | National Academy of Sciences News Release |
| NAS-NRC Release | National Academy of Sciences–National Research Council News Release |

|  |  |
|---|---|
| NAS–NRC–NAE *News Rpt* | National Academy of Sciences–National Research Council–National Academy of Engineering *News Report* |
| NASA Ann | NASA Announcement |
| NASA Gen Mgmt Rev Rpt | NASA Headquarters "General Management Review Report" |
| NASA HHR–39 | NASA Historical Report No. 39 |
| NASA Hist Off | NASA History Office |
| NASA Hq *WB* | NASA Headquarters *Weekly Bulletin* |
| NASA Int Aff | NASA Office of International Affairs |
| NASA *LAR*, XIII/8 | NASA *Legislative Activities Report*, Vol. XIII, No. 8 |
| NASA Leg Off | NASA Office of Legislative Affairs |
| NASA MOR | NASA Headquarters Mission Operations Report, preliminary prelaunch and postlaunch report series; information may be revised and refined before publication |
| NASA prog off | NASA program office (for the program reported) |
| NASA proj off | NASA project office (for the project reported) |
| NASA Release | NASA Headquarters News Release |
| NASA Rpt SRL | NASA report of sounding rocket launching |
| NASA SP–4019 | NASA Special Publication No. 4019 |
| *Natl Obs* | *National Observer* magazine |
| *Nature* | *Nature Physical Science* magazine |
| NBC | National Broadcasting Company |
| NGS Release | National Geographic Society News Release |
| NMI | NASA Management Instruction |
| NN | NASA Notice |
| NOAA Release | National Oceanic and Atmospheric Administration News Release |
| NRL Release | Naval Research Laboratory Release |
| NSC Release | National Space Club News Release |
| NSC *News* | National Space Club *News* |
| NSC *Letter* | National Space Club *Letter* |
| NSF *Highlights* | National Science Foundation's *Science Resources Studies Highlights* |
| NSF Release | National Science Foundation News Release |
| NSTL Release | NASA National Space Technology Laboratories News Release |
| *NY News* | *New York Daily News* newspaper |
| *NYT*, 5:4 | *New York Times* newspaper, section 5, page 4 |
| NYTNS | New York Times News Service |
| *O Sen Star* | *Orlando Sentinel Star* newspaper |
| *Oakland Trib* | *Oakland Tribune* newspaper |
| *Omaha W–H* | *Omaha World-Herald* newspaper |
| ONR *Rev* | Navy's Office of Naval Research *Reviews* |
| *P Bull* | Philadelphia *Evening* and *Sunday Bulletin* newspaper |
| *P Inq* | *Philadelphia Inquirer* newspaper |
| PAO | Public Affairs Office |
| PD | National Archives and Records Service's *Weekly Compilation of Presidential Documents* |
| PIO | Public Information Office |
| PMR *Missile* | USN Pacific Missile Range's *Missile* |
| PMR Release | USN Pacific Missile Range News Release |
| Pres Rpt 74 | *Aeronautics and Space Report of the President: 1974 Activities* |
| SAO Release | Smithsonian Astrophysical Observatory News Release |
| SBD | *Defense/Space Business Daily* newspaper |
| *Sci Amer* | *Scientific American* magazine |
| *Sci & Govt Rpt* | *Science & Government Report*, independent bulletin of science policy |
| SciServ | Science Service News service |
| SD | *Space Digest* magazine |
| *SD Union* | *San Diego Union* newspaper |
| SET *Manpower Comments* | Scientific Manpower Commission's *Scientific, Engineering, Technical Manpower Comments* |

| | |
|---|---|
| *SF* | British Interplanetary Society's *Spaceflight* magazine |
| *SF Chron* | *San Francisco Chronicle* newspaper |
| *SF Exam* | *San Francisco Examiner* newspaper |
| *Sov Aero* | *Soviet Aerospace* newsletter |
| *Sov Rpt* | Center for Foreign Technology's *Soviet Report* (translations) |
| *SP* | *Space Propulsion* newsletter |
| *Spaceport News* | NASA John F. Kennedy Space Center's *Spaceport News* |
| *Spacewarn* | IUWDS World Data Center A for Rockets and Satellites' *Spacewarn Bulletin* |
| SR list | NASA compendium of sounding rocket launches |
| *SSN* | *Soviet Sciences in the News*, publication of Electro-Optical Systems, Inc. |
| *St Louis G–D* | *St. Louis Globe-Democrat* newspaper |
| *St Louis P–D* | *St. Louis Post-Dispatch* newspaper |
| *T–Picayune* | New Orleans *Times-Picayune* newspaper |
| *Tech Rev* | Massachusetts Institute of Technology's *Technology Review* |
| Testimony | Congressional testimony, prepared statement |
| Text | Prepared report or speech text |
| Transcript | Official transcript of news conference or congressional hearing |
| UN Reg | United Nations Public Registry of Space Flight |
| UPI | United Press International news service |
| USGS Release | U.S. Geological Survey News Release |
| USPS Release | U.S. Postal Service News Release |
| *W Post* | *Washington Post* newspaper |
| *W Star-News* | *Washington Star-News* newspaper |
| WFC Release | NASA Wallops Flight Center News Release (after 26 April 1974; formerly Wallops Station News Release) |
| WH Release | White House News Release |
| *WJT* | *World Journal Tribune* newspaper |
| WS Release | NASA Wallops Station News Release (see also WFC) |
| *WSJ* | *Wall Street Journal* newspaper |

# INDEX AND LIST OF ABBREVIATIONS AND ACRONYMS

## A

A-7 (attack fighter aircraft), 91
A-9 (research aircraft), 84
A-10 (close-air-support aircraft), 22, 141–142, 183–184, 225
A-300B (European airbus), 107
AAI Corp., 56–57
AAAS (American Assn. for the Advancement of Science), 45–46
AAS. See American Astronautical Society and American Astronomical Society.
ABM (Antiballistic missile system), 123–124, 215
ABMA (Army Ballistic Missile Agency), 1
Abourezk, Sen. James G., 72
Academy of Sciences, New York. See New York Academy of Science.
Academy of Sciences, Sweden. See Swedish Academy of Sciences.
Academy of Sciences, U.S.S.R. See Soviet Academy of Sciences.
Acapulco, Mexico, 77
Accelerometer, 40, 145
Accident, aircraft, 14, 28, 51, 55, 71, 78, 96, 97, 100–101, 123
ACF. See Air combat fighter.
Acoustic composite nacelle program, 48
ACS. See Attitude control system.
Active control aircraft program, 48
Adams, Capt. Harold B. (USAF), 163
"Advanced Composites" (AIAA presentation), 15
Advanced fighter technology integration (AFTI) aircraft, 76
Advanced medium STOL transport (AMST), 22
*Advanced Nuclear Research* (congressional report), 184
AE (Atmosphere Explorer). See *Explorer 51*.
AEC. See Atomic Energy Commission.
AEC–NASA Space Nuclear Systems Office, 184
AEDC. See Arnold Engineering Development Center.
Aerobee (sounding rocket)
 170, 269, 270
 170A, 270–271
 170A–HE, 270
 200, 2, 3, 264, 265, 267, 269, 270, 271
 200A, 264–265, 266, 267, 268, 270–271
 350, 266–267
Aeroflight simulator, 107

Aeroflot (U.S.S.R. airline), 80
Aeronautical Satellite (Aerosat) system, 77–78, 101, 144, 164, 209
Aeronautics (see also individual aircraft), 140–141, 150
 aerodynamics, 44–45, 73–74, 114, 117, 160, 204
 budget, 21–25, 27, 48, 84, 109
 energy conservation, 25, 29, 44–45, 48, 55–56, 192, 211–212, 224
 research and technology, 44–45, 48, 84, 97, 99–100, 123, 148–149, 154–155, 156, 183, 224–225
*Aeronautics and Space Report of the President: 1973 Activities*, 80
Aeronomy satellite. See *Aeros 1* and *Aeros 2*.
Aeros program (West German–U.S. satellite project), 135
*Aeros 1* (West German aeronomy satellite), 135
*Aeros 2* (Aeros–B), 2, 134–135, 221–222, 227, 242, 257, 259
Aerosat. See Aeronautical Satellite system.
Aerosat Space Segment Program. See Aeronautical Satellite system.
Aerosols, 82–83
Aerospace Corp., 72, 152, 264–265
Aerospace Industries Assn. of America, Inc. (AIA), 3, 10, 19, 94, 119, 189
Aerospace industry, 4–5, 9, 35, 61
 employment, 94, 189
 energy crisis impact, 3, 52, 100
 research, 89–90
 sales, 3, 19, 81, 119, 191
Aerospace Research Center, Italy, 188
Aérospatiale (Société Nationale Industrielle Aérospatiale), 37, 136–137
Aerostat (tethered balloon), 174–175
AFCRL (Air Force Cambridge Research Laboratories), 126
AFGE (American Federation of Government Employees), 115–116
AFML (Air Force Materials Laboratory), 157
Africa, 40–41, 119, 215–216, 254
AFSATCOM (Air Force Satellite Communications System), 21–22
AFSC. See Air Force Systems Command.
AFTI. See Advanced fighter technology integration aircraft.

Agreement. See International cooperation; International cooperation, space; Memorandum of Understanding; and Treaty.
Agreement on Scientific and Technical Cooperation for the Peaceful Uses of Atomic Energy (U.S.-U.S.S.R.), 25
Agricultural Research Service, 86
Agriculture, 62, 86, 105, 109, 132
Agriculture, Dept. of (USDA), 48–49, 65, 90
AIA. See Aerospace Industries Assn. of America, Inc.
AIAA. See American Institute of Aeronautics and Astronautics.
Air combat fighter (ACF; see also YF-16 and YF-17), 22, 94, 116, 183–184
Air Force. See U.S. Air Force.
Air Force Air Weather Service, 120
Air Force Avionics Laboratory, 143
Air Force Cambridge Research Laboratories (AFCRL), 126
Air Force Decoration for Exceptional Service, 157
Air Force Flight Dynamics Laboratory, 76
Air Force Materials Laboratory (AFML), 157
Air Force Satellite Communications (AFSATCOM) System, 21–22
Air Force Systems Command (AFSC; see also Arnold Engineering Development Center and Space Missile Test Center), 7, 10, 12, 91, 143, 144, 145, 149
Air Foundation Certificate of Recognition, 219
Air France, 107
Air Inter (French domestic airline), 38
Air-launched cruise missile, 22
Air Line Pilots Assn. (ALPA), 87
Air pollution (see also Global Air Sampling Program), 12, 29, 38, 86, 91–92, 98, 143, 145–146, 151, 168, 188, 225–226
Air Pollution Control Div., Cleveland, Ohio, 168
Air traffic control (ATC), 12, 15–16, 96, 101, 109, 114, 123, 144, 238
Air Transport Assn. of America (ATA), 99–100, 115–116, 148
Airborne Infrared Observatory (AIRO; C-141 aircraft), 6, 194
Airborne science-shuttle experiment System simulation (ASSESS) program, 177
Airborne warning and control system (AWACS), 60
Airborne weather reconnaissance system (AWRS), 120
Airbus. See A-300B.
Airbus Industries, 107
Aircraft (see also individual aircraft and kinds of aircraft; Accident, aircraft; Noise, aircraft; and Safety), 9, 19, 60, 98, 108, 151, 158–159, 166
  air pollution effects, 12, 213
  budget, 21–22

energy crisis impact, 100
scientific applications, 62, 65, 98, 104–105, 118–119
technology. See Aeronautics.
testing, 141
Aircraft Design Award, 148
Aircraft industry, 12, 44, 119, 138–139, 148, 189
Aircraft survey program, 21
Airlines (see also individual airlines), 44, 51, 68, 71–72, 74, 77–78, 87, 89, 94, 99–100, 115–116, 118, 119, 154–155, 156, 157
AIRO (Airborne Infrared Observatory; C-141 aircraft), 6, 194
Airport (see also individual airports), 96
Aksnes, Dr. Kaare, 167
Alabama Space and Rocket Center, Huntsville, Ala., 28
Alabama, Univ. of, 13
ALADDIN. See Atmospheric Layering and Density Distribution of Ions and Neutrals program.
Alaska, 88, 97, 106–107, 113–114, 130, 167, 186
Albee, Dr. A., 64
Albert, Rep. Carl B., 186–187
Alcyone 1A (rocket motor), 111
Aldrin, Col. Edwin E., Jr. (USAF, Ret.), 135, 138
Alexandria, Va., 164–165
Algeria, 169
Algernon Blair Industrial Contractors, Inc., 31
Ali, Muhammad, 14
Allahbad, India, 63
Alloys, 58, 173, 184
ALPA (Air Line Pilots Assn.), 87
Alpha Orionis. See Betelgeuse.
ALSEP (Apollo lunar surface experiment package), 208–209, 223
Aluminum, 52, 90–91, 100, 173, 210–211, 212
Aluminum oxide, 38
Alyeska Pipeline Service Co., 134
American Airlines, Inc., 74, 120
American Assn. for the Advancement of Science (AAAS), 45–46
American Assn. of Railroads, 86
American Astronautical Society (AAS), 59, 152
American Astronomical Society (AAS), 76, 214
American Can Co., 134
American Federation of Government Employees (AFGE), 115–116
American Geophysical Union, 82, 195
American Institute of Aeronautics and Astronautics (AIAA), 14–15, 99–100, 148, 171, 195
American Physical Society, 90, 92
American Science and Engineering, Inc., 139, 158, 264, 268
American Telephone & Telegraph Co. (AT&T), 3–4, 88, 97, 146–147

Ames Research Center (ARC, NASA)
  aeronautics, 6, 99, 148, 166, 211–212, 273
  applications research, 103
  awards ceremony, 150–151
  biological research, 11
  employment, 220
  establishment, 156
  planetary exploration project, 58–59, 88, 106, 107, 120–121, 153
  space shuttle support, 16, 141, 177
Amino acid, 132
Amir, Salah, 8
Amphibian aircraft, 168
*Amsat Oscar 7* (Orbiting Satellite Carrying Amateur Radio), 200, 201, 221, 227, 250, 257, 259
AMST (Advanced medium STOL transport), 22
Amsterdam, The Netherlands, 178
*Analysis of Federal R&D Funding by Function, Fiscal Years 1969-1975* (NSF report 72-313), 160–161
Anani, Dr. Osama, 26
Anchorage, Alaska, 4
Anders, L/C William A. (USAF, Ret.), 194
Anderson, Dr. David J., 140–141
Anderson, Roy E., 181
Andoeya, Norway, 265, 271
Andover, Me., 3–4
Andrechuk, Dr. Ted, 144–145
Andromeda Space Covers, 168
Anik (Canadian comsat), 169
*Anik 1* (Telesat A), 84
*Anik 2* (Telesat B), 4
Anniversary
  *Apollo 11*, 135–136, 138
  European Space Research Organization, 67
  first manned space flight, 85
  International Telecommunications Satellite Organization, 151–152
  *Skylab 1*, 45, 103
  Soviet Academy of Sciences, 104
*Ans 1* (ANS-A; Netherlands Astronomical Satellite), 2, 159–160, 221–222, 227, 244, 259
ANSPA (Astronomical Netherlands Satellite Program Authority), 160
Antarctica, 94
Antelope Valley Board of Trade, Calif., 189
Antenna, 181
  astronomical research, 14, 45–46, 182
  communications systems, 3–4, 106, 109, 114, 169, 171
  manned spacecraft, 16, 191
  seismology, 163
  solar research, 111, 149, 212
  technology, 47–48
  tracking, 101, 106, 109, 176
Antiballistic missile (ABM), 123–124, 215
Anticyclones, 166–167, 222
Antigua, West Indies, 33, 265
AN/TPN-19 (air traffic control radar system), 12

Apel, Dr. John R., 45, 55
Apollo (program), 45–46, 87, 152, 211
  achievements, 27, 28–29, 136
  completion, 21, 53, 141
  information exchange, 193
  press comment, 31, 69–70, 138, 153
  technology utilization, 10, 13, 191
Apollo (spacecraft), 16–17, 115, 191, 207, 216, 218
*Apollo 11*, 72–73, 135–136, 138, 147, 161
*Apollo 12*, 72–73, 208–209
*Apollo 13*, 35
*Apollo 14*, 72–73, 168, 208–209
*Apollo 15*, 72–73, 149, 215
*Apollo 16*, 28, 72–73, 215
*Apollo 17*, 15, 59, 64, 72–73, 102, 149, 208–209
*Apollo 17: Preliminary Science Report* (NASA SP-330), 72–73
Apollo lunar surface experiment package (ALSEP), 208–209, 223
Apollo-Soyuz Test Project (ASTP), 16, 25, 115, 118, 131, 152, 155, 170, 178, 211, 227, 261
  budget, 21, 41, 84, 204
  communications, 7, 44, 48, 77, 123, 191, 238, 258
  contract, 1, 16, 64, 118, 192
  docking, 147, 164, 172, 191, 232, 243, 252, 262
  emblem selection, 52
  media coverage, 121, 190
  mission preparations, 7, 81–82, 122, 123, 164–165, 192, 193, 207–208, 216
  presidential briefing, 65–66
  press comment, 36, 39, 172
  press conference, 82, 147, 208
  safety, 167
  significance, 124
  space manufacturing, 184
  systems development, 1, 79, 88, 102, 194
  testing, 41, 78, 151, 157, 164, 191, 221, 228, 232, 243
Apollo-Soyuz Test Project Working Group, U.S.-U.S.S.R., 7
Apollo Telescope Mount (ATM), 1, 30, 68, 75, 99–100, 139, 152, 174, 264
Appalachia, 106–107, 129
Applications Technology Satellite (ATS; see also *Ats 1* through *6*), 109, 116, 139
Aquifer, 175–176
Aquila (constellation), 182
Arab–Israeli war, 11, 28, 52
Arab League, 8, 169
ARC. See Ames Research Center.
Arcas (sounding rocket), 93, 265
  Super Arcas, 264, 266–267, 268, 269
Arecibo Observatory, Puerto Rico, 182, 201–202
Argentina, 14
Argon, 26, 68, 70
Ariane (launch vehicle), 189, 196
Ariane Program Board, 189–190
*Ariel 1* (U.K. satellite), 188
*Ariel 4*, 188

Ariel 5 (UK–5), 2, 187–188, 228, 247, 257, 259
ARIES (Astronomical Radio Interferometric Earth Surveying project), 163
Aries (sounding rocket), 57
Arizona, 45
Arizona, Univ. of, 167
Armstrong, Neil A., 135–136, 137, 138, 154
Army. See U.S. Army.
Army Ballistic Missile Agency (ABMA), 1
Army Electronics Command, 110
Arnold Engineering Development Center (AEDC), 39, 55, 163, 167, 202
Artyukhin, Lt. Col. Yuri P. (cosmonaut), 122, 130, 239–240, 241, 262
Asiago Astrophysical Observatory, Italy, 9–10
Assembly of Engineering, 129
Assembly of Parties (INTELSAT), 26
ASSESS (Airborne science–shuttle experiment system simulation program), 177
Asteroid, 37, 45, 177–178
Asteroid Belt, 2, 42, 58–59, 65, 98
ASTP. See Apollo-Soyuz Test Project.
Astrobee (sounding rocket)
  Astrobee D, 270
  Astrobee F, 267
Astronaut, 70–71, 98, 99, 127, 143, 183
  accident, 28
  accomplishments, 7, 99
  awards, 65–66
  Apollo, 15, 59, 135–136, 138, 208–209
  Apollo-Soyuz Test Project, 82, 115, 123, 124, 164–165, 170, 178, 208
  force reduction, 152
  press comment, 51
  press conference, 1, 82
  Skylab, 1, 7, 30, 32, 42–44, 51, 65–66, 67, 153, 157–158, 193, 195
  space flight response, 1, 32, 42–44
  space shuttle, 113
Astronautics Engineer Award (NSC), 58–59
Astronomical Netherlands Satellite Program Authority (ANSPA), 160
Astronomical Radio Interferometric Earth Surveying (ARIES) project, 163
Astronomy (see also individual celestial bodies, observatories, and probes; Telescope; and X-ray telescope), 21, 25, 38, 39, 73, 83, 160, 168, 185, 207, 252, 266
  galactic, 264, 265, 267, 268, 269, 270–271
  interstellar clouds, 132
  radio, 23, 176, 188–189
  solar, 2, 28–29, 32, 57, 67
  x-ray, 125, 133, 159, 160
Astrophysics, 4
AT&T. See American Telephone & Telegraph Co.
ATA. See Air Transport Assn. of America.
ATC. See Air traffic control.

Atlantic Ocean, 33, 40–41, 63–64, 65, 101, 104, 117, 131, 137, 144, 164, 168, 202, 236, 258
Atlas (launch vehicle), 113
  Atlas F, 132–133, 160, 209, 241–242
  Atlas-Centaur, 33–34, 202, 251, 259
ATM. See Apollo Telescope Mount.
Atmosphere (see also individual planets), 16, 23, 24, 102, 149, 153, 166
  balloon observation, 98, 101–102, 185, 197
  chromosphere, 67
  magnetosphere, 77, 111, 177
  pollution, 12, 143, 224
  satellite research, 40, 102, 144, 145, 228, 229–231, 242, 248, 249, 253, 258
  sounding rocket research, 126, 263–271
  stratosphere, 86, 98, 188
  thermosphere, 40, 55, 144
  troposphere, 23
  upper atmosphere, 65, 98, 104, 134–135, 145, 151, 169, 185, 194, 213, 218–219
  U.S.S.R. study (see also Appendix A), 52, 122, 131–132, 194, 195, 215
Atmosphere Explorer (AE). See Explorer 51.
Atmospheric Layering and Density Distribution of Ions and Neutrals (ALADDIN) project, 126, 223, 268
Atmospheric Sciences Laboratory (USA), 197
Atmospheric Variability Experiment (AVE–2), 101–102.
Atomic bomb, 70–71, 125
Atomic Energy Commission (AEC), 22, 23, 66, 72, 80, 83, 87, 107, 184–185, 187, 197
Atrophy, 75–76
ATS (Applications Technology Satellite), 109, 116, 139
Ats 1, 109, 147
Ats 3, 109, 120, 129, 147, 181, 188
Ats 5, 109
Ats 6 (ATS–F), 2, 44, 113–114, 139, 176, 225
  ASTP support, 44, 191
  communications, 116, 165, 221, 225–226
  launch, 108–109, 227, 228, 238, 257, 258
  mission profile, 48, 258
  propulsion technology, 224
  television broadcasting, 48, 106–107, 116, 129
Attitude control system (ACS), 148–149, 158, 183
AU: astronomical unit.
Aurora, 72, 197
Australia, 14, 78, 132, 250
Automotive research, 25–26, 73, 86, 99, 117, 118
Auto-pilot system, 151
Avco Corp., 8
AVE–2 (Atmospheric Variability Experiment), 101–102

Aviation. See Aeronautics, Aircraft, Airlines, Civil aviation, and General aviation.
Avionics, 48, 56, 81, 103, 199
AWACS (Airborne Warning and Control System), 60
Awards
  aeronautics, 14–15, 59, 148
  foreign, 52, 59, 181, 188–189
  government, 157, 219
    NASA, 39, 65–66, 88, 112, 150–151, 171, 195, 203, 213–214
  military, 28, 133, 156, 157, 171
  society, 14–15, 45, 58–59, 95, 118, 137, 148, 157, 169–170, 171, 178, 193, 195, 219
  university, 131, 211
AWRS (Airborne Weather Reconnaissance System), 60

## B

B–1 (advanced strategic bomber), 7, 22, 39, 63, 100, 183–184, 193, 225
B–52 (stratofortress), 7, 52–53, 193, 273–274
Baade, Dr. Walter, 31
BAC. See British Aircraft Corp.
Backfire (U.S.S.R. bomber aircraft), 62–63
Bacteria, 87, 94
Badge of Pilot-Cosmonaut of the U.S.S.R., 181
Bahamas, 174–175
Bali, Indonesia, 100–101
Ballistic missile system, 28
Balloon (see also Lower Atmosphere Composition and Temperature Experiment, and Project da Vinci), 13–14, 40–41, 63, 86, 101–102, 103, 144, 170, 174–175, 185, 226
Balloon-borne ultraviolet stellar spectrometer (BUSS), 185
Baltimore, Md., 19, 42, 134
Bank of America, 134
Barbados, 14
Barking Sands, Hawaii, 270
Barnard (star), 179, 270
Barnes Engineering Co., 134
BART (Bay Area Rapid Transit System), 89–90
Basalt, 170
Bathythermographs, 153
Baton Rouge, La., 83
Bay Area Rapid Transit System (BART), 89–90
Baykonur Cosmodrome, Tyuratam, U.S.S.R., launch (see also Appendix A and Appendix C), 5–6
  Cosmos satellites, 71, 78, 108, 147
  *Intercosmos 11*, 104, 237
  *Luna 22*, 108
  *Luna 23*, 193
  *Molniya 1-S*, 141
  *Salyut 3*, 122

*Salyut 4*, 218
*Soyuz 14*, 122, 130
*Soyuz 15*, 155
*Soyuz 16*, 207–208
*Soyuz 17*, 218
Bean, Capt. Alan L. (USN), 28, 121–122, 193
Beechcraft (aircraft), 65
Beeler, DeElroy E., 176–177
Beirut, Lebanon, 169
Belgium, 54, 67, 101, 138–139, 156
Bell, Larry E., 89
Bendix Corp., 117
Beregovoy, Maj. Gen. Georgy T. (cosmonaut), 218
Bereznyak, Aleksander Y., 131
Bergman, Jules, 51
Bergstrom AFB, Tex., 28
Bering Sea, 102
Berry, Dr. Charles A., 34–35, 43–44, 97
Betelgeuse (Alpha Orionis; star), 214
bev: billion electron volts.
Bhopal, India, 13–14
Bicycle ergometer, 31
Biofeedback, 144–145
Biomedical Application Team (NASA), 95
Birmingham, Ala., 143
Birth control, 105, 109
BIS (British Interplanetary Society), 73, 179
Bisplinghoff, Dr. Raymond L., 136, 182
Bjurstedt, Hilding A., 178
BL Lacertae (quasar), 76
Black Brant (sounding rocket), 72
  Black Brant VC, 264, 265–267, 268, 270–271
Black hole (space phenomenon), 56, 92, 99–100, 133, 171, 182, 187
Blackbird (SR–71; reconnaisance aircraft), 163, 199–200
Blood pressure monitoring system, 144–145, 226
Blume, John A. & Associates, 141
Blumrick, Josef F., 81
BMFT (German Federal Ministry for Research and Technology), 135, 213
Bobko, L/C Karol J. (USAF), 121–122
Boehm, Josef, 1
Boeing 707 (jet passenger transport), 89
Boeing 720, 102
Boeing 727, 17, 38, 99
Boeing 737, 38, 123
Boeing 747, 12, 79, 120, 157, 160, 213, 224
Boeing Co., 1, 19, 35–36, 89–90, 112, 175, 201, 209–210
Bomber aircraft (see also B–1, B–52, Backfire, and Sukhoi), 70, 156, 204
Bombsight, 154–155
Bonn, Univ. of, 268
Booster. See Launch vehicle.
Borman, Col. Frank (USAF, Ret.), 94
Boron, 91
Boron-epoxy composite, 192
Boston, Mass., 19, 89–90, 118
Bow shock, 12–13, 197, 198

Boy Scouts of America, 58
bps: bits per second.
Bradfield (comet), 90
Bragin, Dr. Joseph, 140–141
Brakes, 99
Brand, Vance D., 82, 121–122
Brayton energy conversion system, 29
Brazil, 11, 14, 38, 119, 130, 169, 195, 236–237
Breccia, 72–73
Bremen, West Germany, 199
Brevard County, Fla., 119, 141
Brezhnev, Leonid I., Gen. Sec. Communist Party Central Committee (U.S.S.R.), 25, 123–124, 204
Bridge, Dr. Herbert S., 27
British Aircraft Corp. (BAC), 37, 86–87, 136–137
British Interplanetary Society (BIS), 73, 179
British Labor Party, 136–137
Brooks, Robert A., 136
Brown, Gen. George S. (USAF), 103
Brown, Dr. Robert, 208
Brown Univ., 112
Brussels, Belgium, 99, 123–124
Budget, NASA, 16, 31, 152
  bills introduced, 83–84, 139
  bills passed, 92, 101, 109–110, 118, 124, 149–150, 153
  bills signed, 121, 164
  congressional hearings, 19, 27, 41, 42–44, 46, 47, 48, 54, 55–57, 61–62, 64, 66–67, 68
  press comment, 31, 35, 165, 196
  press conference, 24–25
  reduction, 27, 109–110, 124, 134, 204, 214
  request, 20–25, 27, 73, 176
Bulgaria, 249
Buoys, 119, 120
Burch, Dean, 4
Bureau of Mines, U.S., 54
Burr, Franklin L., Award, 45
Bush, Dr. Vannevar, 125
Bush differential analyzer, 125
Bushuyev, Konstantin D., 81–82, 157, 207–208
BUSS (balloon-borne ultraviolet stellar spectrometer), 185
Butler, T. Melvin, 213–214
Butterfield, Alexander P., 77–78, 101, 144
Bykovsky, Col. Valery F. (cosmonaut), 81–82

## C

C-5A (Galaxy; military cargo aircraft), 14, 165, 192
C-46 (Commando aircraft), 65, 119
C-130 (Hercules; military transport), 192, 199–200
C-141 (research aircraft), 6, 194
CAB (Civil Aeronautics Board), 79
Cadieux, Ambassador Leo (Canada), 144
Cairo, U.A.R., 8
Cal Tech Management Club, 200
Calcium, 157–158
Calcium fluoride, 173
California, 9, 42, 130, 163, 167–168, 169
California Institute of Technology (Cal Tech), 168, 175–176, 177
California, Univ. of (see also Lick Observatory), 115, 267, 270–271
Callisto (Jovian moon), 166–167, 198
Caloris Basin (Mercury), 173
*Calypso* (oceanographic research vessel), 188
Camera (see also Television; Videotelephone; and Vidicon receiving tube), 2, 3, 102, 112, 143, 187
Cameron, Dr. Alastair G. W., 90
Cameron, Dr. Roy E., 94
Campbell, Jack M., 194–195
Campbell, John P., 15
Canada (see also *Anik 2* and Communications Technology Satellite), 4, 38, 39, 71, 77–78, 89, 99–100, 101, 130, 176, 195, 209, 250
Cancer, 86, 153, 200
Canopus (star), 5
Canopus sensor, 231
Cape Canaveral, Fla., 1
Cape Canaveral Air Force Station, 165
Carbon, 27, 167–168
Carbon dioxide, 2, 13, 16, 38, 70, 89, 187
Carbon monoxide, 26
Cardiology, 54, 123–124, 143, 154, 158
Cardiotachometer, 143, 226
Caribbean Sea, 65, 153, 156
Carnarvon Tracking Station, Australia, 77
Carr, Col. Gerald P. (USMC), 1, 7, 30, 31–32, 91, 92, 221
*Carrying the Fire: An Astronaut's Journeys* (book), 147
Case Institute of Applied Sciences, 108
*Casper* (Apollo 16 command module), 28
Castenholtz, Paul D., 15
Castor (rocket engine), 85
Castor II (TX-354-5 rocket motor), 146
Causse, Jean-Pierre, 42, 54
Census, 42
Centaur (launch vehicle upper stage), 178
Centaurus (constellation), 171
Center for Astrophysics, Cambridge, Mass., 133
Central America, 65
Centre National d'Etudes Spatiales (CNES), France, 216
Cerebral palsy, 103
Cernan, Capt. Eugene A. (USN), 115, 121–122
Cesium-137, 190, 247
Chalmers, R. Bruce, 105
Chambers, John A., 69
Chanute, Octave, Award, 148
Charles Stark Draper Laboratory, Inc., 79
Charnell, Robert, 45
Chemistry, 59, 157

Cherry, George W., 44, 56
Chicago, Ill., 89–90, 192
Chicago, Univ. of, 45–46
Childs, Marquis, 3
Chile, 14
China Airlines–Taiwan, 89
China, People's Republic of (P.R.C.), 51–52, 89, 105
China, Republic of (Nationalist), 89
Christensen, Albert F., 95
Chromatography, 168
Chromosphere (earth), 67
Chrysler Corp., 16–17, 117, 118
CIAP (climatic impact assessment program), 86, 265
Cincinnati Electronics, Inc., 110
Circinus X-1 (x-ray source), 133
Civil Aeronautics Board (CAB), 79
Civil aviation, 146, 148, 154–155, 183
Civil Aviation Administration (P.R.C.), 89
Civil Service, 211, 220
Civil Service Commission (CSC), 14, 66, 115–116, 220
Clark, Dr. John C., 110
Clean combustor program, 145–146, 224
Cleary, Charles J., Award (USAF), 157
Clements, William P., Jr., Secretary of Defense, 141–142
Cleveland, Ohio, 168
Climatic impact assessment program (CIAP), 86, 265
Climatology. See Meteorology.
Cloud study (see also Interstellar clouds and Oort cloud), 185, 200–201
  meteorology, 42, 52, 61–62, 104, 119, 249, 253, 258
  planetary, 26, 27, 41–42, 82–83, 168, 208, 222
  U.S.S.R., 92, 131–132, 194, 215, 241
CM (command module; see also Command and service module), 28, 44
CML Satellite Corp., 130
CNES (Centre National d'Etudes Spatiales), France, 216
Coal, 11, 20, 29, 92, 154, 187
Cockpit ground-proximity warning system, 100–101
Cole, Judith A., 102
Collier, Robert J., Trophy, 95
Collins, B/G Michael (USAF, R.), 135, 138, 147
Colombia, 14
Colorado, 70–71, 112
Colorado, Univ. of, 264, 265–266, 270–271
Columbia Univ., 265, 270, 271
Comet (see also Bradfield; Encke; Halley's; and Kohoutek), 9–10, 45, 118–119
Comet Kohoutek Workshop, 118–119
Command and service module (CSM), 41, 82, 151, 165
Command module (CM), 28, 44
Commerce, Dept. of (DOC), 19, 104, 200
Commercial aviation. See Airlines.

Committee for the Reelection of the President, 146
Committee on Energy, 119
Committee on the Peaceful Uses of Outer Space (U.N.), 46, 200
Commonwealth Science and Industrial Research Organization (CSIRO), 171–172
Communications, 7, 15–16, 70, 101, 109, 238
Communications satellite (see also individual programs, satellites, and systems, such as Aeronautical Satellite program, Aerosat, Air Force Satellite Communications System, *Amsat Oscar 7*, *Anik 1*, *Ats 1*, Communications Technology Satellite, Fleet Satellite Communications System, Intelsat, Maritime Communications Satellite System, *Molniya 1-S*, *Oscar 1*, Skynet, *Symphonie 1*, and *Westar 1*), 181, 228
  benefits, 80, 116–117
  budget, 21, 25
  domestic, 4, 37–38, 47–48, 76, 99–100, 130, 146–147
  foreign, 8, 78–79, 165–166, 169, 174, 232, 234, 247
  international agreement, 99
Communications Satellite Act of 1962, 81
Communications Satellite Corp. (ComSatCorp; see also individual Intelsat satellites and Marisat), 3–4, 41–42, 84, 137, 159, 164, 227
COMSAT General Corp., 88, 114, 130, 146–147, 209
Communications Technology Satellite (CTS), 139
Compass Cope program, 151
Composite, 58
Compressor, 44–45
Comptroller General, 126
Computer, 199
  aeronautic applications, 39, 97, 167
  energy research, 111
  medical applications, 54, 68
  pollution research, 168
  spacecraft utilization, 30, 32, 38–39, 57, 81, 105, 186–187
  technology, 62, 131, 143, 177
Computer industry, 137
Computer Sciences Corp. (CSC), 95
ComSatCorp. See Communications Satellite Corp.
Concept verification test (CVT), 19, 177
Concorde (Anglo-French supersonic transport), 12, 38, 64, 86, 118, 136–137, 211–212
Concorde 202, 37
Condon, Dr. Edward U., 70–71
Conference on Scientific Experiments of Skylab (AIAA and American Geophysical Union), 195
Congress, 10, 37, 63, 107, 113, 148, 156, 214, 219
  bills passed, 132, 183–184

Joint Committee on Atomic Energy, 170, 185
Joint Committee on Defense Production, 29
joint conference, 101, 109, 152
Presidential messages, 11, 20–22, 27, 80, 204
report submitted, 51–52, 146–147
Congress, House of Representatives, 28–29, 150
  bills introduced, 78, 83–84
  bills passed, 36, 45, 48, 56, 92, 95, 98, 118, 124, 153
  Budget Committee, 132
  Committee on Appropriations, 124
    Subcommittee on HUD-Space-Science-Veterans, 99–100
    Subcommittee on Transportation, 77–78
  Committee on Merchant Marine and Fisheries, 183
  Committee on Science and Astronautics, 19, 68, 83–84, 91, 92, 95–96, 183
    Subcommittee on Aeronautics and Space Technology, 29, 44, 54, 55–56, 62, 66, 176, 184, 189
    Subcommittee on Energy, 36, 114–115
    Subcommittee on Manned Space Flight, 16, 41, 42–44, 47, 54, 58
    Subcommittee on Space Science and Applications, 25, 47–48, 54–55, 57, 64–65, 117, 182, 203
  Committee on Science and Technology, 183
  Committee Reform Amendments of 1974, 183
  Committee on Un-American Activities, 70–71
  nominations approved and confirmed, 217
  report approved, 48, 118, 153
  resolutions passed, 183
Congress, Senate, 36, 77–78, 99, 124–125, 204, 214
  bills introduced, 44, 62, 139
  bills passed, 61, 101, 106, 149–150, 152, 153
  Budget Committee, 132
  Committee on Aeronautical and Space Sciences, 11, 27–28, 46, 48, 60–61, 66–67, 73–74, 77, 79, 98, 101, 139, 145, 171
    Subcommittee on Aviation, 71
  Committee on Appropriations, 6–7, 149, 150
  Committee on Armed Services, 28, 60
  Committee on Banking, Housing, and Urban Affairs, 106
  Committee on Commerce
    Subcommittee on Aviation, 78
  Committee on Government Operations
    Subcommittee on Reorganization, Research and International Organizations, 83

Committee on Labor and Public Welfare, 106
  nominations submitted, 108, 137
  nominations approved and confirmed, 194, 217
  report, 29, 109, 152, 153
  treaty ratification, 123–124
Conrad, Capt. Charles, Jr. (USN, Ret.), 15, 103, 193, 195
Constellation (aircraft series), 199–200
Constellation. See Corvus and Lacerta.
Construction of facilities, NASA
  budget, 21, 83–84, 121, 144, 150
  earth station, 101
  inflation effects, 209
  simulation facility, 177
  space shuttle, 64, 88, 97, 158, 210–211
  windmill, 6
  x-ray telescope facility, 189, 220
Contract, NASA, 12, 29, 35–36, 80, 200
  aeronautics, 141, 145–146
  Apollo-Soyuz Test Project, 1, 16–17, 79, 102, 117, 118
  automotive research, 26, 86
  computer system, 86
  construction of facilities, 64, 97, 189, 220
  energy research, 187, 192, 204–205
  fire protection, 35
  inflation impact, 209–210
  Large Space Telescope, 201, 220
  launch vehicles, 16–17, 83, 146
  materials research, 184
  satellite, 78, 80, 98, 135, 146, 158
  Skylab, 121
  space probe, 19
  space shuttle, 5, 9, 25, 36, 41–42, 47, 79, 81, 88, 105, 106, 126, 140, 141, 158, 181, 217
  space tug, 140
  study, 12, 121, 165, 205, 217
  support services, 1, 19, 35–36, 62, 78, 93, 95, 115–116, 117, 216
Convair 990 (Galileo; NASA research aircraft), 16, 119
Cooper, Henry S. F., 147
Copernicus (Oao 3), 133, 187
Copper, 58, 166
Cornell Univ., 201–202
Corpus Christi Tracking Station, Tex., 77
Corsair (F-4U; fighter aircraft), 156
Cortright, Dr. Edgar M., 203
Corvus (constellation), 69
Cosmic dust, 212, 253
Cosmic ray, 39–40, 42, 59, 65, 80, 158, 197–198, 212, 253
Cosmic research laboratory, 154
Cosmonaut, 73, 115, 165, 261
  Apollo-Soyuz Test Project, 82, 164–165, 179, 207–208
  awards, 181
  corps increase, 152
  Soyuz missions, 44, 73, 122–124, 139–140, 155, 176, 207, 218, 219, 252
Cosmonautics Day (U.S.S.R.), 85
Cosmos (U.S.S.R. satellite), 193, 222, 228

*Cosmos 605*, 75–76, 190
*Cosmos 626*, 93–94
*Cosmos 628*, 229
*Cosmos 629–632*, 229–230
*Cosmos 633*, 231
*Cosmos 634*, 231
*Cosmos 635*, 232
*Cosmos 636*, 232
*Cosmos 637*, 71, 141, 232
*Cosmos 638*, 78, 147, 222, 225, 232
*Cosmos 639*, 232
*Cosmos 640*, 233
*Cosmos 641–648*, 91, 234–235
*Cosmos 649–651*, 235
*Cosmos 652*, 236
*Cosmos 653*, 236
*Cosmos 654*, 182, 236
*Cosmos 655*, 237
*Cosmos 656*, 108, 237
*Cosmos 657*, 237
*Cosmos 658–661*, 239
*Cosmos 662–665*, 240–241
*Cosmos 666*, 241
*Cosmos 667*, 242
*Cosmos 668*, 242
*Cosmos 669*, 242
*Cosmos 670*, 243
*Cosmos 671*, 243
*Cosmos 672*, 147, 222, 225, 243
*Cosmos 673*, 243
*Cosmos 674*, 244
*Cosmos 675*, 244
*Cosmos 676*, 245
*Cosmos 677–684*, 172–173, 245–246
*Cosmos 685*, 246
*Cosmos 686*, 246
*Cosmos 687*, 246
*Cosmos 688*, 247
*Cosmos 689*, 247
*Cosmos 690*, 190, 247
*Cosmos 691*, 248
*Cosmos 692*, 249
*Cosmos 693*, 249
*Cosmos 694*, 251
*Cosmos 695*, 251
*Cosmos 696*, 252
*Cosmos 697*, 253
*Cosmos 698*, 254
*Cosmos 699*, 254
*Cosmos 700*, 255
*Cosmos 701*, 255
Council for Science and Technology (proposed), 124–125, 129
Council of Economic Advisors, 61
Council of International Cooperation in Space Exploration and Use (U.S.S.R.), 85
Council on Environmental Quality, 61
Cousteau, Jacques-Yves, 188
Crab Nebula, 223, 270, 271
Crater. See Moon.
Cronkite, Walter, 137
Crop damage, 115
Crow, L/G Duward L. (USAF, Ret.), 170
Crowley, John W., 140
Cryo-anchor soil stabilizer, 134

Crystals (see also Silicon), 43, 58, 199, 216
CSC. See Civil Service Commission.
CSIRO (Commonwealth Science and Industrial Research Organization), 171–172
CSM. See Command and service module.
CTS (Communications Technology Satellite), 139
Cuba, 2–3
Cuiba Tracking Station, Brazil, 111
Culbertson, Dr. Phillip E., 47, 113
Cunningham, George, 68
Currie, Dr. Malcolm R., 66–67
Curtin, M/G Robert H. (USAF, Ret.), 54
CVT (Concept verification test), 19, 177
Cyclones, 166–167, 222
Cygnus X–1 (x-ray source), 133, 187, 244, 259
Cyprus, 243
Czechoslovakia, 104, 190, 195, 237, 249

## D

da Vinci, Leonardo, 136
*Daffodil II* (balloon), 13–14
Dakar, Senegal, 104–105
Danish Meteorological Institute, 268–269
Darmstadt, West Germany, 160
Data processing, 9, 38–39, 95, 105, 111, 120, 131
Davis, Dr. William O., 103
Day, LeRoy E., 113
DC–8 (jet transport), 174, 224
DC–9, 38, 160
DC–10, 51, 55, 71, 78
de Florez Training Award, 15
De la Vaulx Medal, 193
de Seversky, Maj. Alexander P. (U.S. Army Air Corps, Ret.), 154–155
De Waard, J., 16
Debus, Dr. Kurt H., 14, 75, 131, 171, 217
Deep Space Network (DSN), 63, 77, 101, 163, 173, 176
Defense, Dept. of (DOD; see also Launch, DOD), 3, 14, 28, 37–38, 41, 100, 130, 132–133, 146, 209
  aeronautics, 148
  budget, 20, 21–22, 183–184
  contract, 31, 190
  employment, 48–49
  environmental research, 119, 264
  funding, 134, 141–142, 143
  space programs, 22, 25, 31, 47, 66–67, 140, 241–242
Defense Meteorological Satellite Program (DMSP), 63, 120, 134, 226, 232, 243
Defense Posture Statement, 51–52
Defense Satellite Communications System (DSCS), 226
Delta (launch vehicle upper stage; see also Thor-Delta), 186, 212
Delta 100 Failure Review Committee, 83
Delta Air Lines, Inc., 74, 83, 87
Delta Project Review Committee, 148

Demin, Col. Lev S. (cosmonaut), 155, 181, 239–240, 244, 262
Democratic National Committee, 146
Denmark, 54, 67, 101, 138–139
*Denpa* (Japanese Radio Explorer satellite), 39–40
Dent, Frederick B., Secretary of Commerce, 130
Desert, 175–176
Detroit, Mich., 175
Deuterium, 27
DIGISAT, 84
Digital albedo horizon sensor, 231
Digital circuitry, 48
Digital data satellite service (DIGISAT), 84
Dirty snowball theory, 9
Disarmament (see also Strategic Arms Limitation Talks), 123–124, 215
Discoverer (satellite), 60
Disher, John H., 47
Disney World, Fla., 141
Distinguished Flying Cross (USAF), 133, 156
Distinguished Service Medal (NACA), 219
Distinguished Service Medal (NASA), 39, 65–66, 88, 150–151, 171, 194–195
Distinguished Service Medal (USN), 28
DMSP. See Defense Meteorological Satellite Program.
DOC. See Commerce, Dept. of.
Docking (see also Apollo-Soyuz Test Project, *Soyuz 14*, and *Soyuz 15*), 85, 123, 139–140, 161, 184
DOD. See Defense, Dept. of.
DOI. See Interior, Dept. of
Domestic communications satellite (see also Anik and Westar), 4
Domestic Council, 124–125
Dopant (electronically active chemical element), 43
Dornbrand, Harry, 194–195
Dornier System GmbH, 210
DOT. See Transportation, Dept. of.
Dow, James E., 137
Dow Jones & Co., Inc., 168
Doyle, Stephen E., 86
"The Dragon's Wings: The China National Aviation Corporation and the Development of Commercial Aviation in China" (history manuscript), 15
Drake, Dr. Frank D., 45–46
Drogue chute, 117
Drone aircraft. See Firebee.
Dryden, Dr. Hugh L., 103
Dryden, Dr. Hugh L., Memorial Fellowship, 59
"The Dryden–Blagonrarov Era of Space Cooperation, 1962–65" (essay), 59
Dryden Research Lecture, 15
DSCS (Defense Satellite Communications System), 226
DSN. See Deep Space Network.
Dubinin, Nikolay, 87–88
Dudley Observatory, 269
Duke, Col. Charles M., Jr. (USAF), 28
Dulles International Airport, Va., 80
Dunne, Dr. James, A., 27
Dyal, Dr. Palmer, 208–209
Dzhezkazgan, U.S.S.R., 123, 207, 241, 252, 262

E

E–Systems, Inc., 3–4
*Eagle* (*Apollo 11* lunar module), 30
Eagleton, Sen. Thomas F., 60
Eaker, Gen. Ira S., 133
Early warning satellite system, 67, 183
Earth
 gravitational field, 98
 radiation, 83, 194, 248
 solar effects, 43–44, 191
 spacecraft observations, 32, 38, 70, 77, 91–92, 100, 109, 114, 122, 126
Earth and ocean physics applications program (EOPAP), 21, 25, 38
Earth Environment and Resources Conference, 168
Earth Observatory Satellite (EOS), 37–38, 78
Earth resources, 4, 9, 28, 32, 42, 46, 55, 80, 86–87, 145, 154, 174, 177–178, 207, 215, 218–219, 225–226, 252
Earth resources experiment package (EREP), 42, 98
Earth Resources Observation Administration (DOI; proposed), 145, 171
Earth resources satellite program, 103, 204
Earth resources survey program, 21, 204
Earth Resources Survey System (proposed), 182
Earth Resources Technology Satellite (ERTS; see also individual satellites such as *Erts 1*), 38–39, 111, 132, 145, 171, 182, 194, 215, 225
Earth station (see also individual stations), 9, 44, 45, 62, 66, 130, 207
 communications applications, 4, 8, 77–79, 81, 86, 88, 99, 106–107, 146–147, 159, 171, 186
 construction, 8, 38, 194
 foreign, 8, 78–79, 81, 111, 139–140, 141
 meteorological applications, 65, 104–105, 110
 reduction, 24
Earthquake, 163, 164, 169
East Germany. See Germany, East.
Eastern Airlines, Inc., 74, 94
Eastern Test Range (ETR), Patrick AFB, Fla. (see also *Ats 6*, *Helios 1*, *Intelsat-IV F8*, *Skynet IIA*, *Skynet IIB*, SMS, *Symphonie 1*, Titan-Centaur, *Westar 1*, and *Westar 2*), 164
Ebert-Fastie spectrometer, 185
Eddy, Dr. John, 67
Edison Electric Institute, 112
Education, 8, 48, 102–103, 106–107, 109, 129, 131, 228, 238, 258
Edwards, John W., 59
Edwards AFB, Calif., 116, 151, 163

EEO. See Equal Employment Opportunity.
Egan, Gov. William A., 4
Eglin AFB, Fla., 101–102
Einstein, Albert, 133
Eisenhower, President Dwight D., 183
Eisenhower, Milton S., Library, 103
El Toro, Calif., 60
Electric field, 77, 190
Electric generator, 173, 175–176
Electric Power Research Institute (EPRI), 211
Electric waves, 39–40
Electricity (see also Thermoelectricity), 34, 63, 73–74, 93–94, 105, 114–115, 186, 187, 204–205, 211
Electrocardiograph, 143
Electrodes, 103
Electromagnetic spectrum, 57
Electronic Systems Center. See International Business Machines Corp.
Electronics, 43, 184
Electrons, 217, 208
Eliot, T. S., 32
Ellington AFB, Tex., 28, 31–32
EME (environmental measurements experiment), 238
Employment, NASA, 6–7, 11, 16, 25, 35–36, 47, 48–49, 66, 102–103, 115–116, 126, 141, 190, 202, 210, 220
Employment, U.S., 131, 141
Encke (comet), 37
Energy. See Electricity, Geothermal energy, Microwave energy, Nuclear energy, Solar energy, Thermal energy, and Wind energy.
Energy conservation
  aeronautics, 25, 29, 44–45, 48, 55–56, 192, 211–212, 224
  automotive, 25–26, 73–74, 118
  NASA plan, 8–9, 73
  national policy, 20, 99, 119
Energy conversion, 73–74, 90–91, 92, 114–115, 130, 167, 187, 192
Energy crisis, 3, 6, 11, 31, 39, 48, 52, 79, 100, 153–154, 175, 177
Energy Emergency Act, 48
Energy Policy Office, 72
Energy Reorganization Act of 1974, 187
Energy research and development (ER &D; see also Solar energy research), 6, 8–9, 29, 34, 61, 99, 105, 160–161, 163–164, 166, 184–185, 211, 220, 226
  budget, 20–21, 23, 73–74, 95, 98, 107, 109, 152
  congressional hearings, 25, 46, 79, 84
  personnel, 15, 194
  press comment, 39, 45, 80, 107
Energy Research and Development Administration (ERDA), 46, 61, 83, 106, 163–164, 187, 194, 220
Energy Resources Council, 187
Engine (see also JT8D, Space shuttle main engine)
  aircraft, 31, 48, 51, 92, 145–146, 193
  rocket, 15, 111, 140, 146

Engineering, 23, 41–42, 45, 102–103, 108, 134
Engineers, 72, 95, 106, 113, 114, 125, 131, 136, 143, 157, 168–169, 179, 187, 191, 212
  awards, 92
  employment, 6–7, 11, 48–49, 66, 190
  energy research, 6, 80
  international cooperation, 7, 16
England. See United Kingdom.
Environment, 17, 23, 38, 48, 73, 84, 98, 160–161, 183, 187
Environmental measurements experiment (EME), 238
Environmental monitoring, 115, 131
Environmental Protection Agency (EPA), 17, 23, 25–26, 54, 117
Environmental warfare, 123–124
EOPAP. See Earth and ocean physics applications program.
EOS (Earth Observatory Satellite), 37–38, 78
EPA. See Environmental Protection Agency.
epndb: effective perceived noise in decibels.
Epoxy, 184, 192
EPRI (Electric Power Research Institute), 211
Equal employment opportunity (EEO), 6–7, 11–12, 34
Equator, 40, 72, 97, 119, 185–186
Equinox, vernal, 65
ER &D. See Energy research and development.
ERDA. See Energy Research and Development Administration.
EREP (earth resources experiment package), 42, 98
Ermenonville Forest, France, 51
ERNO Raumfahrttechnik, GmbH, 54
ERTS. See Earth Resources Technology Satellite.
*Erts 1* (ERTS–A), 25, 98, 111, 115, 132, 225–226
  agricultural applications, 62
  benefits, 32, 80
  international cooperation, 10–11, 38–39, 62, 111, 132, 225–226
  land management, 54, 80, 90, 131
  oceanography, 45
  photographs, 5–6, 32, 90, 100, 130–131, 170, 215
ERTS–B, 21, 25, 111, 132, 225–226
ERTS–C, 98, 109, 124, 150, 171
Ervin, Sen. Sam J., Jr., 44
ESRO. See European Space Research Organization.
Etam, W. Va., 3–4
ETR. See Eastern Test Range.
Europa (Jovian moon), 166–167
Europe, 40–41, 84, 99–100, 108, 116, 118–119, 138–139, 156, 216, 254
European Space Agency, 76

European Space Research Organization (ESRO; see also Maritime Orbital Test Satellite), 9, 160, 169, 210
 anniversary, 67
 comsat program, 77-78, 101, 144, 164, 209
 earth studies, 77, 86-87, 191
 employment, 67
 launch vehicle program, 189-190, 196
 meteorology, 110
 personnel, 42
 planetary exploration, 36-37, 120-121
 reorganization, 76
 Spacelab, 16, 54, 67, 94, 105-106, 113, 181, 199, 223-224, 225
EUV (extreme ultraviolet). See Ultraviolet headings.
EUV spectrometer, 242
ev: electron volt.
EVA. See Extravehicular activity.
Evans, Capt. Ronald E. (USN), 121-122
Exceptional Civilian Service Medal (USA), 171
Exceptional Scientific Achievement Medal (NASA), 194-195, 203
Exceptional Service Medal (NASA), 88, 194-195
Exhibit, 163
Explorer (scientific satellite), 21, 24
*Explorer 1*, 1, 14, 171
*Explorer 42* (*Uhuru*; Small Astronomy Satellite), 133, 187
*Explorer 45* (SSS-A), 114, 177
*Explorer 46* (Meteoroid Technology Satellite), 60, 152-153
*Explorer 50* (IMP-J; Interplanetary Monitoring Platform), 149
*Explorer 51* (AE-C; Atmosphere Explorer), 5-6, 40, 55, 126, 144, 230-231, 266, 271
*Explorer 52*. See *Hawkeye 1*.
External tank (ET), 19, 36, 158, 210-211
Extraterrestrial colonies, 98, 154
Extraterrestial life, 1, 44, 69-70, 94, 97, 135, 201-202
Extravehicular activity (EVA), 28-29, 45, 210-211
Exxon Corp., 147

**F**

F-1 (Mirage 3; French fighter aircraft), 138-139
F-4 (Phantom fighter-bomber aircraft), 94
F-4U (Corsair fighter aircraft), 156
F-5B (fighter-trainer aircraft), 7
F-5E (Tiger II fighter aircraft), 7, 10
F-5F (fighter-trainer aircraft), 10
F-8 (carrier fighter aircraft), 44
F-14 (Tomcat fighter aircraft), 7, 22, 63-64, 100
F-15 (Eagle advanced tactical fighter aircraft), 22, 91, 94, 167, 183-184, 225
F-104 (mach 2 fighter aircraft), 138, 199-200
F-111 (fighter aircraft), 44, 175, 224
FAA. See Federal Aviation Administration.
FAI (Fédération Aéronautique Internationale), 193
FAI Gold Space Medal, 193
Fairbanks Tracking Station, Alaska, 111
Fairchild Republic Co., 76
Fairford, U.K., 37
Falls Church, Va., 95
Fanale, Dr. Fraser P., 208
Farnborough International '74, 163
FCC. See Federal Communications Commission.
FEA. See Federal Energy Administration.
Federal Aviation Administration (FAA), 87, 101, 114, 137, 146, 183, 209
 accident research, 51, 55, 71, 97
 aircraft research, 17, 97, 102, 144, 148, 166
 budget, 22-23, 77-78
 regulations, 78, 79-80, 92-93, 97
Federal Communications Commission (FCC), 84, 88, 113-114, 130, 202, 251
Federal Council for Science and Technology, 60-61
Federal Electric Corp., 117
Federal Energy Administration (FEA), 11, 56, 61, 99, 108
Federal Energy Administration Act of 1974, 99
Federal Energy Office (FEO; see also Federal Energy Administration), 71-72, 80, 88
Federal Housing Administration (FHA), 141
Federal Nonnuclear Energy Research and Development Act of 1974, 220
*Federal Plan for Meteorological Services and Supporting Research FY 1975*, (NOAA proposal), 146
Federal Railroad Administration, 86
Fédération Aéronautique Internationale (FAI), 193
FEO. See Federal Energy Office.
Feoktistov, Dr. Konstantin P. (cosmonaut), 73, 218
Ferrari, Dr. Alfred J., 215
Ferrofluid separator, 166
FG Sagittae (superstar), 143
FHA (Federal Housing Administration), 141
Fifth Annual Lunar Science Conference, 64
Fighter aircraft. See individual aircraft and programs, such as A-7, Air combat fighter, F-4, Mirage 3, P-35, Sukhoi, VFAX, VFX, and YF-16.
Filipchenko, Col. Anatoly V. (cosmonaut), 207, 252, 262
Filton, U.K., 37
Fink, Daniel J., 15
Finke, Robert C., 186
Fire protection, 19, 35, 54, 79-80

Firebee (drone aircraft), 84
Fitzgerald, A. Ernest, 14
Fleet Satellite Communications (FLTSATCOM) System, 67
Fleming, Arthur S., Award, 118, 157
Fletcher, Dr. James C., 17, 35, 53–54, 67, 70, 72, 119, 148, 155, 167, 170, 191, 198, 210, 214
  agreements, 211, 213, 219
  awards, 59
  ceremonies, 65–66, 88, 112, 136, 150–151, 195, 199, 203
  congressional testimony, 27, 46, 60–61, 68, 139, 145
  contract award, 126
  foreign visit, 165, 182
  meeting, 105,
  personnel appointment, 143–144, 150
  press conference, 24–25, 30, 68, 106
  speech, 37, 43, 189, 191–192
Flight Research Center (FRC, NASA)
  aeronautics, 52–53, 84, 151, 160, 175, 273–274
  automotive research, 73, 118
  Flight Activities Office, 158–159
  "Manned Lifting Body Flight Log," 273
  name change (January 1976), 273
  personnel, 176–177, 217, 220
  space shuttle, 189
  symposium, 114
Flight Transportation Laboratory (MIT), 166
Flood, 170
Florida, 5, 131, 167, 214
Florida Acquifer, 131
Florida, Univ. of, 131
FLTSATCOM (Fleet Satellite Communications System), 67
Fluorine, 157
Food supply management, 153–154
*For the Benefit of All Mankind: The Practical Returns from Space* (congressional report), 95–96
Ford, President Gerald R., 146, 156, 199
  agreement, 204
  appointments and nominations, 151, 194
  award, 211
  bills signed, 163–164, 183–184, 187, 192–193, 220
  ceremony, 95
  message to Congress, 204
  press comment, 165
  speech, 175
Ford Motor Co., 156
Forestry, 132
Fort Churchill, Canada, 144, 265–269
Foushee, B. R., 178
France (see also National Research Council and *Symphonie I*), 156, 165
  aeronautics, 12, 38, 64, 81, 107, 136–137, 138–139
  agreements, 101, 213
  space activities, 54, 67, 101, 119, 121, 189–190, 242

Frazier, Joe, 14
FRC. See Flight Research Center.
Friedheim, Jerry W., 14, 42
Fruit fly, 42
Fucino, Italy, 111
Fuel. See individual kinds of fuel, Energy crisis, and Energy conservation.
Fuel cells, 187
Fuel conservation. See Energy conservation.
Fuqua, Rep. Don, 16

## G

Gagarin, Col. Yuri A. (cosmonaut), 36, 85
Gagarin, Yuri, Gold Medal, 193
Galapagos Islands, 238
Galaxy. See C-5A.
Galaxy (see also Milky Way), 76, 133, 160, 214, 223
*Galileo*. See Convair 990.
Gallagher, J. Wes, 190
Galley, Robert, 12
Galveston, Tex., 188
Gamma ray, 68, 80, 90, 108, 158, 190
Ganymede (Jovian moon), 166–167
GAO. See General Accounting Office.
Gardner, Frank, 171–172
Garnet (silicate mineral), 143
GARP. See Global Atmospheric Research Program.
GARP Atlantic Tropical Experiment (GATE), 22, 104–105, 119, 175, 225, 236–237
Garrett Corp., 60, 92
  AiResearch Aviation Co., 145–146
Gas. See Natural gas.
Gas turbine, 25–26, 145–146, 187
Gaseous-core nuclear rocket, 29
Gasoline, 26
GASP (Global Air Sampling Program), 213
Gatch, Thomas L., 40–41
GATE. See GARP Atlantic Tropical Experiment.
Gatos, Dr. Harry C., 42
GAW-1 (general aviation wing), 187
Gazenko, Dr. Oleg G., 75–76, 178
GE. See General Electric Co.
GEC-Marconi (U.K.), 49, 210
Gehrels, Dr. Tom, 82–83
Gemini (program), 28–29, 59, 211
*Gemini 10*, 147
General Accounting Office (GAO), 5, 60, 72, 126
General aviation, 23, 48, 63, 69–70, 96, 97, 99, 183, 189
General Aviation Manufacturers Assn., 189
*General Aviation Programs* (congressional report), 189
General aviation wing. See GAW-1.
General Dynamics Corp., 7–8, 116, 138, 140, 179, 183–184, 190
  Convair Div., 140, 146

General Electric Co. (GE; see also Radio Optical Observatory), 1, 11, 31, 35–36, 78, 147, 165–166, 187, 204–205
General Genetics Institute (U.S.S.R.), 87–88
General Motors Corp., 145–146
General Services Administration (GSA), 36
General Telephone & Electronics Corp. (GTE), 88, 169
Genetics, 87–88
Geneva, Switzerland, 110
Geodetic Explorer. See GEOS–C.
Geology, 122, 143, 170, 241
Geophysical Institute, Peru, 72
GEOS–C (Geodetic Explorer satellite), 2
Geostationary Operational Environmental Satellite (see also Goes 1), 26, 236
Geothermal energy, 13, 24, 175–176
Geothermal Energy Research, Development and Demonstration Act of 1974, 163–164
German Aerospace Show (1974), 94
German Federal Ministry for Research and Technology (BMFT), 135, 213
Germanium selenide, 58
Germany, East, 104, 119, 120, 195, 237, 249
Germany, West (see also Aeros 2, Helios 1, Rocket Research Center, and Symphonie 1), 3, 54, 57, 67, 101, 107, 119, 120, 121, 191, 249–250
Gesellschaft für Weltraumforschung (GfW), 216
GET: ground elapsed time.
ghz: gigahertz (1 billion cycles per second).
Giberson, Walker E., 151
Gibson, Dr. Edward G., 1, 7, 30, 31–32, 91–92, 152
Gibson, Roy, 144
Gilruth, Dr. Robert R., 92
Giscard d'Estaing, President Valery (France), 136–137
Glass, 173, 212
Glenn, Col. John H., Jr. (USMC, Ret.), 99, 199
Glennan, Dr. T. Keith, 219
Global Air Sampling Program (GASP), 213, 224
Global Atmospheric Research Program (GARP; see also GARP Atlantic Tropical Experiment), 21, 23, 224
Glushko, Valentin P., 85
Goddard Award, 15
Goddard, Robert H., Historical Essay Award, 59
Goddard, Robert H., Memorial Dinner, 58
Goddard, Robert H., Memorial Trophy, 58
Goddard, Robert H., Scholarship, 59
Goddard Space Flight Center (GSFC, NASA), 64, 88, 126, 169–170, 227
  Apollo-Soyuz Test Project support, 191
  earth station design, 104–105
  employment, 220
  inflation impact, 209–210
  project management, 104, 146, 160, 200–201, 213
  sounding rocket research, 2, 263, 264–267, 268, 269, 270
  space experimentation, 134–135, 144, 187, 242
  support services, 95
  tracking, 216
Godfrey, Roy E., 69
GOES (Geostationary Operational Environmental Satellite), 26, 236
Goes 1 (GOES–A; SMS–C), 2, 104, 236–237, 258
Gold Star Medal (U.S.S.R. award), 181
Goldsmith, John V., 114–115
Goldstone Tracking Station, Calif., 14, 111, 163
Goldwater, Sen. Barry M., 62, 139
Goodyear Aerospace Corp., 117, 167
GPS. See NAVSTAR Global Positioning System.
Grand Coulee Canyon, 170
Granger, John V., 171
Graphite, 91
Graphite epoxy composite, 210–211
Gravel, Sen. Mike, 4
Gravitational field (earth; see also individual planets), 98
Gravity (see also Weightlessness), 133
Gray, Edward Z., 54
Grazing-incidence telescope, 159
Great Britain. See United Kingdom.
Great red spot (Eye of Jupiter), 82–83, 166, 222
Grechko, Georgy M. (cosmonaut), 218–219, 255
Green Swamp, Fla., 131
Greenbelt Tracking Station, Md., 111
Greenstein, Jesse L., 194–195
Gribbin, John, 169
Grobecker, Dr. Alan J., 86
Groo, Elmer S., 149
Ground station. See Earth station.
Group Achievement Award (NASA), 194–195, 203
Grumman Corp.
  Grumman Aerospace Corp., 7
  Grumman Aircraft Corp., 7–8, 100, 168
GSA (General Services Administration), 36
GSFC. See Goddard Space Flight Center.
GTE (General Telephone & Electronics Corp.), 88, 169
Gubarev, Lt. Col. Aleskey A. (cosmonaut), 218–219, 255
Guggenheim, Daniel, Medal, 137
Guggenheim, Daniel and Florence, International Astronautics Award (1973), 178
Guggenheim Fund for the Promotion of Aeronautics, 156
Guidance and control (see also Navigation), 201

## H

Gyroscope, 5, 183

Hale Observatories, 76, 167, 168
Haley Astronautics Award (AIAA), 195
Hall, Charles F., 150–151
Halley's Comet, 37
Handler, Dr. Philip, 61
Hanover, West Germany, 94
Hardrath, Herbert F., 15
HARM (High-speed antiradiation missile), 108
Harr, Dr. Karl G., Jr., 3
Harris, Mrs. Ruth Bates, 6–7, 11, 150
Harrisburg, Pa., 40–41
Hartford, James J., 99–100
Hartman, Dr. Robert C., 144
Hartsfield, Henry W., 183
Harvard College Observatory, 267, 269
Haughton, Daniel J., 10, 111–112
Hawaii, 12, 21–22, 24, 39, 84, 88, 97, 113–114, 186
Hawaii, Univ. of, 39, 267
Hawk (sounding rocket), 267, 270
Hawker Siddeley Dynamics, Ltd., 49, 58, 210
Hawker Siddeley Group, Ltd., 169
*Hawkeye 1* (*Explorer 52*; scientific satellite), 2, 111, 221–222, 227, 238–239, 257, 259
Hawkins, Willis M., 15, 195
Hawthorne, Calif., 79
HCMM. See Heat-Capacity Mapping Mission.
Health, Education, and Welfare, Dept. of (HEW), 106–107, 134
Health Education Telecommunications (HET) experiment, 129
HEAO. See High Energy Astronomy Observatory.
HEAO–B, 158
Heart-lung machine, 154
Hearth, Donald P., 126
Heat-Capacity Mapping Mission (HCMM), 21, 24, 55, 204
Heat exchangers, 26, 154
Heidelbaugh, Dr. Norman D., 131
Helicopter, 10, 15–16, 51, 81
Helios (solar probe), 21, 33–34, 213, 253, 257, 260
*Helios 1* (Helios–A), 2, 212–213, 221–222, 227, 228, 253
Heliosphere (earth), 197–198
Helium, 27, 68, 144
Helium 3, 179, 185, 197
Hematology, 43–44
Hemke, Dr. Paul E., 108
Henize, Dr. Karl G. (astronaut), 28
*Heos 2* (Highly Eccentric Orbit Satellite), 191
Hercules. See C–130.
Hero of Socialist Labor, 52
Hero of the Soviet Union, 181
Herzberg, Dr. Gerhard, 9

HET (Health Education Telecommunications experiment), 129
Hetherington, Dr. Norris S., 59
HEW. See Health, Education, and Welfare, Dept. of.
Hewish, Dr. Antony, 188–189
Hickam AFB, Hawaii, 188
High Energy Astronomy Observatory (HEAO), 21, 25, 37–38, 80, 99–100, 158, 170, 189
High-speed antiradiation missile (HARM), 108
Highly Eccentric Orbit Satellite (HEOS), 191
Hightstown, N.J., 146
Hill, Louis W., Space Transportation Award, 14, 171
Hinners, Dr. Noel W., 124, 203
History Manuscript Award (AIAA), 15
HL-10 (lifting body), 273
H.M.S. *Beagle*, 30
Hocker, Dr. Alexander, 67, 105,
Holloway, Gen. Bruce K. (USAF, Ret.), 53, 157, 170
Holzer, Dr. Thomas E., 92
Hospitals, 109
Hot Line (U.S.-U.S.S.R.), 81
Hotz, Robert, 183
Housing and Urban Development, Dept. of (HUD), 36, 46, 54, 61, 106, 187, 219, 226
Housing and Urban Development-Space-Science-Veterans Appropriations bill, 149, 152, 153, 164
Houston, Tex., 13, 187, 212
Houston, Univ. of, 59, 269
"Houston, We Have a Problem" (TV film), 35
Hovis, Dr. Warren A., 55
Hubble Constant, 214
HUD. See Housing and Urban Development, Dept. of.
Hudson Bay, 144
Hugh L. Dryden Flight Research Center. See Flight Research Center.
Hughes Aircraft Co., 19, 52, 146–147, 169, 171, 211, 223
Hungary, 195, 249
Huntoon, Dr. Carolyn, 169–170
Huntsville, Ala., 93, 195
Hurricane, 120
Hydraulic pump, 200
Hydrazine, 86
Hydrocarbon, 26
Hydrogen, 2, 3, 5, 9–10, 17, 27, 29, 33–34, 35, 92, 150, 166, 192, 197–198
Hydrogen chloride, 38
Hydrogen cyanide, 77
Hydrogen injection technology, 25–26
Hydrology, 104
Hydroplaning, 75
Hydrothermal Power Co., 175–176
Hypersonic aircraft, 163, 166
Hypersonic technology, 224

# I

IAA (International Academy of Astronautics), 170
IAF. See International Astronautical Federation.
IBM. See International Business Machines Corp.
ICBM. See Intercontinental ballistic missile.
Ice distribution studies, 61–62, 102, 131–132, 170
Idaho, 170
IDCSP (Initial Defense Communications Satellite Program), 204
Illinois, Univ. of, 264, 267, 268–269
ILS (instrument landing system), 144
IMBLMS (integrated medical and behavioral laboratory measurement system), 144–145
IMP-J (*Explorer 50*, Interplanetary Monitoring Platform), 149
Impeachment, 146
Improved Tiros Operational Satellite (ITOS). See individual satellites, such as ITOS-H and *Noaa 2*.
India, 48, 109, 139, 168–169, 221, 238, 258
Indian Ocean, 71–72, 110, 202, 252
Indian Space Research Organization (ISRO), 189–190, 238
Indium-antimonide, 199
Indonesia, 78–79, 169
Inflation, 209
Infrared detectors, 168, 231
Infrared photography, 120–121
Infrared radiation, 91, 109, 177, 216
Infrared radiometer, 98
Infrared spectrometer, 218–219
Infrared telescope, 6, 21, 24, 39, 84, 141
Infrared wavelength, 47–48
Initial Defense Communications Satellite Program (IDCSP), 204
Injun (scientific satellite series; see also *Hawkeye 1*), 111
*Injun 1*, 111
Inscho's Mechanical Contractors, Inc., 220
Institute of Gas Technology, 192
Instituto Nacional de Técnica Aéroespacial (INTA), Spain, 101, 201
Instrument landing system (ILS), 144
INTA (Instituto Nacional de Técnica Aeroespacial, Spain), 101, 201
*Intasat* (Spanish satellite), 200, 201, 221–222, 250–251, 257, 259
Integrated medical and behavioral laboratory measurement system (IMBLMS), 144–145
Integrated propulsion control system (IPCS), 175
INTELSAT. See International Telecommunications Satellite Organization.
Intelsat (comsat series), 2, 3–4, 81, 169, 219–220
Intelsat III, 151–152
Intelsat IV, 14, 151–152
Intelsat IVA, 211
*Intelsat-IV F-2*, 202
*Intelsat-IV F-3*, 202
*Intelsat-IV F-4*, 202
*Intelsat-IV F-5*, 202
*Intelsat-IV F-7*, 146–147, 168, 202
*Intelsat-IV F-8*, 202, 221, 225–226, 227, 251, 257, 259
Interagency cooperation, 38, 126
  AEC
    -DOD, 23
  NASA
    -AEC, 23, 185
    -AEC-DOT, 185
    -AEC-National Geographic Society-USA, 197
    -American Assn. of Railroads-Federal Railroad Admin., 86
    -DOC, 200
    -DOD, 9, 21–22, 66–67
    -DOD-DOT-NAS-NCAR-NOAA-NSF-State Dept., 119
    -DOD-FAA, 148
    -DOI, 54, 73–74
    -DOI-NSF, 187
    -DOT, 54, 118
    -DOT-EPA-FAA, 17, 25
    -DOT-FAA-USN, 166
    -EPA, 25, 54
    -FCC, 202
    -GSA-HUD-NBS-NSF, 36
    -HEW, 129
    -HUD, 54, 187, 219, 226
    -Maritime Admin., 165
    -NOAA, 2, 12, 174, 200, 227, 236
    -NOAA-NESS, 236–237
    -NSF, 6, 73, 94, 105, 204–205
    -Office of Naval Research, 144
    -Smithsonian Institution, 164, 177
    -USAF, 52–53, 126, 163, 167, 192, 223, 224, 252, 273–274
    -USDA, 65
Interagency Panel for Terrestrial Applications of Solar Energy, 204
Intercontinental ballistic missile (ICBM; see also Minuteman and SS-19), 4–5, 14, 22, 155, 192, 204
Intercosmos (U.S.S.R. satellite series), 190, 222, 228
*Intercosmos 11*, 104, 237
*Intercosmos 12*, 195, 249
Intercosmos Council. See Soviet Academy of Sciences.
Interferometry, 214
Interim upper stage (IUS), 25, 66–67, 140, 178, 217
Interior, Dept. of (DOI; see also Earth Resources Observation Administration), 42–43, 54, 73–74, 145, 182, 187
International Academy of Astronautics (IAA), 178
International Aeronautical Federation (IAF), 178

International Assn. of Machinists and Aerospace Workers, 209
International Business Machines Corp. (IBM), 35–36, 81, 93, 105, 114, 140
International Communications Union, 8
International Congress of Electrical and Electronic Communications, 171
International cooperation (see also Disarmament, GARP Atlantic Tropical Experiment, Strategic Arms Limitation Talks, Summit Meeting, and Treaty), 6, 71–72, 105, 126, 132, 143, 200, 202–203, 227–228
France
 -Denmark, 170
 -Netherlands - Spain - U.K. - West Germany, 107
 -U.K., 64, 136–137
Japan
 -P.R.C., 89
U.S.
 -Brazil - East Germany - France - Mexico - Netherlands - U.K. - U.S.S.R. - West Germany, 119, 120
 -Canada, 101
 -Mexico, 62
 -U.S.S.R., 25, 80, 102, 123–124
International cooperation, space (see also Apollo-Soyuz Test Project, ESRO, GARP Atlantic Tropical Experiment, Memorandum of Understanding, and Strategic Arms Limitation Talks), 8, 38, 46, 64, 99, 103, 110, 118–119, 225, 227, 228
Canada
 -ESRO, 209
NASA
 -Canada, 4
 -Canada-France, 39
 -ESRO, 16, 36–37, 54, 77, 120–121, 184, 199, 210, 223–224, 225
 -ESRO-U.K., 77
 -France, 2, 213
 -France-West Germany, 121, 215–216, 227
 -India, 48, 139, 238
 -Iran, 194–195, 225
 -Italy, 2, 40, 111, 187–188, 225, 228, 230–231, 247, 258, 259
 -Japan, 225
 -Netherlands, 2, 159–160, 227, 244, 259
 -Norway, 265, 271
 -Spain, 101, 200, 201, 227, 250–251, 257, 259
 -Sweden, 265
 -U.K., 2, 57–58, 125, 187, 188, 203, 227, 228, 231–232, 247, 257, 258, 259
 -U.S.S.R., 4–5, 193, 225
 -West Germany, 2, 33–34, 57, 134–135, 212–213, 227, 242, 253, 259, 260
U.S.
 -Brazil, 38
 -Canada, 38, 139
 -Canada-ESRO, 77–78, 144
 -Denmark, 268–269
 -Europe, 126–127
 -Italy, 247
 -Peru, 72
 -Spain, 101
 -U.K., 9, 229, 257
 -U.S.S.R., 4–5, 10–11, 118, 121–122, 157, 191, 207–208, 216, 219–220, 225
 -West Germany, 134–135
U.S.S.R.
 -Bulgaria - Czechoslovakia - East Germany - Hungary - Poland - Romania, 249
 -Bulgaria - Czechoslovakia - East Germany - Hungary - Romania, 195
 -Czechoslovakia - East Germany, 104, 237
 -Czechoslovakia - Romania, 190
 -INTELSAT, 81
International Organizations Immunities Act, 9
International Space Hall of Fame, Alamogordo, N. Mex. (proposed), 29
International Sun-Earth Explorers program (ISEE; see also ISEE-A and ISEE-B), 77, 210
International Telecommunications Satellite Organization (INTELSAT; see also Intelsat satellites), 26, 77, 81, 151–152, 159, 169, 202
International Telephone & Telegraph Corp. (ITT), 3–4, 121
International Ultraviolet Explorer (IUE), 77, 209–210
Interplanetary dust, 198, 212
Interstellar clouds, 132
InterTechnology Corp., 19
Io (Jovian moon), 13, 93, 166–167, 208
Ion, 208, 264, 267
Ionization gauge, 145
Ionosphere (earth), 39–40, 72, 134–135, 145, 174, 195, 201, 242, 255, 259, 264, 267, 268–269
Iowa, Univ. of, 111, 238–239, 259
IPCS (integrated propulsion control system), 175
IR. See Infrared headings.
Iran, 7, 194–195, 225
Iridium, 83
Iron, 68, 82, 208–209
ISEE (International Sun-Earth Explorers program), 77, 210
ISEE-A, 210
ISEE-B, 210
Isla de Cedros, Mexico, 171
Isolation garment, 54
Isotope, 87
Israel, 11, 28
ISRO (Indian Space Research Organization), 189–190, 238
Italian Space Commission, 40

Italy (see also San Marco), 2, 26, 38, 54, 67, 101, 111, 173, 187, 194–195, 225, 258
ITOS-G (Improved Tiros Operational Satellite). See Noaa 4.
ITOS-H, 146
ITOS-I, 146
ITOS-J, 146
ITT (International Telephone & Telegraph Corp.), 3–4, 121
IUE (International Ultraviolet Explorer), 77, 209–210
IUS. See Interim upper stage.

## J

Jackson, Nelson P., Aerospace Award, 58–59
Jaffe, Leonard, 38, 46, 106–107, 178
JAL (Japan Airlines Co., Ltd.), 89
Jamaica, 14
Japan, (see also *Denpa*, MU-3C, MU-4C, *Ohsumi, Shinsei,* and *Tansei*), 3, 14, 26, 89, 99–100, 110, 132, 156, 174, 182, 225
Japan Airlines Co., Ltd. (JAL), 89
Japan Rocket Development Assn., 174
Jastrow, Dr. Robert, 116
JATO (Jet-assisted takeoff), 31
Javits, Sen. Jacob K., 4
Jaycees (Washington, D.C.), 118
Jenkins, Dr. Harriet G., 34, 143–144
Jet-assisted takeoff (JATO), 31
Jet Propulsion Laboratory (JPL, Cal Tech), 68, 101, 151
  applications research, 25, 32, 73, 175–176, 177
  space projects, 5, 14, 27, 107, 163, 173, 178–179, 208, 213
Johannesburg Tracking Station, South Africa, 77
Johns Hopkins Univ., 2, 264
Johnson, Clarence L., 199–200
Johnson, Howard W., 199
Johnson, President Lyndon B., 199–200
Johnson, Vincent L., 64, 157
Johnson Space Center (JSC, NASA; see also Shuttle Avionic Integration Laboratory), 13, 15, 59, 65–66, 144–145, 174
  Apollo Spacecraft Program Office, 121–122
  applications research, 62, 65, 187, 212
  ASTP support, 1, 7, 78, 79, 81–82, 102, 164, 165–170, 191, 192, 216
  awards, 88, 131
  Bioscience Payloads Office, 181
  conference, 64, 157–158
  employment, 210
  Flight Operations Aircraft Div., 185
  personnel, 95, 202, 220
  press conference, 30, 147, 155–156, 208
  Skylab Program Office, 203
  space shuttle support, 2, 9, 79, 89, 107, 165–166, 217

Johnston, Richard S., 157–158
Joint Committee on Cooperation in Atomic Energy (U.S.-U.S.S.R.), 25
Joint Spacelab Working Group (NASA–ESRO), 54
Jones, Gen. David C. (USAF), 103
Jones, J. Lloyd, 130
Jordan, 26
JPL. See Jet Propulsion Laboratory.
JSC. See Johnson Space Center.
JT8D (turbofan engine), 51, 145–146, 224
Juneau, Alaska, 4
Juno (launch vehicle), 171
Jupiter (missile), 171
Jupiter (planet; see also Callisto, Great red spot, Io, Mariner Jupiter-Saturn, *Pioneer 10,* and *Pioneer 11*), 13, 14, 32, 36–37, 51, 83, 87, 106, 120–121, 135, 150–151, 153, 169–170, 197–199, 208, 222, 223
Jupiter C (launch vehicle), 171
*The Jupiter Effect* (book), 169
Jupiter orbiter, 99–100
Justice, Dept. of, 169

## K

Kagoshima Space Center, Uchinoura, Japan, 39–40, 182, 230
Kaman Aerospace Corp., 204–205
Kansas, 189
Kansas, Univ. of, 59, 191
Kazakstan, U.S.S.R., 207, 255
Keldysh, Prof. Mstislav V., 165
Kennedy, President John F., 175
Kennedy International Airport, N.Y., 102, 224
Kennedy Space Center (KSC, NASA), 16, 35, 98, 120, 182
  agriculture research, 65
  ASTP support, 41, 88, 117, 151, 165, 194
  ceremonies, 88, 135
  construction of facilities, 64, 97, 210–211, 223–224
  Design Engineering Directorate, 93
  Earth Resources Office, 119
  employment, 35–36, 141, 220
  environmental research, 86, 140
  launch operations, 9, 33–34
  personnel, 171, 217
  Skylab Program Office, 203
  space shuttle support, 12, 64, 66–67, 75, 93
  support services, 16–17, 19, 117
  technology utilization, 100, 167–168
Kentucky, Univ. of, 129
Kenya, 26, 40, 187, 258
Kerr, Dr. Frank, 176
Kerslake, William R., 186
Kerwin, Cdr. Joseph P. (USN), 32, 42, 103, 195
Khozin, Grigory, 172
KIAS: knots indicated air speed.
Kibinge, Ambassador Leonard O., (Kenya), 26

Kidney disease, 157–158
Kilgore, Edwin C., 48, 53, 55–56, 66, 106, 213–214
Kirillin, Vladimir A., 165
Kiruna, Sweden, 265
Kissinger, Dr. Henry A., Secretary of State, 70, 105
Kitt Peak National Observatory, 214
Klebesadel, Dr. R. W., 90
Klimuk, Lt. Col. Pyotr I. (cosmonaut), 255
Kline, Raymond A., 148
Kliore, Dr. Arvydas J., 76
Kohler, Foy D., 172
Kohoutek, Dr. Lubos, 2
Kohoutek (comet), 2, 3, 5, 6, 9–10, 30, 65, 77, 90, 119, 203, 223, 264
Kondo, Dr. Yoji, 185
Kopp, Eugene P., 59
Korea, 14
Korolev, Sergey P., 52
Kourou Space Center, French Guiana, 196, 267
Kowal, Charles T., 167
Kozyrev, Nikolay A., 28,
Kraemer, Robert S., 36–37
Kraft, Dr. Christopher C., Jr., 32, 84–85
Kraft, Dr. Robert F., 143
Krieger, Robert L., 195
KSC. See Kennedy Space Center.
Kubasov, Valery N. (cosmonaut), 81–82

## L

L-1011 (TriStar, jet transport), 10, 111–112
L-band frequency, 77–78, 110, 165
L-band transistor amplifier, 49
LACATE. See Lower Atmospheric Composition and Temperature Experiment.
Lacerta (constellation), 76
Lacey, Lewis L., 184
LAGEOS (Laser Geodynamic Satellite), 164
Lake Victoria, 109, 238
Laminates, 184
LAMPS (Light airborne multipurpose system helicopter), 81
Lancaster, Calif., 189
Land, Elwood W., Jr., 113
Land management, 119, 131, 132
Langley Research Center (LaRC, NASA)
  Administrative Standardization Office, 158–159
  aeronautics, 96, 97, 123, 187, 273
  atmospheric research, 98
  employment, 220
  energy research, 34, 73
  launch vehicle management, 111, 160
  personnel, 213–214
  satellite, 111
LaRC. See Langley Research Center.
Large Radio Astronomy Observatory, 37–38
Large Solar Observatory, 37–38

Large Space Telescope (LST), 21, 37–38, 77, 99–100, 118–119, 124, 127, 150, 201, 220, 223
Larson, Clarence E., 25
Las Cruces, N. Mex., 197
Laser, 47, 102, 121, 131, 140, 151, 154, 177, 184
Laser Geodynamic Satellite (LAGEOS), 164
Launch, DOD (see also Appendix A), 221, 227, 228
  Nts 1, 132–133, 241–242
  Redstone missile, 171
  S3-1 satellite, 145, 194, 228, 248
  unidentified satellite, 36, 63, 194
Launch, NASA, 2, 221–222, 227, 228
  Aeros 2, 134–135, 242, 257, 259
  Amsat Oscar 7, 200, 250, 257, 259
  Ans 1, 159, 244, 257, 259
  Ats 6, 108–109, 238, 257, 258
  failure, 33–34, 222, 229, 257, 258
  Hawkeye 1, 2, 111, 238–239, 257, 259
  Helios 1, 212, 213, 253, 257, 260
  Intasat, 200–201, 250–251, 257, 259
  Intelsat-IV F-8, 202, 251, 257, 259
  Miranda, 57–58, 231–232, 257, 258
  Noaa 4, 200–201, 249–250, 257, 259
  San Marco 4, 40, 230–231, 257, 258
  Skynet IIA, 9, 229, 257, 258
  Skynet IIB, 203, 252, 257, 260
  Sms 1, 103–104, 236–237, 257, 258
  sounding rockets (see also names of individual sounding rockets), 263–271
  Symphonie 1, 215–216, 254, 257, 260
  Westar 1, 85–86, 233, 257, 258
  Westar 2, 185–186, 246, 257, 259
Launch processing system (LPS), 12
Launch vehicle (see also individual launch vehicles), 4–5, 12, 21, 53, 64–65, 78, 113, 116, 135, 168–169
Lava, 82, 215
Law enforcement, 29
Lawrence Livermore Laboratory, 56
Lazarev, Lt. Col. Vasiliy G. (cosmonaut), 255
Lead, 168
Lear 4 mission, 177
Learjet (executive transport aircraft), 160, 177
Leary, Dr. William M., 15
Le Bourget Airfield, 156
Lee, Chester M., 44
Lenin Prize (U.S.S.R.), 52
Leningrad, U.S.S.R., 65, 102, 105–106
Leonov, Lt. Col. Aleksey A. (cosmonaut), 36, 81–82, 178
LeRC. See Lewis Research Center.
LES (Lincoln Experimental Satellite), 23
LES 8, 67
LES 9, 67
Levin, S. Benedict, 171
Levy, Maurice M., 213
Lew, Dr. Hin, 9
Lewis Research Center (LeRC, NASA; see also Plum Brook Station), 210
  aeronautics, 51, 92, 145–146

automotive research, 25
conference, 166
employment, 66, 220
energy research, 29, 34, 61, 73–74, 187, 204–205, 226
environmental research, 168, 213
establishment, 156
launch vehicle management, 33–34, 213
materials research, 173, 174
propulsion technology, 186
Library of Congress, 4
Lick Observatory, 9–10
Life, origin of, 11, 132
Life raft, 151
Life sciences (see also Space biology and medicine), 21, 28–29, 37, 43–44, 97, 131, 134, 140–141
Life support system, 176, 210
Life vest, 151
Lifting body. See HL–10, M2–F3, X–24A, and X–24B.
Light airborne multipurpose system (LAMPS) helicopter, 81
"Light Blue." See *Tansei 2*.
*Light Heart* (balloon), 40–41
Lighter-than-air vehicles, 166
Lighter-than-Air Workshop, 166
*Lightning*. See P–38.
Lightning, 151
Lightweight fighter prototype program, 31, 79, 91, 94, 116
Lilly, William E., 24, 209
Lincoln Experimental Satellite (LES), 23, 67
Lindbergh, B/G Charles A. (U.S. Army Air Corps, Ret.), 156–157
Lindstrom, Robert E., 69
Liquid-fueled rocket, 52
Liquid-oxygen pressurization system, 148–149
Liquid oxygen and hydrogen rocket engine, 15
Lockheed Aircraft Corp., 7–8, 10, 15–16, 111–112, 130, 151, 163, 192, 220
  Lockheed–California Co., 168
  Lockheed Missiles & Space Co., 201, 266–267
  Lockheed Propulsion Co., 5, 36, 126
Loening, Albert P., 168
Loening amphibian aircraft, 168
Logan International Airport, Boston, Mass., 118
Loki (sounding rocket), 65
London, England, 10, 51, 101, 107, 163
Long-haul technology, 48
Long-range radio-control system, 151
Loomis, Henry, 106–107
Lord, Douglas R., 54
Lorenzini, Dino A., 15
Los Alamos Scientific Laboratory, 270
Los Angeles, Calif., 134, 148, 163, 164, 259
Lousma, L/C Jack R. (USMC), 28, 121–122
Love, Maj. Michael V., 274
Lovelace, Dr. Alan M., 15, 157

Lovell, Capt. James A., Jr. (USN, Ret.), 35
Low, Dr. George M., 8–9, 11, 25, 28, 53, 135, 190, 200, 211
Low-emission combustor, 26
Lower Atmospheric Composition and Temperature Experiment (LACATE), 98, 226
Lower mesosphere (earth), 266
Loweth, Hugh T., 182
Lox (liquid oxygen). See Oxygen.
LPS (launch processing system), 12
LRV (lunar roving vehicle), 37
LST. See Large Space Telescope.
Lubbock, Tex., 197
Lucas, Dr. William R., 53, 120, 199
Luftwaffe, 156
Luna (U.S.S.R. lunar probe), 222, 228
*Luna 16*, 72–73
*Luna 22*, 108, 223, 237
*Luna 23*, 193–194, 223, 248
Lunar agriculture, 154
Lunar data analysis and synthesis program, 223
Lunar exploration (see also individual manned missions, such as *Apollo 11* and *Luna 16*), 21, 32, 36, 37, 81, 85, 132, 135–136, 154, 161
*Lunar Orbiter 5*, 215
Lunar orbiter program, 217
Lunar roving vehicle (LRV), 37
Lunar sample analysis, 21
Lunar Science Conference, Fifth Annual, 64
Lunar tide, 45
Lundin, Bruce T., 26, 66
Lunney, Glynn S., 82, 118, 157, 207–208
L'vov, Ukraine, 81
Lyman-alpha radiation, 3, 5
Lynn, James T., Secretary of Housing and Urban Development, 219

# M

M2–F3 (lifting body), 273
M–X (Japanese launch vehicle), 174
McCall, Robert T., 45
McConnell, Dr. Dudley G., 6–7, 11–12, 34, 143–144
McCormack, Rep. Mike, 46
MacDonald Observatory, Fort Davis, Tex., 45
McDonnell Douglas Corp., 51, 64, 71, 76, 81, 83, 134, 140, 196, 205
  McDonnell Douglas Astronautics Co., 7–8, 41–42, 217
McFall, Russell W., 134
McLucas, John L., Secretary of the Air Force, 183
Madrid Space Communications Complex, Spain, 101
MAF. See Michoud Assembly Facility.
Magnesium, 100
Magnet, 29
"Magnetic domain bubbles," 143

Magnetic field (earth), 48, 77, 103, 111, 114
Magnetic storm, 40, 114
Magnetohydrodynamics, 187
Magnetometer, 54, 208–209, 253
Magnetopause, 12–13
Magnetosphere (earth). See Atmosphere.
Magnetotail, 149
Makarov, Oleg G. (cosmonaut), 255
Malaysia, 78–79, 169
Malkin, Dr. Myron S., 113
Malloy, Dr. James A., Jr., 59
MAMA (Mobile automatic metabolic analyzer), 13
Manching, West Germany, 173
Manke, John A., 52–53, 274
Manned space flight (see also Apollo, Apollo-Soyuz Test Project, Appendix A, Appendix C, Astronaut, Cosmonaut, Gemini, Mercury, Salyut, Skylab, Soyuz, Space biology and medicine, Space shuttle, and Spacelab), 137, 221, 228
  anniversary, 85, 137–138
  award, 137
  budget, 21, 24, 27, 84, 109, 160–161
  congressional consideration, 28, 41, 42–44, 47
  man-hours in space, 261
  medical aspects, 34–35, 42, 43–44, 97
  press comment, 31, 32, 51, 69–70, 85
  technology utilization, 95
  U.S.S.R., 152, 161
Manned Space Flight Network (MSFN), 176
Mapping, 38–39, 42, 55, 61–62, 65, 114, 130, 168, 169–170, 215
Maran, Dr. Stephen P., 2
Marathon Construction Co., 136
Marine Corps Air Station, Calif., 60
Marine resources, 131–132
Mariner (spacecraft), 23, 34, 54
*Mariner 4*, 87
*Mariner 9*, 70
*Mariner 10* (Venus-Mercury probe), 14, 72, 130, 151
  data transmission, 5, 68, 71–72, 82, 173
  mission profile, 5, 26–27, 63, 222–223
  press comment, 37, 75
  press conference, 16, 27, 69
Mariner Jupiter-Saturn 1977 (MJS '77) mission, 21, 23, 25, 212, 223
Mariner Jupiter-Uranus 1979 mission, 99–100
Marisat (maritime comsat), 2, 67, 114
Maritime Administration, 165
Maritime communications satellite (Marisat) system, 2, 67, 114
Maritime industry, 147, 238
Maritime Orbital Test Satellite (MAROTS), 49, 210
Mark 500 (nuclear warhead), 151
MAROTS. See Maritime Orbital Test Satellite.

Mars (planet; see also individual probes, such as *Mars 3*, and Viking program), 13, 32, 33, 34–35, 36–37, 43, 70, 87, 94, 112, 135, 154, 169–170
*Mars 3*, (U.S.S.R. Mars probe), 60
*Mars 4*, 33, 60, 70
*Mars 5*, 33, 60, 70
*Mars 6*, 33, 59, 60, 70
*Mars 7*, 33, 59–60, 70
Marshall Space Flight Center (MSFC, NASA), 19, 41–42, 88, 119, 174
  aeronautics, 102
  Apollo-Soyuz Test Project support, 118, 216
  astronomy, 93, 158, 170
  conference, 118–119, 164, 184
  construction of facilities, 21, 176–177, 189, 220
  employment, 25, 90–91, 115–116, 210, 220
  energy research, 6, 73, 91, 111, 114, 175, 192, 226
  meteorology, 101–102
  personnel, 1, 53–54, 69, 120, 199
  Skylab, 58, 80, 139, 203
  space shuttle support, 36, 121, 140, 150, 184
  space transportation system, 7–8, 34, 47, 75, 88, 103, 217
  Spacelab support, 177, 184, 185, 205
  technology utilization, 10, 13, 143, 167
  transportation research, 86
Martin Marietta Corp., 1, 7–8, 86, 140, 158, 165, 172, 201, 220
Maryland, 130
Maryland, Univ. of, 45–46
Mascon, 72–73, 215
Mass spectrometer, 13, 145, 231, 242
Massachusetts, 168
Massachusetts Institute of Technology (MIT; see also Flight Transportation Laboratory), 126, 199, 270–271
Materials research, 58, 91, 214
  adhesives, 184
  aircraft, 48, 91, 99, 192
  alloys, 173, 184
  coatings, 111, 216
  crystals, 43, 115, 199
  solar energy research, 90–91, 105, 178
Mathews, Charles W., 47–48, 61–62, 182
Mattingly, Cdr. Thomas K. II, (USN), 28
Mauna Kea, Hawaii, 24, 39, 84
Max Planck Institute for Extraterrestrial Physics, West Germany, 191
Mayall Telescope, 214
Mayflower Hotel, Washington, D.C., 211
MCI Communications Corp., 130
Mecca, William R., 209–210
Medal of Freedom, 31
Medal of Honor, 156
Medicine (see also Space biology and medicine)
  aviation, 22
  space manufacturing, 43
  technology utilization, 29, 54, 68, 95, 100, 143, 144–145, 226

telecommunications, 48, 106–107, 109, 139, 228, 238, 258
Mediterranean Sea, 63–64
Memorandum of Understanding, 58, 77–78, 101, 111, 144, 160, 188, 194–195, 204, 209, 211, 213
Mercure (European short-range transport), 38
Mercury (metal), 34
Mercury (planet; see also *Mariner 10*), 14, 26, 29, 36–37, 87, 222
Mercury (program), 28, 211
Mercury Redstone 2, 136
MESH consortium, 210
Messerschmitt-Boelkow-Blohm GmbH, 54, 113, 169
Messier 13 (star cluster), 201–202
Messmer, Premier Pierre (France), 38
Meteor, 237
Meteor (U.S.S.R. meteorological satellite), 120, 222, 228
*Meteor 16*, 52, 231
*Meteor 17*, 92, 235
*Meteor 18*, 131–132, 241
*Meteor 19*, 194, 248
*Meteor 20*, 215, 253
Meteorite, 11, 208, 215
Meteoroid, 60, 65, 152–153
Meteoroid detector, 42, 60
Meteoroid Technology Satellite (MTS). See *Explorer 46*.
Meteorological satellite (see also individual satellites, systems, and programs, such as Defense Meteorological Satellite Program, *Goes 1*, ITOS–H, *Meteor 16*, National Operational Meteorological Satellite System, Nimbus, *Noaa 2*, SEASAT, *Sms 1*, and Tiros–N), 28, 85, 101–102, 110, 154, 174, 228
Meteorology (see also Atmospheric Layering and Density Distribution of Ions and Neutrals project, Climatic Impact Assessment Program, Global Atmospheric Research Program, GARP Atlantic Tropical Experiment, and Weather and Climate program), 24, 109, 132, 183, 266–267
 budget, 21, 109
 international cooperation, 105
 observation studies, 65, 101–102, 188, 213, 236–237, 253
 program proposal, 146
 weather forecasting, 22, 154, 215, 231, 235, 248
 weather modification, 41–42, 143
Methane, 98
Methyl cyanide, 77, 118–119
Methylamine, 132
Metzenbaum, Sen. Howard M., 99, 199
mev: million electron volts.
Mexican National Commission for Outer Space, 62
Mexico, 14, 42, 62, 119
Mexico City, 171
mhz: megahertz (1 million cycles per second). .

Miami, Fla., 174–175
Miami, Univ. of, 140
Michigan, Univ. of, 266, 268–269
Michoud Assembly Facility (MAF, NASA), 10, 19, 88, 158
Microdensitometer, 100
Microdot, 58
Microelectronics, 48
Micrometeorite, 195, 249
Microphone, 150
Microphone meteoroid sensors, 60
Microwave energy, 115, 178–179
Microwave frequency, 56, 61–62, 73
Microwave radiometry, 102
Middle East War. See Arab-Israeli War.
Midway Island, 14
Milky Way (galaxy), 76, 90, 132, 201–202, 214
Miller, G. William, 111–112
Mini-Sniffer (radio-controlled aircraft), 151
Mining, 129
Ministry of Defence, U.K., 203
Minneapolis, Minn., 147
Minnesota, Univ. of, 266–267
Minorities, 6–7, 11, 102, 126, 202
Minuteman (intercontinental ballistic missile), 22, 57
Minuteman I, 192
Mirage 3, (F–1; French supersonic fighter aircraft), 138–139
*Miranda* (UK–X4; U.K. satellite), 2, 57–58, 221–222, 227, 231–232, 257, 259
Mirror, 189
MIRV. See Multiple independently targetable reentry vehicles.
Missile (see also individual missiles, kinds of missiles, and missile systems), 4–5, 22, 28, 51–52, 62–63, 70, 81, 151, 155, 157, 163, 193
Missile Definition Group, 120–121
Mississippi Test Facility (MTF, NASA; see also National Space Technology Laboratories), 10, 31
Missouri, Univ. at Rolla, 136
MIT. See Massachusetts Institute of Technology.
MIUS (modular integrated utility system), 187
MJS'77. See Mariner Jupiter-Saturn 1977 mission.
Mlavsky, Dr. A. I., 105
Mobile automatic metabolic analyzer (MAMA), 13
Modular integrated utility system (MIUS), 187
Molecular science, 118–119, 154
Molniya (U.S.S.R. comsat series), 81, 219–220, 222, 228
*Molniya I–27*, 88–89, 233–234
*Molniya I–28*, 192, 247
*Molniya I–S*, 141, 243
*Molniya II–9*, 93, 235
*Molniya II–10*, 139, 242
*Molniya II–11*, 217, 254
*Molniya III–1*, 202–203, 251

Monaco, 59
Monopropellant, 31
Montana, 170
Monterey, Calif., 166
Moon (see also lunar headings, individual lunar missions, Sea of Crisis, and Taurus-Littrow), 13, 32, 36, 87, 98, 208–209, 212, 215
Moonquakes, 208–209
Moorer, Adm. Thomas H. (USN), 103
Morgantown, W.Va., 89–90
Moritz, Bernard, 53
Morocco, 41
Moroz, Vasily I., 70
Morrison, Dr. Dennis R., 181
Morrison, Dr. Phillip, 90
Morse, Samuel F. B., 134
Mosaic, 90
Moscow, U.S.S.R., 4–5, 14, 70, 80, 81, 122, 123–124, 157, 164, 165, 191, 219–220
Moss, Sen. Frank E., 15, 62, 73, 139, 204
Mossinghoff, Gerald J., 1, 102
Motion sickness, 42,
Mount Palomar, Calif., 167
Mount Wilson, Calif., 168
MRCA (multirole combat aircraft), 173
MSFC. See Marshall Space Flight Center.
MSFN (Manned Space Flight Network), 176
MS-T1 *(Tansei 1)*, 39–40
MS-T2. See *Tansei 2.*
MTF (Mississippi Test Facility), 10, 31, 119
MTS (Meteoroid Technology Satellite). See *Explorer 46.*
MU–3C (Japanese launch vehicle), 39–40
MU–4S, 39–40
Mulligan, Dr. J. H., Jr., 129
Mulready, Richard C., 15
Multiple independently targetable reentry vehicle (MIRV), 14, 22, 42, 70, 140
Multirole combat aircraft (MRCA), 173
Multispectral-image-analysis system, 98
Multispectral scanner, 38–39, 65, 90
Munich, West Germany, 191
Murray, Dr. Bruce C., 82
Musgrave, Dr. Franklin Story, 181
Mutch, Dr. Thomas A., 112
Myers, Dale D., 39, 41, 47, 88, 101

# N

N–2 (Japanese launch vehicle), 174
NAA (National Aeronautic Assn.), 95, 173
NAACP (National Assn. for the Advancement of Colored People), 202
NACA. See National Advisory Committee for Aeronautics.
NAE. See National Academy of Engineering.
Nani, India, 63
NAS. See National Academy of Sciences.
NASA. See National Aeronautics and Space Administration.
NASA Aircraft Office, 158–159
NASA Astronaut Office, 149
*NASA Energy-Related Research and Development* (report), 73–74
NASA Headquarters, 89, 90, 119, 120, 148–149, 181, 184, 211, 213, 220
  Apollo Program Office, 217
  personnel, 95, 149, 213–214
  Skylab Program Office, 203
NASA Office for Center Operations, 53, 106, 148, 158–159
NASA Office of Aeronautics and Space Technology (OAST), 48, 53, 56, 66, 106, 130, 157, 213–214
NASA Office of Applications (OA), 6, 104, 143–144, 145
NASA Office of Automatic Data Processing Management, 148
NASA Office of DOD and Interagency Affairs, 157, 170
NASA Office of Earth Resources Survey Systems (proposed), 145, 171
NASA Office of Energy Programs, 102, 175, 226
NASA Office of Equal Opportunity Programs, 6–7, 11–12, 34, 143–144, 150
NASA Office of General Council, 1
NASA Office of Legislative Affairs, 1, 102
NASA Office of Manned Space Flight (OMSF), 39, 47, 101, 136, 211
NASA Office of Personnel, 126
NASA Office of Planetary Programs, 36–37
NASA Office of Public Affairs for Community and Human Relations, 150
NASA Office of Space Science (OSS), 33–34, 57, 111, 124, 157, 160, 169–170
NASA Office of Technology Applications (proposed), 60–61
NASA Office of Tracking and Data Acquisition (OTDA), 62
NASA Office of User Affairs, 95
NASA Spacelab Program Office, 54
NASA–University Conference on Aeronautics, 191
Nash, Carl E., 117
NATE (neutral-atmosphere temperature experiment), 242
National Academy of Engineering (NAE), 86, 97, 129
National Academy of Sciences (NAS), 86, 92, 119, 124–125, 126–127, 129, 140–141, 176
National Advisory Committee for Aeronautics (NACA), 70–71, 93, 156, 176–177, 211, 219
National Aeronautic Assn. (NAA), 95, 173
National Aeronautics and Space Act of 1958, 60–61, 62, 78, 163–164

National Aeronautics and Space Administration (NASA; see also individual Centers, offices, and programs; Awards; Budget, NASA; Construction of facilities, NASA; Contract, NASA; Employment, NASA; Interagency cooperation; International cooperation; International cooperation, space; Launch, NASA; and Personnel, NASA), 8–9, 177–178, 227
National aerospace fellowship program, 102–103
National Assn. for the Advancement of Colored People (NAACP), 202
National Astronomy and Ionosphere Center, 201–202
National Bureau of Standards (NBS), 36, 70–71
National Center for Atmospheric Research (NCAR), 119, 185
National Civil Service League Career Service Award, 157, 169–170
National Civil Service League Special Achievement Award, 169–170
National Commission for Outer Space, Mexico, 62
National Defense Research Council, 70–71
National Environmental Satellite Service (NESS, NOAA), 200–201, 249, 259
National Geographic Society, 45, 183, 197
National Maritime Research Center, 165
National Oceanic and Atmospheric Administration (NOAA; see also National Environmental Satellite Service, National Operational Meteorological Satellite System, and *Noaa* series)
 budget, 22
 cooperation, 2, 12, 28, 66, 119, 146, 200, 213, 249
 *Meteorological Services Research for 1975*, 146
 research, 45, 114, 213
 satellites, 103–104, 114, 146, 183, 200, 249, 258
 solar observatory, 174
National Operational Meteorological Satellite System (NOMSS), 146, 236–237, 249
National Radio Astronomical Observatory, 118–119
National Research Council, Canada, 9–10, 39
National Research Council, France, 39
National Research Council, U.S., 129
National Resources Council (proposed), 60–61
National Science Foundation (NSF), 119, 134, 201–202
 budget, 6, 23, 164
 conference, 175–176
 interagency cooperation, 6, 94, 204
 personnel, 136, 182
 report, 48–49, 72, 143, 160–161, 190

solar energy research, 19, 36, 46, 56–57, 105, 163–164, 187, 204–205
 wind energy research, 6, 34, 63, 66, 73
National Scientific Balloon Facility, 185
National Security Council, 124–125
National Severe Storms Laboratory, 101–102
National Solar Energy Program, 204
National Space Club (NSC), 37, 58, 87
National Space Week, 132, 135–136
National Space Technology Laboratories (NSTL, NASA), 119
National Transportation Safety Board, 51, 71
National Weather Service, 101–102
Nationalist China. See China, Republic of.
"The Nation's Energy Future" (AEC report), 107
NATO. See North Atlantic Treaty Organization.
Natural gas, 11, 20, 129, 154
Naugle, Dr. John E., 53, 57, 77, 87, 112–113, 124, 169–170
Naumova, Yelena, 171
Navajo (Piper aircraft), 97
Naval Research Laboratory (NRL), 3, 57, 121, 132–133, 160, 241–242, 264, 269, 270
Naval Weapons Center, 151
Navigation, 38, 55, 63, 67, 109, 121, 176, 249–251, 253
Navigation satellite (see also NAVSTAR Global Positioning System), 21–22, 37, 209
Navigation System Using Time and Ranging (NAVSTAR). See NAVSTAR Global Positioning System.
Navigational Technology Satellite (NTS), 209
NAVSTAR Global Positioning System (GPS), 21–22, 67, 121, 132–133, 160, 190, 209, 226, 228, 241–242
Navy. See U.S. Navy.
Navy Astronaut Wings, 28
Navy Distinguished Service Medal, 28
NBS (National Bureau of Standards), 36, 70–71
NCAR (National Center for Atmospheric Research), 119, 185
Nebula. See Orion.
NEC (Nuclear Energy Commission), 83
Neon, 27, 68
Neoprene, 91
Neptune (planet), 37, 45, 135, 222
NERVA (nuclear engine for rocket vehicle application program), 184
NESS. See National Environmental Satellite Service.
Netherlands (see also *Ans 1*), 16, 54, 67, 101, 107, 119, 138–139, 159–160, 244
Netherlands Astronomical Satellite (ANS). See *Ans 1*.
Neutral-atmosphere temperature experiment (NATE), 242

Neutral mass spectrometer experiment, 231
Neutron, 168
Neutron star, 182
New Jersey, 168
New Mexico, 29
New York Academy of Sciences, 56
New York, N.Y., 4, 8, 13, 80, 134, 156, 163, 164, 165
Newfoundland Tracking Station, Canada, 77
Ney, Dr. Edward P., 90
NHK Technical Research Laboratory, 165–166
Nicaragua, 14
Nickel, 58, 173
Nickel-cadmium battery, 250–251
Nickel-iron, 98
Nike-Apache (sounding rocket), 264, 265, 266, 267, 268–269
Nike-Cajun, 266
Nike-Iroquois, 126
Nike-Malemute, 267, 270
Nike-Tomahawk, 264–265, 267, 268–269, 270–271
Nimbus (meteorological satellite), 146
*Nimbus 5*, 21, 120, 188
Nimbus-F, 2, 21, 188
Nimbus-G, 98, 188, 204, 226
"The 1973 NASA Payload Model" (study), 126
Nippon Electric Co., 169
Nitric acid, 98
Nitric oxide, 188, 265, 271
Nitrogen, 5, 13, 40, 79–80, 151, 157–158
Nitrogen dioxide, 98, 266
Nitrogen oxide, 26, 86
Nitrous oxide, 98
Nixon, President Richard M., 9, 65–66, 72, 130, 136, 143
 agreements, 25
 appointments and nominations by, 88, 103, 108, 137
 bills signed, 99, 121, 132
 budget request, 20
 committee establishment, 119
 messages and statements, 11, 20, 30, 72, 80
 press comment, 129, 153–154
 reports, 72, 107, 124–125, 146–147
 resignation, 146
 resolutions, 132
 speeches, 65–66
 summit meeting, 123–124, 241
NOA: new obligational authority.
NOAA. See National Oceanic and Atmospheric Administration.
*Noaa 2* (ITOS-D; meteorological satellite), 62, 120, 200
*Noaa 3* (ITOS-F), 62, 120, 183, 200
*Noaa 4* (ITOS-G), 2, 200, 221–222, 225–226, 227, 249–251, 257, 259
Nobel Prize, 188–189
Noise, aircraft, 17, 23, 25, 48, 51, 84, 99–100, 123, 174, 189, 224
Noise reduction, 150

Nome, Alaska, 4
NOMSS. See National Operational Meteorological Satellite System.
Noordwijk, The Netherlands, 199
NORAD. See North American Air Defense Command.
Nordberg, Dr. William, 54
North America, 65
North American Air Defense Command (NORAD), 2–3, 174
North Atlantic Treaty Organization (NATO), 123–124, 138–139
North Carolina, 13
Northrop Corp., 7, 10, 79, 84, 116, 138, 146, 183–184
Norway, 138–139, 210, 265, 271
Nott, Julian, 13–14
Nova, 31
NRL. See Naval Research Laboratory.
NSC. See National Space Club.
NSF. See National Science Foundation.
NSTL (National Space Technology Laboratories), 119
NTS (Navigational Technology Satellite), 209
*Nts 1* (Timation 3A), 132–133, 228, 241–242
Nu-life (medical device), 68
Nuclear energy, 11, 14, 20, 21, 23, 25, 29, 56, 84, 107, 118, 143, 154, 182, 184–185
Nuclear Energy Commission (NEC), 83
Nuclear engine for rocket vehicle application (NERVA program), 184
Nuclear microbomb, 179
Nuclear-powered shuttle, 115
Nuclear reactor, 20, 23, 25, 29, 56, 105–106, 107, 184–185
Nuclear Regulatory Commission, 187, 194
Nuclear Test Ban Treaty (1963), 90, 179
Nuclear weapons, 105, 123–124, 151, 177–178, 185
Nutrition, 131
Nylon, 91

# O

O-model (Japanese launch vehicle), 174
OA. See NASA Office of Applications.
OAO (Orbiting Astronomical Observatory), 21
*Oao 3* (*Copernicus*), 133, 187
Oasis experiments, 44, 218–219
OAST. See NASA Office of Aeronautics and Space Technology.
Observatory (see also individual observatories), 169–170, 174
Oceanography, 38, 188
 ocean dynamics, 24, 45, 119, 131, 153
 satellite applications, 102, 104, 182
Oceans and International Environmental and Scientific Affairs (State Dept.), 194
O'Dell, Dr. Charles R., 118–119
Oertel, Dr. Goetz K., 43–44

Office of Management and Budget (OMB), 24, 27, 28, 80
Office of Naval Research (ONR), 144
Office of Science and Technology (OST, President's), 125
Office of Telecommunications Policy (White House), 86
Ogiso, Ambassador Motoo (Japan), 26
Ogorodnikov, Kiriol, 171
Ohio, 182, 199
*Ohsumi* (Japanese satellite), 39–40
Oil, 10, 54, 129, 130, 132, 154
Oil shortage, 20, 45
O'Leary, P. J., 178
OMB. See Office of Management and Budget.
Omega Centauri (star cluster), 171
Omega-inertial navigation system, 120
Omegatron, 40
Omnidirectional wheel, 81
OMSF. See NASA Office of Manned Space Flight.
O'Neill, Dr. Gerald K., 98
ONR (Office of Naval Research), 144
Oort cloud, 90
OOS (orbit-to-orbit stage). See Interim upper stage.
Operation Kohoutek. See Kohoutek.
Optical Society of America, 90, 92
Orbit-to-orbit stage (OOS). See Interim upper stage.
Orbita network (U.S.S.R. comsat system), 88–89, 93, 139, 202–203, 217, 233–234, 235, 242, 247, 254
Orbital Workshop (OWS). See *Skylab 1*.
Orbiter, shuttle, 47, 79, 107
Orbiting Astronomical Observatory (OAO; see also *Oao 3*), 21
Orbiting Explorers, 21
Orbiting Satellite Carrying Amateur Radio (OSCAR). See *Amsat Oscar 7*, *Oscar 1*, and *Oscar 6*.
Orbiting Solar Observatory (OSO; see also OSO-1), 21
Orbiting Test Satellite (OTS), 49, 169
Order of Lenin (U.S.S.R.), 52, 181
Order of the Red Banner of Labor (U.S.S.R.), 52
Ordway, Frederick, I., III, 15
Orfila, Ambassador Alejandro (Argentina), 26
Orion (constellation), 90, 168–169, 214, 247, 270
*Orion* (NASA barge), 88
*Orion 2* (space station telescope), 44, 73, 239–240
Orly Airport, France, 51
Oro, Dr. John, 140–141
*Oscar 1* (Orbiting Satellite Carrying Amateur Radio), 201
*Oscar 6*, 201
*Oscar 7*. See *Amsat Oscar 7*.
OSO (Orbiting Solar Observatory), 21
OSO-1, 25, 77
OSS. See NASA Office of Space Science.
Osseo, Minn., 19

OST (Office of Science and Technology, President's), 125
OTDA. See NASA Office of Tracking and Data Acquisition.
OTS (Orbiting Test Satellite), 49, 169
Outer Planet Probe Technology Workshop, 106
Outer Space Act of 1967, 46
"Outlook for Aeronautics" (study), 148, 191–192
"Outlook for Space" (study), 126, 177–178
Outstanding Leadership Medal (NASA), 203, 213–214
Overmyer, L/C Robert F. (USMC), 121–122
OWS (Orbital Workshop). See *Skylab 1*.
Oxygen, 2, 3, 13, 15, 27, 33–34, 35, 75–76, 89, 148–149, 150
Ozone, 12, 59, 98, 188, 265, 266

## P

P-35 (fighter aircraft), 154–155
P-38 (*Lightning*; fighter aircraft), 199–200
P-80 (jet fighter), 199–200
Pacific Ocean, 9, 14, 42, 202, 217–218, 236–237, 251, 258
Paige, Dr. Lowell J., 182
Paine, Dr. Thomas O., 136, 194
Palestine, Tex., 185
Palmdale, Calif., 7, 193
Pan American World Airways, Inc., 12, 74, 79, 100–101, 157
Parachute, 65, 117, 176, 210
Paris, France, 12, 36–37, 51, 71, 78, 80, 105–106, 107, 113, 118, 136–137, 144, 156, 189
Parker, Dr. Robert A., 149, 203
Patent, 91, 95–96
Patterson, Dr. William J., 184
PCOM (Pioneer command system), 107
Peenemuende, West Germany, 1
Peking, P.R.C., 89
Pendray, G. Edward, Award, 15
Pennsylvania State Univ., 264, 267, 268, 270
Pentagon, 28
People's Republic of China (P.R.C.), 51–52, 89, 105
Pequet, Henri, 63
Permafrost, 94, 134
Pershing (missile), 171
Personnel, NASA
  appointment, 1, 15, 34, 47, 53–54, 69, 95, 97, 101, 102, 106, 120, 124, 130, 136, 143–144, 148, 150, 157, 158–159, 170, 199, 213–214, 217
  awards and honors, 14, 15, 28, 39, 59, 65–66, 88, 95, 112, 148, 169–170, 171, 193, 195, 203, 211, 213–214
  dismissal, 6–7
  resignation, 39, 176–177
  retirement, 136, 157, 228, 170, 171, 213–214, 217, 220

Peru, 72
Petrone, Dr. Rocco A., 13, 15, 53–54, 68, 120, 182, 203
Petrosyants, Andronik M., 25
Petrov, Dr. Boris N., 85, 165
Phantom fighter bomber aircraft. See F-4.
Philadelphia, Pa., 168
Philippines, 14, 78–79, 169
Phoenix (air-to-air missile), 7
Phoenix, Ariz., 42
Phosphorus, 157–158
Photography (see also Infrared photography and Ultraviolet photography), 5–6, 32, 38, 68, 69, 100, 101, 104–105, 119, 121, 173
Photospheric field, 67
Photovoltaic technology, 114–115
Physical therapy, 103
Physics, 21, 38
Physiology, 43–44, 178
Pickering, Dr. William H., 26
Pilot, 52–53, 63, 96, 118, 148, 151, 152, 154–155, 156–157
Pilotless Aircraft Research Center. See Wallops Flight Center.
Pioneer (program), 14, 21, 24, 29, 121
Pioneer (spacecraft), 13, 19, 120–121
*Pioneer 10* (Pioneer-F; interplanetary probe), 58–59, 88, 107, 120–121, 150–151
  mission status, 2, 42, 65, 197
  photographs, 82–83
  press comment, 3, 13
  results, 12–13, 76, 82–83, 93, 166–167, 199, 222
*Pioneer 11* (*Pioneer–Saturn*; Pioneer-G), 2, 42, 65, 88, 107, 153, 166–167, 197–199, 210, 222
Pioneer command (PCOM) system, 107
*Pioneer–Saturn*. See *Pioneer 11*.
Pioneer–Venus mission, 19, 21, 24, 204, 223
Piper Seneca (research aircraft), 187
Pittsburgh, Univ. of, 265–266, 268
Plagemann, Stephen, 169
Planetary evolution, 118–119
Planetary exploration (see also individual planets and probes), 21, 24, 32, 34, 36–38, 44, 87, 106, 154, 179, 214, 222–223
Planning Research Corp., 93
Plasma study, 27, 29, 77, 114, 191, 238–239
Plesetsk, U.S.S.R., launch (see also Appendix A)
  Cosmos satellites, 91, 147, 172–173, 190
  *Intercosmos 12*, 195, 249
  *Meteor 16*, 7, 231
  *Meteor 17*, 92, 235
  *Meteor 18*, 241
  *Meteor 19*, 194, 248
  *Meteor 20*, 215, 253
  *Molniya I-27*, 88–89, 233, 234
  *Molniya I-28*, 192, 247

*Molniya II-9*, 93, 235
*Molniya II-10*, 139
*Molniya II-11*, 217, 254
*Molniya III-1*, 202–203, 251
Plum Brook Station, Ohio (LeRC), 6, 66, 92, 204–205
Pluto (planet; see also *Pioneer 10*), 2, 69–70, 135, 222
pndb: perceived noise in decibels.
Podgorny, Presidium Chairman Nikolay V. (U.S.S.R.), 181
Pogue, Col. William R. (USAF), 1, 7, 30, 31–32, 91–92, 221
Poker Flats, Alaska, 264–265, 266–267
Poland, 249
Polar caps, 111, 238–239
Polarimetry, 198
Polaris (submarine), 93–94
Pole, Felix, 93–94
Pollution (see also Air pollution; Atmosphere, pollution; Noise, aircraft; and Water pollution), 91–92, 117, 119, 126
Pollution control (see also Air pollution; Climatic impact assessment program; and Water pollution), 10, 11, 20, 21, 23, 25–26, 226
Polonium 210, 93–94
Polyethylene, 144
Polymer, 99
Polyurethane, 91
Popovich, Col. Pavel R. (cosmonaut), 122, 130, 181, 239–240, 241, 262
Poritzky, Siegbert B., 99–100
Poseidon (missile), 140
Postal Service. See U.S. Postal Service.
Potassium, 73–74, 187
Pozinsky, Norman, 195
Pratt & Whitney Div. See United Aircraft Corp.
P.R.C. See China, People's Republic of.
Presidential Medal for Merit, 219
Press comment
  Apollo program, 31, 69–70, 138, 153
  Apollo-Soyuz Test Project, 36, 39, 172
  *Ats 6* (ATS-F), 116
  energy research and development, 39, 45, 80, 107
  international cooperation, space, 103, 105
  manned space flight, 31, 32, 51, 69–70, 85, 115
  *Mariner 10*, 37, 75
  NASA budget, 31, 35, 165, 196
  *Pioneer 10*, 3, 13
  science and technology, 85, 105, 129, 153–154
  Skylab program, 8, 31, 32, 34–36, 39, 87
  space program, national, 31–32, 51, 69–70, 165, 183
Press conference
  Apollo-Soyuz Test Project, 147, 208
  *Ats 6*, 106–107
  DOD, 14
  Lockheed Aircraft Co., 10

*Mariner 10*, 16, 27, 69
NASA budget, 24–25
Skylab program, 1, 30
space shuttle, 24, 112–113
*Soyuz 15*, 155–156
Prince Albert Tracking Station, Canada, 111
Prince Ranier III (Monaco), 59
Princeton Univ. (see also Woodrow Wilson School of Public and International Affairs), 98
Probe (see also individual probes such as *Helios 1*, *Luna 16*, *Mariner 10*, *Mars 3*, and *Pioneer 10*), 14, 126
Project Able, 167
Project Airstream, 185
Project ALADDIN. See Atmospheric Layering and Density Distribution of Ions and Neutrals.
Project Cyclops, 45–46
Project da Vinci (atmospheric research balloon flight), 197
Project Daedalus, 179
Project Far Side, 103
Project Independence, 20, 153
Project Interdependence (proposed), 175
Propane, 58
Propulsion technology (see also Nuclear energy), 25, 34, 44–45, 51, 56, 61, 73, 99–100, 109, 117, 161, 179, 184, 185, 186, 224
Prosthetics, 226
Protocol on Cooperation in Controlled Thermonuclear Fusion and Plasma Physics Research, 25
Protons, 197, 208
Proximity fuse, 125
Proxmire, Sen. William, 6–7, 167
Public Affairs Agreement, U.S.–U.S.S.R., 191
Public Service Award (NASA), 39
Puerto Rico, 11–12, 14, 40–41, 88, 186, 201–202
Pulsar, 176, 182, 188–189, 223
Puppis A (supernova remnant), 125

## Q

Quasars (see also individual quasars, such as BL lacertae), 76, 93, 133, 163, 176
Quiet propulsive lift technology program, 27

## R

Radar (see also Scanning laser radar), 12, 14, 41, 60, 67, 72, 93–94, 119, 125, 151, 182, 215
Radar satellite, 86–87, 93–94

Radiation (see also Cosmic ray, Gamma ray, Infrared radiation, Lyman-alpha radiation, Solar radiation, Ultraviolet radiation and X-ray), 16, 42, 60, 77, 83, 98, 103, 109, 119, 121, 122–123, 140, 143, 153, 159, 166–167, 213
Radiation, Inc., 134
Radio, 27, 41, 70, 77, 147, 178–179
astronomical probe, 80, 201–202
balloon transmission, 174–175
interference, 145, 198
planetary emission, 45–46, 197
satellite transmission, 48, 59, 68, 106–107, 139, 192, 202–203, 217, 243, 247, 254
stellar emission, 163, 182
Radio Amateur Satellite Corp. (see also *Amsat Oscar 7*), 201, 250
Radio-Optical Observatory, 181
Radio Solar Telescope Network (RSTN), 149
Radio waves, 160, 212
Radioactive material, 170
Radioactivity, 87, 92–93, 208–209
Radioastronomy, 23, 176, 188–189
Radioisotope thermoelectric generator (RTG), 23, 29, 67
Radiometer, 98, 200–201, 249–250
Radionuclides, 170
Radiotelescope, 154, 160, 176, 188–189
Railroad (see also Federal Railroad Administration), 20
Raines, Martin L., 195
Rambaut, Dr. Paul G., 131
Ramjet engine, 92
Rancho Los Amigos Hospital, Downey, Calif., 167–168
R&D. See Research and development.
R&PM. See Research and program management.
Rankine topping cycle, 187
RANN. See Research Applied to National Needs.
Rasool, Dr. S. Ichtiaque, 16
Ray, Dr. Dixy Lee, 194
Raytheon Corp., 12
RCA Corp., 8, 64, 134
RCA Alaska Communications, Inc., 4
RCA Astro Electronics Div., 146
RCA Global Communications, Inc., 3–4
RDT&E. See Research, development, test, and evaluation.
Reconnaissance aircraft (see also SR–71 and U–2), 151
Reconnaissance satellite, 183, 243
Rectenna receiving system, 178–179
Redstone (missile), 171
Redstone Arsenal, Ala., 102
Reed, Sylvanus Albert, Award, 15
Reeder, John P., 15
Reentry vehicle nosetip test program (USAF), 8, 11
Refanned engine program, 224
Reflector satellites, 167
Regional Dissemination Centers (NASA), 95–96, 226

Relativity, theory of, 45, 133
Remote sensing (see also Sensor), 11, 42, 46, 55, 62, 65, 98, 111, 115, 140
Remotely piloted research vehicle (RPRV), 48, 91
Remotely piloted vehicle (RPV), 151
Rensselaer Polytechnic Institute, 108
Republic Aircraft Corp., 154–155
Research aircraft. See individual aircraft, such as A–9, C–141, Convair 990, Piper Seneca, WB–57F, and X–1.
Research and development (R&D; see also Energy research and development), 4–5, 23, 47–48, 126, 153–154, 183
  budget, FY 1975, 20–21, 23, 83–84, 121, 124, 134, 143, 149–150, 160–161
*Research and Development in Industry 1972* (NSF 74-312), 143
Research and program management (R&PM), 21, 83–84, 121, 124, 149–150
Research Applied to National Needs (RANN), 23, 204–205
Research, development, test, and evaluation (RDT&E), 21–22
Reynolds, Smith and Hills, 64
Rice Univ., 266
Rio Grande River, 42
Riser, Charles R., 168
Riyad, Mahud, 8
RL–10 (engine), 140
Roberts, Dr. Leonard, 148
Robledo de Chavela Tracking Station, Spain, 101
Robotics, 177
Rockefeller, John D., II, 211
Rockefeller, Vice President Nelson A., 151, 217
Rockefeller Center, N.Y., 8
Rockefeller Public Services Award for Administration, 211
Rocket Research Center, West Germany, 1
Rocket technology, 52, 103, 124, 125
Rockwell International Corp., 7, 39, 63, 76, 88, 100, 121, 160, 168, 181, 193, 194, 209
Rocky Mountains, 106–107
Rohr Industries Inc., 89–90
Romania, 190, 195, 249
Romatowski, Raymond G., 213–214
Rome, Univ. of. See Aerospace Research Center.
Roosevelt Committee on Uranium Research, 70–71
Rosen, Dr. Harold A., 15
Rosman, N.C., 129, 191
Ross Island, Antarctica, 94
Royal Air Force (U.K.), 13–14
Royal Norwegian Council for Scientific and Industrial Research, 265, 271
RPRV (remotely piloted research vehicle), 151
RPV (remotely piloted vehicle), 151
RSTN (Radio Solar Telescope Network), 149
RTG. See Radioisotope thermoelectric generator.
Rukavishnikov, Nikolay N. (cosmonaut), 207, 252, 262
Russia. See U.S.S.R.
Ryan Aeronautical Div. See Teledyne, Inc.
Ryan monoplane, 156
Ryle, Sir Martin J., 188–189

S

S–2 (antisubmarine aircraft), 43
S–3A (Viking antisubmarine aircraft), 43
*S3–1* (USAF atmospheric research satellite), 145, 194, 228, 248
Safety
  aircraft, 23, 48, 78, 79–80, 92–93, 100–101, 160, 166, 224
  mine, 29, 54, 92
Safety valves, 10
Sagan, Dr. Carl E., 70
Sagdeyev, Roald, 33
Sagitta (constellation), 143
Sagittarius B (interstellar cloud), 132
Sahara Desert, 41
SAIL (Shuttle Avionic Integration Laboratory), 2
Saint-Gobain-Pont-à-Moussan, 42
Sakharov, Andrei D., 154
SALT (Strategic Arms Limitation Talks), 42, 70, 204
Salt, 208
Salt lenses, 216
Salyut (U.S.S.R. scientific space station), 193, 222
*Salyut 1*, 123, 219
*Salyut 2*, 4, 123, 219
*Salyut 3*, 122–123, 130, 139–140, 155, 156, 164, 176, 181, 217–218, 219, 221, 228, 239–240, 241, 244, 255, 261, 262
*Salyut 4*, 217–218, 219, 221, 228, 255
SAM–D (surface-to-air missile), 9
SAMSO (Space and Missile Systems Organization), 34, 160, 203
San Andreas Fault, 163
San Antonio, Tex., 95
San Clemente, Calif., 136
San Diego, Calif., 31–32
San Francisco, Calif., 4, 45–46, 89–90
San Marco (Italian launch facility), 187, 188, 222, 230–231, 247, 259
*San Marco 1* (NASA–Italy cooperative satellite), 40
*San Marco 2*, 40
*San Marco 3*, 40
*San Marco 4* (San Marco C2), 2, 40, 222, 228, 230–231, 257, 258
S&AD (Science and Applications Directorate). See NASA Astronaut Office.
Sandage, Dr. Allen R., 214
Sandia Laboratories, 57
Santa Ana, Calif., 115
*Santa Maria* (ship), 30

SAO (Smithsonian Astrophysical Observatory), 139
Sarafanov, Lt. Col. Gennady V. (cosmonaut), 155, 181, 239–240, 244, 262
Sary Shagan, U.S.S.R., 215
SAS (Small Astronomy Satellite). See Explorer 42.
Satellite power stations, 115
Satellite Tracking and Data Acquisition Network. See Space Tracking and Data Acquisition Network.
Satellite warning system, 21–22
Saturn (launch vehicle), 33–34, 184
    Saturn IB, 16–17, 41, 88, 113, 118, 151, 216
    Saturn V, 41, 119, 167
Saturn (planet; see also Mariner Jupiter-Saturn 77 Mission, *Pioneer 10*, *Pioneer 11*, and Titan), 37, 88, 106, 166–167, 198, 222
Saudia Arabia, 7, 169
Sawhill, John C., 88, 108, 130
Saxbe, William B., 199
Scanning electrostatic analyzer, 26
Scanning laser doppler system, 102
Scanning laser radar, 184
Scanning radiometer, 249–250
Scheer, Julian, 138
Schenectady, N.Y., 181
Scherer, Capt. Lee R. (USN, Ret.), 195, 217
Schlesinger, Dr. James R., Secretary of Defense, 9, 28, 51–52, 94, 193
Schmidt telescope, 167
Schmitt, Dr. Harrison H., 15, 79, 102, 115
Schneider, William C., 30, 32, 39, 42, 58–59, 95
Schweickart, Russell L., 95
Schwenk, Francis C., 29
Science, 20, 23, 47, 85, 105, 124–125, 129, 153–154, 200
Science Adviser (President's), 23, 125, 176
Science Research Council (U.K.), 187, 188
Scientific-Atlanta, Inc., 114
Scientists, 119, 122–124, 126–127, 140–141, 151, 153, 169, 214, 225
    astronomy, 3, 59, 65, 76, 118–119, 132, 133, 143, 160, 169–170, 171–172, 176, 182, 208, 214, 223
    awards, 92, 95, 188–189, 203
    conference, 64, 98, 102
    death, 125
    earth study, 11, 104–105, 114–115, 119, 168, 170, 191
    employment, 6–7, 11, 48–49, 80, 190
    energy research, 114–115
    foreign, 4–5, 72, 87–88, 108, 225
    manned space flight, 7, 16, 31, 32, 41, 113
    planetary exploration, 14, 68, 106, 112, 120–121, 154, 166, 179
    propulsion technology, 186
Scientist-astronaut, 121–122, 149, 152

Scout (launch vehicle), 40, 93, 134–135, 159, 160, 188, 213, 230–231, 244, 247, 258, 259
    Scout D, 57–58, 231, 242, 258, 259
    Scout E, 111, 238–239, 259
Screwworm, 62
Sea of Crisis (Mare Crisium), 194, 248
Seamans, Dr. Robert C., Jr., 61, 97, 129, 194
Seaplane, 154–155
SEASAT (Specialized Experimental Applications Satellite), 24, 124, 150
SEASAT-A, 21, 24, 55, 204
Seelye Stevenson Value and Knecht Co., 97
Seismography, 104
Seismometer, 208–209
Senegal, 175
Sensor, 57–58, 60, 103, 119, 120, 134, 141, 152–153, 182, 188, 231
SEP (Société Européenne de Propulsion), 189–190
SEPS (Solar-electric propulsion stage), 34, 184
Sergeyev, Georgy, 36
*Sert 2* (Space Electric Rocket Test satellite), 186
Service Cross of the German Eagle, 156
Sethna, H. N., 105
Sevastyanov, Vitaly I. (cosmonaut), 255
Seversky Aircraft Corp. See Republic Aircraft Corp.
Sewell, Charles A., 155
Shale, 154
Shanghai, P.R.C., 89
Shapland, Dr. D. J., 16
Shapley, Willis H., 6–7, 220
Shatalov, Maj. Gen. Vladimir A. (cosmonaut), 78, 85, 122, 147, 155–156, 208
Sheldon, Dr. Charles S., II, 4–5
Shepard, R/A Alan B., Jr. (USN, Ret.), 95, 136, 137, 168, 171
Sheremetyevo Airport, Moscow, 80
*Shinsei* (Japanese satellite), 39–40
Ships, 30, 40–41, 52, 102, 104–105, 114, 119, 147, 165
Short takeoff and landing aircraft (STOL), 23
Shramo, Daniel J., 178
Shuttle Avionic Integration Laboratory (SAIL), 2
Sidney, Australia, 193
Silicates, 82
Silicon, 28
Silicon carbide whiskers, 58
Silver, 58
Simon, William E., 88
Singer Co., 217
Sirius (star), 244, 259
6th Aircraft Design, Flight Test, and Operations Meeting, 99–100, 148
Skyhook program, 144
Skylab (program), 13, 16–17, 21, 34–35, 41, 47, 53, 64, 98, 118–119, 149, 152, 184, 223

achievements, 13, 17, 27, 42–44, 58, 67, 80, 87, 221
awards, 59, 88, 95, 131, 169–170, 183, 203
budget, 21, 186–187
commemoration, 45, 103
conference, 152, 157–158
congressional consideration, 19, 91–92
contract, 1, 139
employment, 25, 141
information exchange, 193
photographs, 131
press comment, 31, 32, 36, 39, 87
significance, 17, 28–29, 30, 32
technology utilization, 134, 143, 144–145
tracking and data acquisition, 24, 44, 77, 102
*Skylab 1* (Orbital Workshop; see also Apollo Telescope Mount), 1, 8, 28–29, 32, 42–43, 45, 80, 91–92, 98, 103, 153, 183, 199
*Skylab 2* (first manned mission), 32, 43–44, 45, 153, 183, 193, 195
*Skylab 3* (second manned mission), 7, 43–44, 153, 183, 193
*Skylab 4* (third manned mission; see also Apollo Telescope Mount), 14, 25, 35–36, 152, 216, 228, 264
achievements, 7, 8, 28–29, 30, 153, 221
awards, 65–66
medical aspects, 31, 32, 43–44
press comment, 8, 31, 32, 51
press conference, 1, 30
tracking and data acquisition, 40
Skylab Life Sciences Symposium, 157–158
Skylark (U.K. sounding rocket), 125
Skynet (U.K. comsat), 2
*Skynet A*, 9. 182, 204
*Skynet B*, 9, 182, 204
Skynet II comsat program, 9
*Skynet IIA*, 9, 64, 83, 158, 182, 203, 221, 227, 229, 252, 257, 258
*Skynet IIB*, 182, 203–204, 227, 252, 257, 260
SL-1. See *Skylab 1*.
SL-2. See *Skylab 2*.
SL-3. See *Skylab 3*.
SL-4. See *Skylab 4*.
Slayton, Maj. Donald K. (USAF, Ret.), 82, 121–122
Sloop, John L., 15
SLV-3 (launch vehicle), 189–190
SM (service module). See Command and service module.
Small Astronomy Satellite (SAS). See *Explorer 42*.
Small tactical aerial mobility platform (STAMP), 60
Smith, A. M. O., 148
Smith, Dr. Malcolm C., Jr., 131
Smith, Richard G., 199
Smithsonian Astrophysical Observatory (SAO), 139
Smithsonian Institution, 136, 164, 177

SMM (Solar Maximum Mission), 57
SMS (Synchronous Meteorological Satellite), 2
*Sms 1* (SMS-A; Synchronous Meterorological Satellite), 21, 103–104, 119, 176, 221, 225–226, 227, 236–237, 257, 258
SMS-B, 21
SMS-C. See *Goes 1*.
Smylie, Robert E., 130
Snow, 92, 194, 215, 253
Société Nationale Industrielle Aérospatiale (Aérospatiale), 37, 136–137
Sociéta per Azioni per lé Communicazioni Spaziali (Telespazio), 111
Société Européenne de Propulsion (SEP), 189–190
Society of Automotive Engineers, 137
Sodium, 208
Soendre Stroemfjord, Greenland, 268–269
Soil Cartographic Division (Agriculture, Dept. of), 90
Soil studies, 62, 65
Solar astronomy (see also Helios), 2, 28–29, 32, 57, 67
Solar conversion devices
array, 34, 58, 67, 186, 249–250, 253
battery, 176, 218
cell, 33–34, 57–58, 73, 105, 114–115, 224, 231, 250–251
panel, 8, 32, 45, 63, 122, 217–218
Solar electric propulsion stage (SEPS), 34, 184
Solar energy (see also Solar heating and cooling), 8, 12–13, 72, 83, 105, 118
applications, 43, 56–57, 98, 111, 114–115, 152–153, 201
environmental effects, 40, 107
research and development, 6, 17, 20, 29, 36, 45, 46, 73, 106, 114, 149, 152, 192–193, 204, 219
Solar Energy Coordination and Management Project, 192–193
Solar Energy Research, Development, and Demonstration Act of 1974, 192–193
Solar heating and cooling, 6, 8, 19, 45, 56–57, 68, 73, 91, 111, 112, 114–115, 118, 226
Solar Heating and Cooling Demonstration Act of 1974, 36, 45, 46, 61, 95, 106, 163–164, 175, 219, 226
Solar Maximum Mission (SMM satellite), 57
Solar phenomena, 28
corona, 43–44, 67
flares, 57, 90, 104, 115, 177
tide, 45
wind, 5, 9–10, 13, 27, 43–44, 60, 111, 197–198, 212, 253
Solar physics, 57, 212, 253, 260, 264–265, 267, 269, 270, 271
Solar proton monitor, 200–201, 249–250
Solar radiation, 9–10, 12, 60, 76, 77, 104, 114, 134–135, 139, 144, 145, 174, 212, 237, 242
Solar simulation system, 66

Solar system, 2, 13, 37, 42, 43, 51, 58–59, 64, 70, 72, 82, 118–119, 136, 166–167, 169, 179, 222
Solar telescope, 218–219, 255
Solid-fueled rocket booster (SRB), 36, 38, 150, 210–211, 212
Solid-fueled rocket motor (SRM), 5, 33–34, 36, 41, 55, 106, 210–211, 212
Solid-state technology, 47–48, 143, 144
Sommer, Robert W., 158–159
Sounding rocket (see also Aerobee, Aries, Black Brant, Loki, Nike-Iroquois, Skylark, and Super Loki), 47, 57, 67, 103, 112–113, 118–119, 126, 223, 226, 263–271
South America, 65, 89, 156, 216, 238, 254
South Atlantic Anomaly, 159
Southern California, Univ. of (USC), 152, 271
Southwest Research Institute, San Antonio, Tex., 95
Soviet Academy of Sciences (see also Council of International Cooperation in Space Exploration and Use, and General Genetics Institute), 52, 104, 157, 165, 190, 193, 225
Soviet Mission Control Center, 157, 191
Soviet Union. See U.S.S.R.
Soyuz (U.S.S.R. spacecraft; see also Apollo-Soyuz Test Project), 108, 115, 121, 157, 167, 191, 193, 208, 216, 218, 222, 228
*Soyuz 4*, 161
*Soyuz 5*, 161
*Soyuz 10*, 123, 219
*Soyuz 11*, 73, 123, 167, 219
*Soyuz 12*, 4, 73, 178
*Soyuz 13*, 4, 44, 73, 178
*Soyuz 14*, 122–123, 130, 139–140, 155, 156, 164, 176, 178, 181, 207–208, 217–218, 221, 228, 241, 261, 262
*Soyuz 15*, 155–156, 164, 167, 176, 181, 207–208, 217–218, 221, 228, 239–240, 244, 261, 262
*Soyuz 16*, 207–208, 221, 225, 228, 252, 261, 262
*Soyuz 17*, 218–219, 228, 255
*Soyuz 18*, 255
*Soyuz 20*, 255
Spaatz, Gen. Carl A., 133
"Space: A Report to the Stockholders" (TV documentary), 137
Space and Missiles Systems Organization (SAMSO, USAF), 34, 160, 203
Space and Rocket Center, Huntsville, Ala., 28
Space biology and medicine, 4, 28–29, 30, 41, 73, 97, 154, 207, 239–240, 252
 animal experiments, 44, 75–76, 190, 193, 247
 flight simulation, 181
 human research, 31–32, 34–35, 42–44, 103, 122, 123, 155, 157–158, 178, 218–219, 241, 252
Space debris, 2–3, 82, 174

Space Detection and Tracking System (SPADATS), 67
Space Electric Rocket Test Satellite (*Sert 2*), 186
Space manufacturing, 21, 32, 38, 43, 58, 122, 154, 178, 184, 199, 241
Space Missile Test Center, 209
Space Plasma High Voltage Interaction Experiment (SPHINX) spacecraft, 2, 33–34, 222, 258
Space Power Facility (SPF), 66
Space program, national (see also Budget, NASA), 37–38, 87, 126, 165, 227, 228
 applications technology, 81, 95–96, 106–107, 137, 200
 benefits, 84–85
 press comment, 31, 32, 35, 51, 69–70, 103, 165, 183
Space science, 21, 24, 57, 214, 222–223, 241
Space Science Award (AIAA), 14
Space shuttle (see also External tank, Interim upper stage, Solid-fueled rocket booster and motor, Space shuttle main engine, Space shuttle orbiter, Space tug, and Spacelab), 16, 19, 30, 53, 57, 65–67, 99–100, 115, 131, 149, 171, 173
 aircraft use, 120
 budget, 17, 21, 24, 27, 35, 41, 47, 98, 109, 113, 160–161, 165, 195, 214
 capabilities, 21–22, 32, 37–38, 57, 62, 78, 98, 112–113, 124, 126–127, 167
 construction of facilities, 21, 31, 64, 97, 210
 development status, 47, 223–224
 fuel crisis impact, 100
 international cooperation, 37, 38, 181, 184
 landing site, 189
 press comment, 39, 69–70, 103, 131
 study, 121, 126, 140, 217
 system development, 9, 12, 41–42, 56, 81, 89, 93, 105, 113, 117, 141, 174, 184, 185, 199, 201, 210
 testing, 55, 119, 120, 150, 151
 tracking and data acquisition, 24, 47
 use, 21–22, 37–38, 47, 48, 66–67
Space Shuttle Cost and Review Committee, 210
Space shuttle main engine (SSME), 31, 47, 84, 88, 150
Space Shuttle Operational Management Assessement Team, 195
Space shuttle orbiter, 36, 47, 55, 79, 107, 120, 141, 165, 168, 181, 210, 217
Space station (see also Salyut, and Skylab), 32, 47, 73, 85, 98, 103
"Space Stations, Present and Future" (IAF conference), 178
Space Tracking and Data Acquisition Network (STADAN), 176
Space transportation system (STS), 19, 34, 47, 205, 217
Space tug, 7–8, 19, 25, 34, 56, 66–67, 103, 126, 140, 178, 184, 217

Space weapons, 183
Space Week (National Space Week), 132, 135-136
Spacecraft Design Award (AIAA), 15
Spaceflight Tracking and Data Network (STDN), 40, 77, 176, 226
Spacelab, 42, 47, 126, 185
  budget, 19, 54
  capabilities, 38, 127, 184
  contract, 113, 141, 217
  development status, 223-224
  international cooperation. See European Space Research Organization, Spacelab.
  laboratory, 16, 41-42, 94
  testing, 176-177, 181
Spacelab Operations Working Group, 184
Spacelab Payload Utilization and Planning Study, 205
Spacelab Preliminary Requirements Review, 199
"Spaceship Earth-A New Perspective" (AIAA conference), 14
Spacesuit, 154
SPADATS (Space Detection and Tracking System), 67
Spain (see also *Intasat*), 41, 54, 67, 101, 107, 200, 201
Span-load tailoring, 160
Spanish Foreign Legion, 41
Special Energy Bill, 149
Special Task Force on Energy Conservation, 8-9
Specialized Experimental Applications Satellite (SEASAT), 24, 124, 150
Spectrogram, 29, 44
Spectrometer (see also Balloon-borne ultraviolet stellar spectrometer, Ebert-Fastie Spectrometer, EUV spectrometer, Infrared spectrometer, Mass spectrometer, Ultraviolet spectrometer, and X-ray spectrometer), 2, 40, 126, 145, 172, 255
Sperry, Lawrence, Award, 15
Sperry Support Services, 143
SPF (Space Power Facility), 66
SPHINX. See Space Plasma High Voltage Interaction Experiment spacecraft.
Spicules, 67
Spin tunnel, 96
*Spirit of St. Louis* (aircraft), 30
Spokane River, 170
Sprankle, Roger S., 175-176
Spread F (ionospheric disturbance), 72
SR-71 (Blackbird; reconnaissance aircraft), 163, 199-200
SRB. See Solid-fueled rocket booster.
SRC (Science Research Council, U.K.), 187, 188
SRM. See Solid-fueled rocket motor.
SS-19 (U.S.S.R. ICBM), 14
SSM (Support systems module), 201
SSME. See Space shuttle main engine.
SSS (Small Scientific Satellite). See *Explorer 45*.
SST (supersonic transport), 153

SSX-16 (U.S.S.R. missile), 42
SSX-18, 42
SSX-19, 42
STADAN (Space Tracking and Data Acquisition Network), 176
Stafford, B/G Thomas P. (USAF), 82, 121-122, 152, 178
STAMP (Small tactical aerial mobility platform), 60
Star (see also Barnard, Betelgeuse, Canopus, Neutron star, Omega Centauri, Pulsar, Quasar, Supernova, and 31 Crater), 44, 45, 73, 90, 118-119, 133, 143, 159, 160, 168, 176, 177, 185, 214
Star City. See Zvezdny Gorodok, U.S.S.R.
STAR consortium, 210
Star-pointing-and-control system, 185
Starr, Dr. Chauncey, 211
State, Dept. of, 119, 194
STDN. See Spaceflight Tracking and Data Network.
Steam, 73-74, 187
Steel, 58
Stethoscope, 54
Stevens, Sen. Ted, 4
Stevenson, Robert E., 153
Stever, Dr. H. Guyford, 46, 60-61, 112, 125, 182
Steward Observatory, 167
Stewart, Frank M., 69
Stockholm, Sweden, 188-189
STOL (short takeoff and landing) aircraft, 22, 23
Stone, Donald, 184
Storms, 62, 120
Strategic Arms Limitation Talks (SALT), 42, 70, 204
Stratford, Conn., 51
Stratofortress. See B-52.
Stratosphere (earth), 86, 98, 188, 266
Stream Lite-1 Million (spotlight), 8
Streamlight, Inc., 8
Strip mining, 91-92
STS. See Space transportation system.
Submarine, 43, 62-63, 140
Submarine-launched cruise missile, 22
Subsonic aircraft, 148
Sukhoi (U.S.S.R. fighter-bomber aircraft), 62-63
Sulfur, 93, 208
Sulfur dioxide, 86
Sulfuric acid, 27
SUMMA (superconductive magnetic apparatus), 29
Summit meeting, 123-124, 136-137, 241
Sun (see also solar headings), 28-29, 36-37, 43-44, 67, 118, 143, 166-167, 169, 176
Sundstrand Data Control, Inc., 100-101
Super Arcas (sounding rocket), 264, 266-267, 268, 269
Super Loki (sounding rocket), 65
Superconductive magnetic apparatus (SUMMA), 29
Supercritical wing, 44, 48, 68, 112, 148, 173, 187, 224

Supernova, 31, 90, 125
Supersonic aircraft (see also names of individual aircraft), 2-3, 104, 112, 114, 166, 167, 175, 176-177
Supersonic transport (SST; see also Concorde and Tu-144), 153
Support systems module (SSM), 201
Surface surveillance program, 36, 230
Sweden, 54, 67, 101
Swedish Academy of Sciences, 188-189
Swedish Space Corp., 265
Swing wing, 7, 39, 193, 211-212
Switzerland, 54, 67, 101
*Symphonie 1* (Symphonie-A; French-West German comsat), 2, 215-216, 221, 227, 254, 257, 260
Synchronous Meteorological Satellite (SMS; see also *Goes 1*, *Sms 1*, and SMS-B), 2
Systems Requirements Review, 203
Szostak, Rosemarie, 59

## T

T-38 (trainer aircraft), 28
Tactical fighter aircraft. See F-5E, F-15, etc.
Taft, Robert, Jr., 199
Taiwan. See China, Republic of (Nationalist).
Tamman, Gustav, 214
Tanegashima Space Center, Japan, 182
Tanker/missile carrier aircraft, 22
*Tansei 1* (MS-T1, Japanese satellite), 39-40
*Tansei 2* ("Light Blue"; MS-T2), 39-40, 222, 228, 230
Tape recorder, 38, 143, 259
Tashkent, U.S.S.R., 193
Task Force on Remote Sensing (U.N.), 46
Taurus-Littrow region (lunar landing site), 72-73
Taylor Valley, Antarctica, 94
TC-1. See Titan-Centaur.
TCV (Terminal-configured vehicle), 123
TDRSS. See Tracking and Data Relay Satellite System.
Teague, Rep. Olin E., 15, 28-29, 58, 78
Technology, 20-21, 23, 25, 28-29, 38, 47-48, 105, 112-113, 123-125, 129, 135, 146, 153, 177-178, 200
"Technology in the Service of Man" (LeRC review), 166
Technology Resources Survey and Applications Act, 60-61
Technology utilization, space, 10, 24, 56, 95-96, 99-100, 134, 143, 166, 187
  budget, FY 1975, 21, 27
  medicine, 13, 54, 68, 95, 143, 144-145, 167-168, 212, 226
Teflon, 91
Tehran, Iran, 194-195
Teledyne, Inc., 151
Telegraph, 88-89, 93, 139, 151-152, 254

Telemetry, 131, 152-153, 185
Telemetry transmitter, 103
Telephone, 78-79, 88-89, 93, 116-117, 139, 151-152, 171, 216, 251, 254
Telesat Canada, 4, 84, 169
Telescope (see also Apollo Telescope Mount, Grazing-incidence telescope, Infrared telescope, Large Space Telescope, Mayall telescope, *Orion 2*, Radio telescope, Schmidt telescope, Solar telescope, Ultraviolet telescope, Wolter Type I Telescope, and X-ray telescope), 9-10, 16, 37, 39, 69, 77, 154, 168, 171-172, 185
Telespazio (Sociéta per Azioni per lé Communicazioni Spaziali), 111
Teletype, 216, 218
Teletypewriter, 147
Television, 32, 48, 59, 102, 116, 117, 119, 122, 137, 143, 174-175, 191
  satellite transmission, 54, 63, 68, 69, 78-79, 84, 86, 106-107, 129, 139, 147, 151-152, 154, 165-166, 186, 192, 202-203, 215, 217
  space exploration, 5, 27, 59, 101, 108, 121, 173, 207, 237
Telex, 151-152
Terminal-configured vehicle (TCV) program, 123
Terry, Sydney L., 117
Tethered Communications (TCom. Inc.). See Westinghouse Corp.
Texas, 164, 171
Texas, Univ. of, 268-269
Texas A&M Univ., 188
Texas Instruments, Inc., 108
Textron, Inc., 111-112
Thailand, 14
Thermal energy, 92, 194, 215, 231, 235
Thermal mapper, 55
Thermal stress analysis, 114
Thermionic converter, 29, 56
Thermionic technology, 184-185
Thermodynamics, 117
Thermoelectricity, 93-94
Thermonuclear energy, 154
Thermosphere (earth), 40, 55, 144, 230-231
Thiokol Corp., 5, 36, 41, 106, 126, 146, 229
31 Crater (star), 69
Thole, John M., 157, 195
Thompson, Dr. Floyd L., 137
Thompson, James R., 69
Thomson-CSF, 49, 86-87, 169
Thor-Burner II (launch vehicle), 63, 232, 243
Thor-Delta (launch vehicle), 9, 64, 83, 85-86, 103-104, 109, 146, 148-149, 158, 174, 182, 185-186, 196, 200, 201, 203, 204, 229, 233, 236, 246, 249, 252, 254, 258, 259, 260
Thorium, 68
Thornton, Dr. W. E., 158
3C 268 (galaxy), 160
Thrust-vector-control system (TVC), 39-40

Tiger II (F-5E), 7, 10
Tikhonravov, Prof. Mikhail, 52
Timation 3A. See *Nts 1*.
Timonium, Md., 56-57
Tires, 99
Tiros-N (meteorological satellite), 21, 204, 213
Titan III (launch vehicle), 1
   Titan IIIB, 239
   Titan IIIB-Agena, 36, 230, 243
   Titan IIIC, 108-109, 139, 238, 258
   Titan IIID, 233, 248
   Titan IIIE, 33-34
   Titan IIIE-Centaur, 212, 253, 257, 258, 260
Titan (Saturn moon), 106, 198
Titan-Centaur (launch vehicle), 2, 33, 203, 222
Titanium, 99, 166
Toksoz, Dr. M. Nafi, 208-209
Tokyo, Japan, 89
Tokyo, Univ. of, 39-40
Tomcat. See F-14 aircraft.
Tracking and data acquisition (see also Earth station, Spaceflight Tracking and Data Network, and Tracking and Data Relay Satellite System), 144, 157, 160, 173, 176, 185, 188, 204, 213, 223, 226, 242
   Apollo-Soyuz Test Project, 7, 207, 216
   budget, 21
   international cooperation, 135
   space shuttle, 21, 40, 47
Tracking and Data Relay Satellite System (TDRSS), 24, 62, 66, 78, 127, 176, 226
Tracking station. See Earth station and individual tracking stations.
Trainer aircraft. See F-5B, F-5F, and T-38.
Trans World Airlines, Inc. (TWA), 74, 79, 156
Transceiver, 110
Transponder, 78-79, 84, 86, 169, 181, 202, 251
Transport aircraft (see also Supersonic transport aircraft), 25
Transportation, 29, 73, 84
Transportation, Dept. of (DOT; see also Transportation Systems Center), 17, 22-23, 25, 54, 79, 86, 117, 118, 119, 166, 185
*Transportation of Radioactive Material by Passenger Aircraft* (report), 170
Transportation Systems Center (DOT), 118
Traveling wave tube, 47-48
Treaty, 11, 123-124, 200
Trident (undersea long-range missile system; ULMS), 22, 140
Trident I (missile), 140, 151
TriStar (jet transport). See L-1011.
Troposphere (earth), 23
Trubshaw, Brian, 37
Trucks, 99
Truman, President Harry S., 133
Truszynski, Gerald M., 62, 77

TRW Inc., 58-59, 78, 158
Tselinograd, Kazakhstan, U.S.S.R., 219, 262
Tsunamic data, 104
Tu-144 (U.S.S.R. supersonic transport), 86, 211-212
Tucson, Ariz., 219
Tungsten, 173
Turbine discs, 26
Turbine engine, 26, 48
Turbine wheel cooling, 26
Turbofan engine (see also JT8D), 193
Turbogenerator, 63, 73
Turboprop engines, 145-146
Turcat, André, 118
Turkey, 243
Turkish Airlines, Inc., 51, 55, 71, 78
TVC (thrust-vector-control system), 39-40
TWA. See Trans World Airlines, Inc.
TX-354-5 (Castor II rocket motor), 146
Tyuratam, U.S.S.R. See Baykonur Cosmodrome.

## U

U-2 (reconnaissance aircraft), 131, 188, 199-200
UAL. See United Air Lines.
Uchinoura Space Center, Japan, 182
UDMH (unsymmetrical dimethylhydrazine rocket fuel), 200
UFO (unidentified flying object), 70-71
*Uhuru* (Small Astronomy Satellite; *Explorer 42*), 133, 187
U.K. See United Kingdom.
UK-5. See *Ariel 5*.
UK-X4. See *Miranda*.
Ukrainian Academy of Sciences, 178
ULMS (undersea long-range missile system; Trident), 22, 140
Ultraviolet photography, 3, 26, 27, 44, 120-121, 266-267
Ultraviolet radiation, 2, 12, 77, 86, 104, 134-135, 159, 173, 242, 265-266, 270
Ultraviolet scanner, 69
Ultraviolet spectrometer, 5
Ultraviolet telescope, 159
Ultraviolet wavelength, 185
U.N. See United Nations.
Undersea long-range missile system (ULMS; Trident), 22, 140
Underwood-Prescott Memorial Award, 131
UNESCO (United Nations Educational, Scientific, and Cultural Organization), 8, 99
Unidentified flying object (UFO), 70-71
Unidentified satellite, DOD, 36, 63, 194, 230, 232, 233, 239, 243, 248-249
Unifi, Inc., 13
Union of Soviet Socialist Republics. See U.S.S.R.
United Air Lines, Inc. (UAL), 12, 17, 74, 174, 213, 224
United Aircraft Corp., 156

Pratt & Whitney Div., 51, 140, 145–146
Sikorsky Aircraft Div., 51
United Kingdom (U.K.; see also *Ariel 5*, Concorde, *Miranda*, Royal Air Force, Skylark, and Skynet), 14, 16, 38, 54, 58, 67, 77, 81, 101, 107, 119, 136–137, 156, 173, 176
United Nations (U.N.; see also Committee on the Peaceful Uses of Outer Space, Task Force on Remote Sensing, and World Meteorological Organization), 10–11, 26, 105
United Nations Educational, Scientific, and Cultural Organization (UNESCO), 8, 99
United States (U.S.; see also Aerospace industry; Hot line; International cooperation; International cooperation, space; Space program, national; and Strategic Arms Limitation Talks), 12, 97, 104, 116, 118, 139, 156, 176, 177, 186, 195, 236
  defense, 28, 51–52, 138–139, 155, 204
  foreign policy, 63, 105
  trade, 7, 11, 28, 81
*United States and Soviet Progress in Space: Summary Data through 1972 and a Forward Look* (74-35SP), 4
Universe, 132, 136, 141, 160, 214
Universities (see also individual universities), 6, 102–103, 105, 126, 134
Unsymmetrical dimethylhydrazine (UDMH) rocket fuel, 200
Upper atmosphere (earth), 65, 98, 104, 134–135, 145, 151, 169, 185, 194, 213, 218–219, 264, 265–267, 268–269, 270–271
Uranium, 23, 56, 68
Uranium hexafluoride, 29
Uranus (planet; see also *Pioneer 10*), 37, 106, 222
U.S. Air Force (USAF; see also individual bases, centers, and commands; Air Force Air Weather Service; Air Force Avionics Laboratory; Air Force Cambridge Research Laboratories; Air Force Materials Laboratory; Atmospheric Layering and Density Distribution of Ions and Neutrals program; Initial Defense Communications Satellite Program; Launch, DOD; and Space and Missiles Systems Office), 9, 66, 70–71, 72, 101–102, 138, 163, 192, 200, 209, 224
  Aeronautical Systems Div., 146
  aeronautics (see also X-24 and individual aircraft, such as YF–12 and B–1), 31, 52–53, 144, 151, 273
  budget, 21–22
  employment, 131, 141
  personnel, 13, 170
  space program participation, 1, 8, 11, 33–34, 90, 121, 134, 140, 183, 229, 252, 273

U.S. Army (USA; see also Army Ballistic Missile Agency, Army Electronics Command, and Atmospheric Sciences Laboratory), 31
U.S. Army Air Corps, 156
U.S. Bureau of Mines, 54
U.S. Geological Survey (USGS), 10, 32, 170
U.S. Information Agency, 59
U.S. Marine Corps (USMC; see also Marine Corps Air Station), 60
U.S. Navy (USN; see also individual aircraft; LAMPS helicopter; and Launch, DOD), 22, 28, 36, 43, 52, 62–63, 66, 108, 136, 151, 166
  ocean surveillance program, 36, 230
  Sea Systems Div., 140
  Weather Reconnaissance Squadron Four, 153
U.S. Postal Service, 45, 103, 233
*U.S.-Soviet Cooperation in Space* (book), 172
USA. See U.S. Army.
USAF. See U.S. Air Force.
USC. See Southern California, Univ. of.
USDA See Agriculture, Dept. of.
USGS. See U.S. Geological Survey.
USMC. See U.S. Marine Corps.
USN. See U.S. Navy.
USNS *Vanguard*, 119
U.S.S. *Enterprise*, 63–64
U.S.S. *New Orleans*, 30, 31
U.S.S.R. (Union of Soviet Socialist Republics; see also Aeroflot Soviet Air Lines; Apollo-Soyuz Test Project; International cooperation; International cooperation, space; Orbita network; Soviet headings; and Strategic Arms Limitation Talks), 10–11, 14, 15–16, 63, 105–106, 165
  communications, 172–173
  disarmament, 204, 215, 241
  missile (see also individual missiles, such as SS–19 and SSX–16), 4–5, 51
  satellites and spacecraft (see also individual satellites and spacecraft, such as Cosmos, Intercosmos, Luna, Mars, Meteor, Molniya, Salyut, and Soyuz), 110, 227–228
  space program, 4–5, 16, 36, 38, 43, 78, 85, 93–94, 103, 115, 139–140, 152, 183, 221, 227
U.S.S.R. State Committee on Science and Technology, 112
Utah, 5
Utah State Univ., 43
UV. See Ultraviolet headings.

# V

VAB (Vehicle Assembly Building). See Kennedy Space Center.
Valve Engineering Co., 83
Van Allen, Dr. James A., 83
Van Allen radiation belts, 143, 159–160

Van Biesbroeck, Dr. George A., 45
Vandenberg AFB, Calif. (see also Western Test Range), 1, 36, 63, 66–67, 127, 132–133, 194
Vanik, Rep. Charles A., 117
Variable-geometry wing, 173
VDS (Viking Dynamic Simulator), 33
Vehicle Assembly Building (VAB). See Kennedy Space Center.
Vela (nuclear detection satellite), 90
Venezuela, 14
Venus (planet; see also *Mariner 10*, and Pioneer–Venus), 5, 16, 32, 36–37, 87, 99–100
Vertical or short takeoff and landing aircraft. See V/STOL.
Vertical takeoff and landing aircraft. See VTOL.
Vertical-temperature-profile radiometer (VTPR), 200–201, 249–250
Very-high-resolution radiometer (VHRR), 200–201, 249–250
VFAX fighter program, 183–184
VFW–Fokker/ERNO Raumfahrttechnik, GmbH, 113, 223–224
VFX (fighter aircraft), 22
VHRR (Very-high-resolution radiometer), 200–201
Victory, Dr. John F. 219
Video-telephone, 154
Videotape, 106–107
Vidicon receiving tube, 143
Vietnam, 20, 174–175
Viking (program), 21, 57, 77, 87, 99–100, 124, 150, 203, 222–223
Viking (spacecraft), 23, 33, 57, 172, 222
 lander, 112, 186–187, 203, 222–223
 orbiter, 186–187, 203, 222–223
Viking (S3-A antisubmarine aircraft), 43
Viking Dynamic Simulator (VDS), 33
Viking rocket, 189–190
Vikram Sarabhi, India, 168–169
Vinyl cyanide, 171–172
Virginia, 165
Vitasign Attendant Monitor, 95
Vladivostok, U.S.S.R., 204
Volcano, 68, 82
Volusia County, Fla., 65
von Braun, Dr. Wernher, 1, 14, 154, 171
von Kármán, Theodore, Memorial Lecture, 15
*Vostok 4* (U.S.S.R. manned spacecraft), 122
V/STOL (vertical or short takeoff and landing) aircraft, 183–184
VTOL (vertical takeoff and landing) aircraft, 23
VTPR (vertical-temperature-profile radiometer), 200–201, 249–250

## W

Waddy, Joseph C., 115–116
Wagon Mound, N. Mex., 197
Wake vortex, 99–100, 102, 160

Waldmann, Raymond J., 26
Wallops Flight Center (WFC, NASA), 93, 126, 188, 220, 264, 267, 268–269, 270, 271
Warner, John W., Secretary of the Navy, 28
Warning system, 149
Warrenton, Va., 19
Warsaw Pact countries, 71
Washington, 170
Washington Cathedral, Washington, D.C., 136
Washington, D.C., 4, 14, 25, 26, 37, 42, 58, 59, 82, 83, 88, 94, 95, 106, 112–113, 118, 136, 146, 156, 181, 195, 202
Water pollution, 10, 29, 91–92, 188, 225–226
Water quality monitoring, 140
Water resources management, 32, 119, 131, 132
Water vapor, 38, 98, 102
Watergate controversy, 31, 146, 153–154
Waterton, Colo., 1
WB–57F (research aircraft), 185
WC–130 (research aircraft), 120
Weather. See Meteorology.
Weather and climate program, 21, 61–62
Webb, James E., 138
Weightlessness, 169–170
 animal and plant experiments, 75–76, 87–88, 190, 247
 human response, 31, 32, 34–35, 42–44, 91, 157–158, 178, 218–219
 space manufacturing, 43, 178, 184
Weinberger, Casper W., Secretary of Health, Education, and Welfare, 106–107
Weitz, Capt. Paul J. (USN), 195
West Germany. See Germany, West.
Westar (Western Union comsat), 2
*Westar 1* (Westar-A), 64, 85–86, 116, 164, 171, 186, 221, 225–226, 227–228, 233, 257, 258
*Westar 2* (Westar-B), 97, 110, 113–114, 185–186, 221, 225–226, 227, 246, 257, 259
Westar Satellite Communications System, 186
Westerbork, The Netherlands, 160
Western Test Range (WTR; see also Vandenberg AFB, Calif.), launch
 *Aeros 2*, 135, 242, 260
 *Ans 1*, 159, 244
 *Hawkeye 1*, 111, 238–239, 259
 *Miranda*, 57–58, 231, 258
 *Nts 1*, 241–242
 *Noaa 4*, 200, 249, 259
 unidentified satellite, DOD, 230, 232, 233, 239, 243, 248–249
Western Union International Corp., 3–4
 Western Union Telegraph Co. See Westar satellites.
Westinghouse Electric Corp., 102, 134, 174–175, 187
WFC. See Wallops Flight Center.
Whedon, Dr. G. D., 157–158

Wheelchair, 81
Whipple, Dr. Eldon C., Jr., 92
Whipple, Dr. Fred L., 9–10, 77
Whirling chair, 31
Whitcomb, Dr. Richard T., 112, 148, 173
White, Dr. Robert M., 211, 213
White, Gen. Thomas D., Space Trophy, 183
White hole, 133
White House (see also Office of Science and Technology, and Office of Telecommunications Policy), 48, 153–154, 164–165, 214
White Sands Missile Range (WSMR), N. Mex., 2, 3, 98, 264–265, 266–267, 268, 269, 270, 271-257
Wilford, John N., 59
Wilson, Prime Minister Harold S. (U.K.), 136–137
Wind energy, 6, 29, 34, 66, 73, 115, 120, 126, 168, 169, 226
Wind tunnel, 8–9, 39, 55, 92, 114, 141, 160, 163, 167, 202, 211–212
Windmills, 6, 34, 63, 204–205
Winfield, Rodney, 136
Wing, aircraft (see also GAW-1, Supercritical wing, Swing wing, and Variable-geometry wing), 7, 48, 91, 112, 160, 192, 211–212
Winnewisser, Gisbert, 171–172
"Winning the Initiative: NASA and the U.S. Space Science Program, 1958-60" (essay), 59
Winter, Dr. David L., 97
Wisconsin, Univ. of, 269, 270–271
Wolfe, Dr. John H., 14
Wolter Type I telescope, 125
Women, 6–7, 11–12, 102–103, 169–170, 171, 202
Wood, Dr. Lowell, 56
Woodling, Carroll, H., 15
Woods, David, 161
Woods Hole, Mass., 87, 126–127
Working group. See Apollo-Soyuz Test Project and Committee on the Peaceful Uses of Outer Space.
Working Group on Space Biology and Medicine, U.S.-U.S.S.R., 193
World Data Center, 120
World Energy Conference, 174
World Intellectual Property Organization, 99
World language, 117
World Meteorological Organization, U.N., 110, 119
World War II, 70–71, 125, 133, 154–155, 156
*World Weather Program, Plan for Fiscal Year 1975*, 143
Wright Brothers Lecture, 148
Wright Brothers Memorial Trophy, 173, 219
WSMR. See White Sands Missile Range.
WTR. See Western Test Range.
Wurzburg, Germany, 171–172
Wydeven, Dr. Theodore, Jr., 216

## X

X-1 (rocket research aircraft), 176–177
X-15, 176–177
X-24A (lifting body), 163, 273
X-24B, 52–53, 163, 224, 273–274
X-24C, 163
X-ray, 67, 80, 90, 100, 104, 133, 139, 158, 174, 187, 212, 228, 237, 244, 269, 271
X-ray astronomy, 125, 133, 159–160, 259, 266–267, 270
X-ray spectrometer, 172
X-ray telescope, 21, 37–38, 125, 154, 158, 189, 218–219
XB-70 (experimental mach 3 aircraft), 176–177
Xenon, 8, 68

## Y

Yardley, John F., 101, 210
Yates, Iva C., 184
YCH-53E (helicopter), 51
Yeliseyev, Dr. Aleksey S., 81–82
YF-12 (research aircraft), 114
YF-12 Flight Loads Symposium, 114
YF-16 (lightweight fighter aircraft), 79, 91, 94, 116, 138, 146, 183–184
YF-17, 31, 79, 94, 116, 138, 146, 183–184
YJ-101 (aircraft engine), 31
Young, Capt. John W. (USN), 28, 95

## Z

Zaire, 14
Zarb, Frank G., 171
Zero-gravity atmospheric-cloud-physics experiment laboratory, 41–42, 185
Zinc, 166
Zvezdny Gorodok (Star City), U.S.S.R., 121–122, 123, 152, 208
Zwicky, Dr. Fritz, 31

# ERRATA

## In Earlier Volumes of Astronautics and Aeronautics

*Astronautics and Aeronautics, 1963*

p. 189: Delete second item from bottom (USAF unidentified satellite launch). Insert "from Point Arguello, Calif.," after "back" in third line from bottom of page.

p. 389: In second item from the top, *October 16* (twin Vela launch), substitute identification "Tetrahedral Research Satellite (TRS II)" and weight of "3 lbs." for "Pygmy" and "2½ lbs." Add "Pres. Rpt. on Space, 1963, 1/27/64" to sources at the end of the item. (See correction for p. 391.)

p. 391: Delete third item from top, DOD launch. (See information added to p. 389, above. The launch from Cape Canaveral was late on Oct. 16 by Eastern Standard Time and was on Oct. 17 by Greenwich Mean Time.)

p. 509, Appendix A, launch table: In Launch Date column, change "Oct. 17" to "Oct. 16." For that date, also change first "DOD Spacecraft" to "Vela I" and second "DOD Spacecraft" to "Vela II."

p. 560, Index, left column: After "M–2," change "291" to "292."

*Astronautics and Aeronautics, 1964*

p. 457: Appendix A, launch table: In Launch Date column, change last entry from "Oct. 10" to "Oct. 9*" (with an asterisk).

p. 487, Index, right column: After "Fire 1," change "483" to "466."

p. 500, Index, left column: After "Mars, unmanned," delete "470–471, 474–475" and substitute "443, 444."

p. 502, Index, left column: After "Moon, photographs," change "460" to "360."

*Astronautics and Aeronautics, 1965*

p. 591, Appendix A, launch table: In Aug. 13 entry, insert new name: *Spasurad 1*, 1965–65G, with 736-mi apogee, 680-mi perigee, 90.0° inclination, and 108.1-min period; still in orbit.

For last SURCAL, substitute international designation "1965K" for "1965L" and substitute orbital parameters 736-mi apogee, 680-mi perigee, 90.0° inclination, and 108.1-min period. Still in orbit.

p. 654, Index, left column: After "Moon, mapping," change "495" to "496."

*Astronautics and Aeronautics, 1966*

p. 257: In 11th line from top of page (third item), change "Space Tracking and Data . . ." to "Satellite Tracking and Data. . . ."

p. 354: In 13th line from top of page, correct the last name of M/G Vincent G. "Houston" to "Huston."

p. 455, Index, left column: Correct "Maxwell, B/G Jewell C. (USAF, Ret.)" to read: "Maxwell, M/G Jewell C. (USAF)." Gen. Maxwell was promoted in August 1966. He was not retired.

p. 468, Index, right column: Change "Space Tracking and Data Acquisition Network" to "Satellite Tracking and," etc., and move to p. 465, bottom of left column.

p. 469, Index, left column: After "STADAN" change "Space" to "Satellite."

p. 470, Index, left column: After "SURVEYOR I, launch," change "197–198" to "196–197."
After "SYNCOM II" change "133–134" to "134–135."

p. 474, Index, left column: In top line, after "Venus (planet)," change "142" to "141."

## Astronautics and Aeronautics, 1967

p. 154: In third paragraph, correct "B/G J. C. Maxwell" to read, "M/G J. C. Maxwell (USAF)."

p. 169: In *May 31* USAF launch, third item from bottom, change first sentence to begin: "USAF launched 9 satellites. . . ."

p. 196: In first *June 29* item, change first sentence to read: "USAF launched two satellites, *EGRS–9* and *Aurora I*, on one Thor-Burner II booster from WTR."

p. 203: In last item, change "B/G J. C. Maxwell (USAF, Ret.)" to read: "M/G J. C. Maxwell (USAF)."

p. 293: In next to last item, make second sentence read: ". . . Williams was eighth astronaut to be killed in an accident and fourth [instead of third] to die in a T–38 crash. . . ." Delete "one other has died in an air crash." (This was the third T–38 crash, but the fourth death.)

p. 299: In top item, substitute the following orbital parameters: 308-km (214-mi) apogee, 196-km (124-mi) perigee, 89.7-min period, and 65.6° inclination.

p. 332: In sixth line from top, delete "first." (See July 25 item.)

p. 404, Appendix A, launch table: For *Surveyor III*, change international designation to "1967–35A."

p. 410, Appendix A, launch table: For May 31 DOD spacecraft, change first word in Remarks column to "Nine."

p. 419, Appendix A, launch table: For *Surveyor V*, change international designation to "1967–84A." In last line of Remarks column, change date to "12/16/67."

p. 422, Appendix A, launch table: In first entry, Oct. 11, substitute the 214-mi apogee, 214-mi perigee, 89.7-min period, and 65.6° inclination.

p. 425, Appendix A: Change *Surveyor VI* launch vehicle to "Atlas-Centaur."

p. 434, Appendix B: Change Surveyor VI launch vehicle to "Atlas-Centaur."

p. 446, Index, right column: Change Cernan's middle initial to "A."

p. 448, Index, right column: After "Conrad, LCdr. Charles" delete the "E."

p. 449, Index, left column: After "Cooper, Maj." transpose name and initial to read "L. Gordon."

p. 453, Index, right column: In top line, insert a "t" to make the surname "Feoktistov."

p. 458, Index, left column: After "Jaffe, Dr. Leonard D.," change "307" to "377."

p. 460, Index, left column: Substitute the following for the two entries below "Launch Complex 34":
    Launch Complex 37, 81
    Launch Complex 39, 338, 339

p. 463, Index, left column: After "Maxwell, M/G," delete "Ret."

p. 471, Index, left column: After "President's Science Advisory Committee," change "45" to "43–45."

p. 484, Index, right column: After "Webb—Continued, space program," change "301" to "300."

p. 485, Index, right column: After "X–15 . . . , record," change "344" to "345."

## Astronautics and Aeronautics, 1968

p. 15: In second item (AIAA's Goddard Award), second line, change "Warsham" to "Worsham."

p. 317: In last item (X–15 flight), third line, change date to "1959."

p. 381, Index, left column: Change Aldrin's rank to "Col."

p. 391, Index, left column: After "Disarmament," change "160" to "159, 161."

p. 392, Index, left column: After "Electronics Research Center," change "164" to "165."

p. 409, Index, left column: Delete "Naval Research Laboratory (NRL), 184."
In right column: Insert "184" in page numbers for top item.

p. 412, Index, left column: After "Paine, Dr. Thomas O.," change "366" to "336." Also insert "245."

p. 425, Index, left column: Change, "Warsham" to "Worsham" and move to alphabetical position on p. 426, right column.

## Astronautics and Aeronautics, 1969

p. 21: In third paragraph of *Jan. 21* item, delete the second "n" from "Sonnett"—to read "Dr. Charles P. Sonett."

p. 93: In photo caption, next to last line, change "Mariner IV" to "Mariner VI" (transposition).

p. 106: In last item, second line, delete "near polar."

p. 230: In second paragraph, second line, change "(20–30 m)" to "(2–3 m)."

p. 247: In third paragraph, fourth line, change "5 million" to "50 million."

p. 261: In fourth item (Lick Observatory), make note that fourth line from this item should be inserted after second line of third item.

p. 318: In third paragraph, third line, insert omitted words after "our little 8,000-mile-diameter":

    earth throughout our vast 8-billion-mile-diameter

and pick up again with "Solar System."

p. 349: In second item (Silverstein), second paragraph, fourth line, change "NASA" to "NACA." And in fifth paragraph, first line, insert "NACA and " after "served."

p. 421: In *During 1969*, third paragraph of item, third line, change date to "1965." (*Mariner IV* was launched in 1964 but reached Mars in 1965.)

p. 466, Appendix C: In March 3 entry, Crew column, change spelling of "Schweikart" to "Schweickart."

p. 469, Index, right column: In Oct. 12 entry, Crew column, change spelling of "Garbatko" to "Gorbatko."

p. 494, Index, right column: Delete "FOBS" entry, top two lines.

p. 522, Index, left column: Change "Sonnett" to "Sonett."

p. 524, Index, left column: After "Space Shuttle, reusable," change "23" to "25–26."

p. 533, Index, right column: After "Zähringer," change "Joseph" to "Josef."

*Astronautics and Aeronautics, 1970*

p. 95: In top line, change "12 200-km" to "12 200-m."

p. 199: In *June 11* item (HL–10 flight), fifth line, change "mach 0.07" to "mach 0.7."

p. 262: In top item, ninth line, change "7.6-cm" to "76-cm."

p. 348: In top item, ninth line, insert "Space" before "Nuclear Systems Office."

p. 351: In third item, change dateline from *"October 29"* to *"October 28"* (item should be combined with *October 28* item on p. 349). Put *October 28* date on fourth item on p. 251 (Frutkin briefing).

*Astronautics and Aeronautics, 1971*

p. 202: In second paragraph from bottom, fifth line of paragraph, change "British Astronautical " to "British Interplanetary."

p. 250: In fifth paragraph from the top, first line, change "Space Nuclear Propulsion Office" to "Space Nuclear Systems Office." (Office name had been changed.)

p. 419, Index, left column: Change "AEC–NASA Space Nuclear Propulsion Office" to "AEC–NASA Space Nuclear Systems Office."

p. 427, Index, left column: Delete "British Astronomical Society" and add "202" to page numbers for "British Interplanetary Society" below.

p. 467, Index, right column: Change "Technology unitization" to "Technology utilization."

*Astronautics and Aeronautics, 1972*

p. 207: In *May 29* item, first line, change "Secretary General" to "General Secretary."

p. 555, Index, left column: After "Powell, Cecil," change "379" to "380."

*Astronautics and Aeronautics, 1973*

p. 35: In second item, end of third line, change "350" to "3.5."

p. 285: In last item (X–24B flight), first line, change "fifth captive" to "fifth glide." In second line, change "attached to" to "launched from" a B–52 aircraft.

p. 416, Index, right column: Change Bobko's rank to "L/C."

p. 437, Index, right column: After "ITOS–E," delete "See *Noaa 3*" and add page numbers "1, 209, 212." Insert new entry: "ITOS–F. See *Noaa 3*." (ITOS–E was intended to become *Noaa 3*, but failed to reach orbit; it was replaced by ITOS–F.)

p. 453, Index, right column: Change the *Noaa 3* item to read: "*Noaa 3* (ITOS–F; see also ITOS–E), 312, 352, 358."

# NASA HISTORICAL PUBLICATIONS

## HISTORIES

- Frank W. Anderson, Jr., *Orders of Magnitude: A History of NACA and NASA, 1915–1976*, NASA SP–4403, 1976, GPO.\*
- William R. Corliss, *NASA Sounding Rockets, 1958–1968: A Historical Summary*, NASA SP–4401, 1971, NTIS.\*\*
- Constance McL. Green and Milton Lomask, *Vanguard—A History*, NASA SP–4202, 1970, GPO; also Washington: Smithsonian Institution Press, 1971.
- Barton C. Hacker and James M. Grimwood, *On the Shoulders of Titans: A History of Project Gemini*, NASA SP–4303, in press.
- Edwin P. Hartman, *Adventures in Research: A History of the Ames Research Center, 1940–1965*, NASA SP–4302, 1970, NTIS.
- Mae Mills Link, *Space Medicine in Project Mercury*, NASA SP–4003, 1965, NTIS.
- Alfred Rosenthal, *Venture into Space: Early Years of Goddard Space Flight Center*, NASA SP–4301, 1968, NTIS.
- Robert L. Rosholt, *An Administrative History of NASA, 1958–1963*, NASA SP–4101, 1966, NTIS.
- Loyd S. Swenson, James M. Grimwood, and Charles C. Alexander, *This New Ocean: A History of Project Mercury*, NASA SP–4201, 1966, NTIS.

## REFERENCE WORKS

- *The Apollo Spacecraft: A Chronology*, NASA SP–4009: Volume I, 1969, NTIS. Volume II, 1973, GPO. Volume III, 1976, GPO. Volume IV, in press.
- *Astronautics and Aeronautics: A Chronology of Science, Technology, and Policy*, annual volumes from 1961, with an earlier summary volume, *Aeronautics and Astronautics, 1915–1960*. Early volumes available from NTIS; recent volumes from GPO. *Astronautics and Aeronautics, 1973*, NASA SP–401, 1975.
- Katherine M. Dickson (Library of Congress), *History of Aeronautics and Astronautics: A Preliminary Bibliography*, NASA HHR–29, NTIS.
- *Documents in the History of NASA: An Anthology*, NASA HHR–43, August 1975, NASA History Office.
- *Project Gemini Technology and Operations: A Chronology*, NASA SP–4002, 1969, NTIS.
- *Project Mercury: A Chronology*, NASA SP–4001, 1963, NTIS.
- *Project Ranger: A Chronology*, JPL/HR–2, 1971, NTIS.
- *Skylab: Preliminary Chronology*, NASA HHN–130, May 1973, NTIS.
- Jane Van Nimmen and Leonard C. Bruno, *NASA Historical Data Book, 1958–1968*, Vol. I, *NASA Resources*, NASA SP–4012, 1976, NTIS.
- Helen T. Wells, Susan H. Whiteley, and Carrie E. Karegeannes, *Origins of NASA Names*, NASA SP–4402, 1976, GPO.

---

\*GPO: Available from Superintendent of Documents, Government Printing Office, Washington, D.C. 20402.
\*\*NTIS: Available from National Technical Information Service, Springfield, Virginia 22161.

www.ingramcontent.com/pod-product-compliance
Lightning Source LLC
Chambersburg PA
CBHW081613200526
45167CB00020B/3620